The Singular Universe and the Reality of Time

A Proposal in Natural Philosophy

Cosmology is in crisis. The more we discover, the more puzzling the universe appears to be. How and why are the laws of nature what they are?

A philosopher and a physicist, world-renowned for their radical ideas in their fields, argue for a revolution. To keep cosmology scientific, we must replace the old view in which the universe is governed by immutable laws by a new one in which laws evolve. Then we can hope to explain them.

The revolution that Roberto Mangabeira Unger and Lee Smolin propose relies on three central ideas. There is only one universe at a time. Time is real: everything in the structure and regularities of nature changes sooner or later. Mathematics, which has trouble with time, is not the oracle of nature and the prophet of science; it is simply a tool with great power and immense limitations. The argument is readily accessible to non-scientists as well as to the physicists and cosmologists whom it challenges.

ROBERTO MANGABEIRA UNGER is a philosopher, social and legal theorist, and politician. His engagement with cosmology and natural philosophy in this book deepens and generalizes ideas that he has developed in *False Necessity*, *The Self Awakened*, and *The Religion of the Future*, among other writings.

LEE SMOLIN is a theoretical physicist who has made important contributions to quantum gravity. Born in New York City, he was educated at Hampshire College and Harvard University. He is a founding member of the Perimeter Institute for Theoretical Physics. His earlier books explore philosophical issues raised by contemporary physics and cosmology: *Life of the Cosmos*, *Three Roads to Quantum Gravity*, *The Trouble with Physics*, and *Time Reborn*.

Unger and Smolin have been collaborating for eight years on the project that this work brings to fruition.

The Singular Universe and the Reality of Time

A Proposal in Natural Philosophy

ROBERTO MANGABEIRA UNGER
AND
LEE SMOLIN

CAMBRIDGE
UNIVERSITY PRESS

CAMBRIDGE
UNIVERSITY PRESS

University Printing House, Cambridge CB2 8BS, United Kingdom

Cambridge University Press is part of the University of Cambridge.

It furthers the University's mission by disseminating knowledge in the pursuit of education, learning and research at the highest international levels of excellence.

www.cambridge.org
Information on this title: www.cambridge.org/9781107074064

First published 2015

Printed in the United States of America by Sheridan Books, Inc

A catalogue record for this publication is available from the British Library

Library of Congress Cataloging in Publication data
Unger, Roberto Mangabeira.
The singular universe and the reality of time : a proposal in natural philosophy / Roberto Mangabeira Unger, Lee Smolin.
pages cm
ISBN 978-1-107-07406-4
1. Cosmology. I. Title.
BD511.U54 2014
113–dc23
2014016833

ISBN 978-1-107-07406-4 Hardback

Contents

The nature and scope of this work

Roberto Mangabeira Unger and Lee Smolin

To think of the universe as a whole rather than of something within the universe is one of the two most ambitious tasks that thought can undertake. Nothing matches it in ambition other than our attempts to form a view of ourselves. In addressing this topic, we soon reach the limits of what we know and even of what we can ever hope to know. We press science to the point at which it passes into philosophy and philosophy to the point at which it easily deceives itself into claiming powers that it lacks.

Yet we cannot cast this topic aside. First, we cannot avoid it because we are driven to understand whatever we can about our place in the world, even if what we do know, or might discover, represents only a small and superficial part of the enigmas of nature. Second, we should not seek to escape it because no one can develop and defend ideas about parts of natural reality without making assumptions, even if they remain inexplicit, about nature as a whole. Third, we need not turn away from it because among the greatest and most startling discoveries of science in recent times are discoveries about the universe and its history. The most important such discovery is that the universe has a history. Part of the task is to distinguish what science has actually found out about the world from the metaphysical commitments for which the findings of science are often mistaken.

* * *

In this book, we deal with this subject directly. Three ideas are central to our argument.

The first idea is the singular existence of the universe. (We use singular here in the sense of unique, not in the sense in which relativists use it to mean a singularity at which the curvature of spacetime and other quantities become infinite. In fact we later argue that the universe cannot be singular in that sense.) There is only one universe at a time, with the qualifications that we discuss. The most important thing about the natural world is that it is what it is and not something else. This idea contradicts the notion of a multiverse – of a plurality of simultaneously existing universes – which has sometimes been used to disguise certain explanatory failures of contemporary physics as explanatory successes.

The second idea is the inclusive reality of time. Time is real. Indeed, it is the most real feature of the world, by which we mean that it is the aspect of nature of which we have most reason to say that it does not emerge from any other aspect. Time does not emerge from space, although space may emerge from time.

That time is inclusive as well as real means that nothing in nature lasts forever. Everything changes sooner or later, including change itself. The laws of nature are not exempt from this impermanence. By implying the mutability of the laws of nature, the idea of the inclusive reality of time contradicts a dominant interpretation of what the physics and cosmology of the last hundred years teach us about the workings of nature.

Twentieth-century science overthrew the conception of an invariant background in space and time to the events and phenomena of nature. Einstein's greatest accomplishment in inventing general relativity was to replace Newton's absolute space and time with a conception of spacetime that is both relational and dynamical. When he did so, however, he reaffirmed the notion of an immutable framework of natural laws. We have ordinarily expected such timeless laws to supply warrants to our practice of causal explanation. If the laws of nature change, how can we hope to establish scientific inquiry on a secure basis? A major concern of this book is to propose answers to this question.

Now, however, we have grounds to overthrow the view that was reaffirmed when belief in an invariant background of space and time was abandoned. Unless we accomplish this second overturning we cannot do justice to the most important discovery made by the cosmology of the twentieth century: the discovery that the universe, and everything in it, has a history. The prevailing accounts tell that history against a background of immutable laws of nature. We argue that there is more reason to read that history as including the evolution of the laws themselves. History then subjects the laws as well as everything else to the effects of time.

If time is inclusively real in cosmology, which has the whole universe for its subject matter, it must be inclusively real in every department of science and in every piece of nature.

The third idea is the selective realism of mathematics. (We use realism here in the sense of relation to the one real natural world, in opposition to what is often described as mathematical Platonism: a belief in the real existence, apart from nature, of mathematical entities.) Now dominant conceptions of what the most basic natural science is and can become have been formed in the context of beliefs about mathematics and of its relation to both science and nature. The laws of nature, which it has been the supreme object of science to discern, are supposed to be written in the language of mathematics.

We cannot give an adequate account of the singular existence of the universe and of the inclusive reality of time without developing and vindicating a certain view of mathematics. Mathematics has two subject matters: nature (viewed in its most general aspects) and itself. It begins in an exploration of the most general relations in the world, abstracted from time and of phenomenal particularity, but it soon escapes the confines of our perceptual experience. It invents new concepts and new ways of connecting them, inspired by its previous ideas as well as by the riddles of natural science.

Our mathematical inventions offer us no shortcut to timeless truth either about nature or about some special realm of mathematical objects outside nature. They have no prophetic role, notwithstanding the vast power and prestige of mathematics. They may or may not be useful. They never replace the work of scientific discovery and of imagination. The effectiveness of mathematics in natural science is reasonable because it is limited and relative.

The singular existence of the universe, the inclusive reality of time, implying the mutability of the laws of nature, and the selective realism of mathematics all have justifications of their own. However, they are more than a collection of separate and loosely related propositions. The more deeply we understand them, and appreciate the reasons for holding them to be true, the more clearly do we come to recognize their many and intimate relations to one another. They represent three sides of the same comprehensive view. They support and refine one another. It is only when we appreciate their connections that we can grasp just how much they require us to break with certain ideas that continue to enjoy wide influence both within and outside science.

* * *

This work deals with foundational problems in basic science. It proposes a reinterpretation of some of the most important discoveries of twentieth-century cosmology and physics, the historical character of the universe first among them. The reinterpretation has consequences for the future agenda of these sciences. It seeks to distinguish what we in fact know – the hard empirical residue of scientific discovery – from the lens of assumptions through which we are accustomed to see the larger significance of these factual findings.

The history of physics and cosmology has been in large part the history of a marriage between two sources of inspiration. One source is our probing of the manifest world, through observation and experiment, conditioned by our success at inventing and deploying equipment that enables us to extend or exceed our powers of perception. The other source is a vision of reality at the center of which there often stands an ontological program: a view of the kinds of things that there ultimately are and of the ways in which they connect. Such were the ontological programs associated with the science of Aristotle, of Newton, and of Einstein.

It will sometimes happen that no fundamental progress can be achieved in science without dissolving this marriage between the empirical residue and the philosophical gloss. Once the marriage is dissolved, it becomes possible to see the discoveries of science with new eyes. It is never possible, however, to do so without changing some of our beliefs about how nature works.

Two large philosophical traditions inform the ideas of this book. They can be placed under two labels: the relational approach to nature and the priority of becoming over being. In this work, we make no attempt to justify them as philosophical conceptions outside the scientific contexts in which we make use of them. The case for them here lies in the insights that they together make possible.

The relational idea is that we should understand time and space as orderings of events or phenomena rather than as entities in themselves. More generally, it is the view that within a network of causal connections, extending outward to a causally connected universe, everything influences everything else through causal links. In understanding the operation of nature, this relational structure matters more than any of its parts. Its parts matter, and exert their effects, by virtue of the role that they perform within the relational network to which they belong.

In the history of physics and of natural philosophy the two chief statements of the relational view have been those formulated by Gottfried Leibniz in the late seventeenth century and by Ernst Mach in the late nineteenth century. A complication of our argument is that neither of these versions of the relational approach is wholly adequate to our purpose. We must therefore develop another version along the way.

A second philosophical inspiration of this book is less easy to associate with a single doctrine, a ready-made description, or a few names. It is the tradition of thought that affirms the primacy of becoming over being, of process over structure, and therefore as well of time over space. It insists on the impermanence of everything that exists. On this view, the rudimentary constituents of nature, described by particle physics, are impermanent. So, too, are the laws of nature, expressed in the language of mathematics, which it has been the chief ambition of modern science to establish.

The present quest for a grand theory of everything – of the fundamental forces and fields in nature – goes forward on the basis of viewing these law-like regularities and elementary constituents as if they were forever. As a result, we argue, it fails fully to appreciate the most important cosmological discovery: that the universe has a history. Cosmology must be a historical science if it is to be a science at all: a historical science first, a structural science only second, not the other way around.

In the history of Western philosophy, the line of thought that affirms the impermanence of structure has spoken in the voices of thinkers as different as Heraclitus, Hegel, Bergson, and Whitehead. Among the philosophical schools of other civilizations, notably of ancient India, it represented the hegemonic metaphysic.

Although it is not a view that has ever enjoyed commanding influence over the physics that Galileo and Newton inaugurated, it plays a major part in the life sciences as well as in the study of society and of human history. The structures investigated by the naturalist, the historian, or the social scientist may be enduring. No one, however, thinks of them as eternal. Moreover, insofar as there are regularities or laws that govern their workings, they evolve together with the phenomena that they govern.

The philosophical ideas that have guided and interpreted the program of modern physics have traditionally regarded this lack of eternal structures and laws in the life sciences and in the study of human affairs as a sign of the derivative or precarious character of those disciplines. The gold standard of scientific inquiry continues to be supplied, in the eyes of this tradition, by a way of thinking that treats impermanence, and thus time itself, as threats to the achievement of our most far-reaching explanatory endeavors.

One of our aims in this book is to show that the idea of the primacy of becoming over being deserves to hold in cosmology a place no less central than the one that it occupies in the supposedly less happy and less basic sciences. If it is entitled to this role in cosmology, which is the science of the whole universe and its history, it must merit it as well in physics, which studies pieces of the universe and moments of its history.

Among the implications of this philosophical conception, and of the idea of the inclusive reality of time, is the thesis that the new can emerge and does emerge during the evolution of the universe. The new is not simply a possible state of affairs, prefigured by eternal laws of nature. It is not simply waiting to fulfill the conditions that, according to such laws, allow it to move from possibility to actuality. The new represents a change in the workings of nature. Such change embraces the regularities – that is to say, the laws – as well as the states of affairs.

The emergence of the new is a repeated event in the history of the universe. It continues, under novel forms and constraints, in our own experience: the appearance of mind and the exercise of our human power to accelerate the production of novelty in the universe. Our science and our mathematics rank among the most notable instances of the exercise of this power.

The relational approach to space, time, and other physical properties and the primacy of becoming over being each solve a problem that the other leaves unsolved. Timeless versions of relational spacetime leave inexplicable basic features of nature such as the choice of laws and of initial conditions. Our best hope of explaining these enigmas is to put the laws of nature under the dominion of time: to hypothesize that they are mutable and that they have become what they are by evolving in real time. On the other hand, the priority of becoming over being has often been affirmed against the backdrop of an absolute rather than a relational view of time. The result may be to substitute a mystical notion for a scientific program by invoking an external force or entity that produces becoming in an otherwise passive universe. Only when we understand becoming from the perspective of relational time can we subject it to a dynamics that is internal to the universe. Only then can we lay it open to explanation by the methods of science.

The development of our three central ideas, in the spirit of these two traditions of thought, defines a position that can be labeled temporal naturalism. This position in turn informs an approach to the central problems and future agenda of cosmology.

* * *

The discourse in which we present our argument invokes, and seeks to reinvent, the vanished genre of natural philosophy.

This book is not an essay in popular science: the presentation of contemporary scientific developments to a broad readership. We hope that it will be accessible to readers who come to it from many different backgrounds, not just to cosmologists and physicists. Nowhere, however, have we deliberately compromised the formulation of the ideas to make them more accessible. The limitations of our arguments are those that are imposed by the limits of our understanding; they do not result from deliberate simplification.

In the absence of an established discourse of natural philosophy, scientists have often used the presentation of ideas to a general educated public as a device by which to address one another with regard to the foundational matters that they cannot readily explore in their technical writings. Here, however, we set our hands to natural philosophy directly, not under the mask of popularization.

The discourse of this book is also to be distinguished from the philosophy of science as that discipline is now ordinarily practiced. The work of the philosophy of science is to argue about the meaning, implications, and assumptions of present or past scientific ideas. It offers a view of part of science, from outside or above it, not an intervention within science that seeks to criticize and redirect it. It foreswears revisionist intentions.

The proximate subject matter of the philosophy of science is science. The proximate subject matter of natural philosophy is nature. Unlike the philosophy of science, natural philosophy shares its subject matter with science.

A natural-philosophical argument about the universe and its history is not simply or chiefly an argument about cosmology. It is a cosmological argument. It intervenes, and takes a position, in the cosmological debates with which it deals. It does so on the basis of ideas and considerations both internal to contemporary science and external to it. It tries to describe and to explore a broader range of intellectual options than is represented in the contemporary practice of the fields that it addresses. Its goals are frankly revisionist: to propose and defend a redirection of cosmology that has implications for the path that physics can and should take.

In all these respects, the discourse of this book resembles nothing so much as what was known, up to the middle of the nineteenth century, as natural philosophy. The trouble is that, despite occasional and exceptional efforts by individual scientists and philosophers, natural philosophy has long ceased to exist as a recognized genre. (A major exception to its near-disappearance in the intervening period was the work of Ernst Mach at the turn of the twentieth century, together with the way in which Albert Einstein made use of Mach's ideas. Another exception was the natural-philosophical writing of Mach's contemporary, Henri Poincaré. To this day, biology has benefited from a long line of natural philosophers, many of them active scientists.) The duo of popular science and philosophy of science has usurped the place of natural philosophy.

Here we seek to breathe new life and form into this defunct way of thinking and writing. It is impossible to do justice to the intellectual difficulties and opportunities that we explore without defying the limits of the established technical discourse of cosmology and physics. Neither, however, can we advance the agenda that we set for ourselves without engaging these disciplines on their own terms as well as on terms that remain foreign to them.

The reasons to cross, in both directions, the frontier between science and philosophy, go beyond the practical need to find broader sources of inspiration when confronted with perplexities that established scientific ideas may be insufficient to overcome. These reasons have to do, as well, with an ideal of scientific inquiry and with a conception of the mind.

Science is corrupted when it abandons the discipline of empirical validation or disconfirmation. It is also weakened when it mistakes its assumptions for facts and its ready-made philosophy for the way things are. The dialectic between openness to the promptings of experience and openness to the surprises of the imagination is the vital requirement of its progress. When "normal science" begins to take on some of the characteristics of "revolutionary science" – the science of "paradigm change" – what results is a higher, more powerful practice of scientific work. Natural philosophy can be an ally of science in this effort to raise the sights and to enhance the powers of scientific thinking.

It is an effort that can succeed because the mind is what it is. We can always see and discover more than any set of methods and presuppositions, in any discipline, can prospectively. Vision exceeds method, and reshapes practice and discourse, according to its needs.

Natural philosophy, however, cannot be at the beginning of the twenty-first century what it was at the beginning of the nineteenth. It must turn into something else. Rather than providing a theory of this something else, we here offer an example of it.

* * *

Each of us presents separately the whole argument of this book, recording, each in his own way, the product of eight years of collaboration and discussion. One of us renders our joint argument as a systematic view in natural philosophy. The other expresses it in a version that, without ceasing to be natural philosophy, comes closer to the debates and theories of cosmology and physics today. He states it in the context of problems and ideas immediately familiar to contemporary cosmologists and physicists. He explores its implications for their present and future work.

The two of us agree about the overall direction and the central claims of the argument. We do not, however, agree about all the matters on which we touch. Some of the differences between us are minor. Others are substantial. Whether small or large, these differences serve as a salutary reminder that there are many ways to develop the same general view. We list and explore these disagreements in a note at the end of this book.

* * *

Our subject of study is the universe and its history. Our negative thesis is that the ways of thinking about the universe and its history that now enjoy the widest influence within cosmology fail adequately to convey the significance of what cosmology has found out about the world. They provide a flawed basis for its future development. Our affirmative thesis is that the intellectual instruments are already at hand to develop another and better way of thinking about these issues. This alternative is incompatible with commonly held views about the plurality of universes, the emergent or illusory nature of time, and the power of mathematics to serve science as its oracle and prophet.

The subject matter could not be more fundamental. Nothing can be properly compared to it other than our study of ourselves. Cosmology is not just one more specialized science. It is the study of the universe as a whole, beyond which, for science, there lies nothing.

All our ideas about parts of nature will be influenced, whether knowingly or not, by our assumptions about the whole universe. Contemporary physics and cosmology have repeatedly inverted this principle: they have tried to apply to the study of the universe and of its history procedures that are useful only when applied to the study of local phenomena. This inversion has led them into some of their gravest mistakes.

The science of cosmology, by which we mean the scientific study of the universe as a whole, cannot be just the physics of local or small phenomena, scaled up to the largest scales, as it usually has been. For reasons that we consider, physics has been the study of subsystems of the universe. This approach is incapable of providing answers to the central questions of cosmology, such as the nature of time and space and the origins or explanations of the laws and initial conditions of the universe. To answer these questions scientifically, with hypotheses open to empirical confirmation or falsification, requires a new approach, based on new principles and enlisting new methods. Our aim is to develop methods and principles adequate to a science of cosmology that is not simply a scaled-up version of contemporary physics. To develop them, we take as points of departure three conceptions: the singular existence of the universe, the inclusive reality of time, and the selective realism of mathematics.

Part I **Roberto Mangabeira Unger**

1 The science of the one universe in time

THE SINGULAR EXISTENCE OF THE UNIVERSE

This book develops three connected ideas about the nature of the universe and of our relation to it. The first idea is that there is only one universe at a time. The second idea is that time is real and inclusive. Nothing, including the laws of nature, stands outside time. The third idea is that mathematics has this one real, time-drenched world as its subject matter, from a vantage point abstracting from both time and phenomenal particularity.

On the view defined by these three ideas, the universe is all that exists. That there is only one universe at a time justifies using the terms universe and world interchangeably. If there were a plurality of universes, the world would be that plurality. The singular universe must, however, be distinguished from the observable universe, for our universe may be much larger than the part of it that we can observe. In this book, we use the words cosmology and cosmological to designate what pertains to the universe as a whole, not just to its observable portion. Observational astronomy has continued, in recent decades, to make remarkable discoveries about the observable universe. Cosmology, however, risks losing its way. The arguments of this work are cosmological: they concern the whole of the universe and the way to think about it.

Each of the three central ideas developed in this book has implications for how we interpret what science, especially in physics and cosmology, has already discovered about the world and for how we view what science can and should do next. It has consequences as well for our view of the place of these scientific discoveries in our self-understanding.

The first idea is the solitary existence of the universe. We have reason to believe in the existence of only one universe at a time, the universe in which we find ourselves. Nothing science has discovered up to now justifies the belief that our universe is only one of many, although the universe may well have predecessors. The multiplication of universes in contemporary cosmology has not resulted from any empirical discovery or inference from observation; it has been the outcome of an attempt to convert, through this fabrication, an explanatory failure into an explanatory success. The explanatory failure is the compatibility of a prevalent view of how nature works at the level of its elementary constituents with many states of nature other than the one that we observe. (Today, in the early twenty-first century, string theory, with its prodigious surfeit of alternative consistent versions, almost all of them not realized in the observed universe, provides the most striking example of such underdetermination of phenomena by prevailing theories.) The conversion of failure into success proceeds by the simple expedient of supposing that for each version or interpretation of the theory in favor there is a corresponding universe in which what it says is true.

If these unobserved universes were held to be merely possible, the question would arise why only one of the possible universes in fact exists. Therefore, the most radical form of the conversion of failure into success consists in claiming that these other universes are more than merely possible; they are actual, even though we have no evidence of their existence (the multiverse idea).

The most widely accepted causal hypothesis today to explain the genesis of such a multiverse is "eternal inflation," postulating the creation of an infinite number of universes formed as bubbles from phase transitions on an eternally inflating medium. Within string theory, it is plausible to believe that such bubble universes are described by laws, chosen by a stochastic process from the immense range of theories that are compatible with the string-theoretical approach. The retrospective teleology of the "strong anthropic principle," according to which the criterion of selection of the laws in our universe is that they make possible our human life and consciousness, closes the circle of prestidigitation.

The sleight of hand represented by this combination of ideas amounts to an ominous turn in the history of science. It is a turn away from some of the methods, standards, and presuppositions that have guided and disciplined science until relatively recently.

Although the opposing idea, of the singular existence of the universe, may appear self-evident to some scientists and to many non-scientists, it raises a problem of the first order. Individual being, wrote Aristotle, is ineffable. We can provide law-like explanations of recurrent phenomena in parts of the universe. But how can there be a law-like explanation of the universe as a whole if the universe is one of a kind? How can we offer such an account if we are not entitled to represent and to explain our world as one of many possible or even actual worlds? The theory of the universe would have to be the theory of an individual entity. For such a theory the history of science offers no model.

THE INCLUSIVE REALITY OF TIME

The second idea defended and developed in this book is that time is inclusively real. According to this thesis, nothing in this singular universe of ours remains outside time.

The reality of time may seem an empty truism. In fact, it is a revolutionary proposition. It contradicts not only certain speculative doctrines that openly affirm the illusory character of time, but also ideas about causation and scientific explanation that may seem beyond reproach and doubt.

When the idea of the reality of time is combined with the idea of the unique existence of the observed universe, it results in the view that this one world of ours and every piece of it have a history. Everything changes sooner or later.

Recognition of the reality of time gives rise to a philosophical conundrum about causation. If time were not real, there could be no causal relations for the reason that there would be no before (the cause) and after (the effect). Causes and consequences would be simultaneous. They would therefore be unreal or mean something different from what we take them to mean. Nothing would distinguish causal connections, which are time-bound, from logical or mathematical relations of implication, which stand outside time. What we, in causal language, call causes and effects would in fact be aspects of a relational grid in a timeless reality.

If, however, everything is time-bound, that principle must apply as well to the laws, symmetries, and constants of nature. There are then no timeless regularities capable of underwriting our causal judgments. Change changes. It is not just the phenomena that change; so do the regularities: the laws, symmetries, and supposed constants of nature.

Our conventional picture of causation must be confused. For we seem to believe, on the evidence of the way in which we use our causal language, outside science as well as within it, that time is real, but not too real. It must be somewhat real; otherwise there would be no causal connections at all. It must not, however, be so real that our causal judgments are all adrift on a sea of changing laws.

In this book we argue that the evidence of science – the deliverances of the science of today, viewed in the light of its recent history – does not entitle us to circumscribe the reality or the reach of time. Our causal judgments cannot indeed be anchored in immutable laws and symmetries. That need not mean, however, that we stand condemned to explanatory impotence. Causal explanation, properly reinterpreted and redirected, can survive the overcoming of our equivocations about the reality of time. It can make peace with the view that time is real and that nothing remains beyond its reach.

This intellectual program brings us face to face with a further riddle, a puzzle that comes into sight when we begin to take seriously the notion that the laws of nature, as well as its other regularities – symmetries and supposed constants – are within time, and therefore susceptible to change, rather than outside time, and therefore changeless. We seem faced with an unacceptable choice between two troubling positions.

One position is that higher-order or meta-laws govern the change of the laws and other regularities of nature. In this event, however, the problem presented by the time dependence of the laws is simply pushed to the next level. Either such higher-order laws are themselves within time and liable to change, or they are timeless and changeless. Nothing fundamental would have shifted in the structure of the problem.

The other possibility is that no such higher-order laws exist. Then our causal judgments would remain bereft of any apparent basis. The change of laws would seem an enigma for which no adequate explanation can exist: change requires causal explanation, and causal explanation must in turn be warranted, or so it is traditionally believed, by laws and symmetries of nature.

We consider ways out of this dilemma. One of them plays an especially large role in our argument, as it has in the development of the life and earth sciences and of social and historical study, although not of physics. According to this view, the laws, symmetries, and supposed constants change together with the phenomena. Causal connections are, on this view, a primitive feature of nature. In our cooled-down universe, they recur over a discriminate structure of natural phenomena, which is to say that they exhibit law-like form. In other, extreme states of nature, however, those that occurred in the very early history of the universe, they may be, or have been, lawless.

The idea that the laws of nature are susceptible to change and that the laws may develop coevally with the phenomena that we take them to govern may be puzzling: for the reasons that I have suggested, it renders unstable the laws of nature that we habitually take as warrants of causal explanation. However, it is neither nonsensical nor unprecedented. We are accustomed to invoke it in the life sciences as well as in social and historical study. It saves us from needing to appeal to speculative metaphysical conjectures, such as the notion of a multitude of unobservable worlds.

The conjecture of the mutability of the laws of nature seems to give rise to insuperable paradoxes. The impression of paradox, however, begins to dissolve once we turn on its head the conventional picture of the relation between laws and causal connections, and recognize that the former may derive from the latter rather than the other way around. This idea may lead us to think in a new light of a broad range of familiar and intractable facts. Among these facts are the unexplained values of the universal constants of nature, especially of those constants that we do not and cannot use as conventional units of measurement and that are, for this reason, conventionally called dimensionless. Their seemingly arbitrary values may be the result of earlier states of the universe and of the operation of laws or symmetries different from those that now hold. They may be vestigial forms of a suppressed and forgotten history: testimonials to a vanished world – the one real world earlier on.

* * *

A simple way to grasp what is at stake for science in the idea of the inclusive reality of time and of its corollary, the conjecture of the mutability of the laws, symmetries, and supposed constants of nature, is to ask the question: Where do these regularities come from? Because the laws and symmetries of nature, as we now understand them, fail to account uniquely for the initial conditions of the universe, we need to ask as well a second question: Where do these initial conditions come from? (The mysterious constants of nature help describe these conditions. They do not explain them. On the contrary, they require explanation, which the established laws and symmetries fail to provide. Thus, even though we can count the constants, together with the laws and symmetries, as regularities, we cannot expect them to help explain the initial conditions of the universe. They form, from the outset, part of the problem rather than part of the solution.)

There are, broadly, three ways to approach these questions.

A first approach is to say that the laws and symmetries comprise an immutable framework of natural events. They are what they are. If they fail to apply to the very earliest moments of the universe or to explain its initial conditions that must be only because our knowledge of the laws and symmetries remains incipient and incomplete. It is this first approach that, at least until recently, has been ascendant in the history of physics, from Newton to Einstein. It represents part of the intellectual backdrop to the major discoveries of twentieth-century cosmology.

An objection to this approach is that what we already know about the very early universe suggests that the laws and symmetries as we now formulate them could not have held in the extreme conditions that existed then and that may exist again later in the history of the universe: for example, in the interior of black holes. How can we speak of laws and symmetries if, in such extreme states, there is no discriminate structure – no stable repertory of different kinds of things, such as those described by the standard model of particle physics, interacting in ways that laws and symmetries can capture?

Another objection is that the initial conditions of the universe, and therefore its subsequent evolution, seem extremely unlikely if nature is indeed constituted as our laws and symmetries say that it is. Under this first approach, the initial conditions of the universe remain unexplained by the laws and symmetries. The laws and symmetries seem applicable only to a universe that has already organized itself in the ways that are characteristic of the cooled-down universe.

It is true that so long as it resists the temptation to succumb to a rationalist metaphysics science can never show that the universe had to be what it has become. Science must in the end recognize what I here call the facticity of the universe: that it just happens to be what it is rather than something else. The problem with the first approach, however, is that it may prematurely and unnecessarily narrow the field open to causal inquiry. It may mistake nature for a subset of natural processes. It may codify as laws and symmetries how nature works in these familiar variations: those that prevail in the relatively cold and differentiated universe in which we find ourselves.

It is one thing to respect the inability of science to show that the universe must be what it is. It is another thing to reduce science to a body of precise laws, symmetries, and constants that are unable to account either for themselves or for the initial conditions of the universe.

A second approach to the question of where the laws and initial conditions of our universe come from is to take this universe as only one of a multitude. The chief object of explanatory ambition shifts, under this approach, from the laws, symmetries, and constants that happen to prevail in this universe of ours to the laws and mathematical conceptions governing the multitude of universes, of which ours would be only one. It will soon appear that there is hardly any difference in this view between laws and mathematical notions.

The effective laws that have up until now been the chief object of science become, under this approach, simply a variant among many sets of higher-order laws applying to the crowd of universes. The strangeness of the initial conditions of our universe can be discounted as a trait of a universe that happens to be an outlier in the crowd. The detachment of the higher-order laws from the realities of the universe that we observe, and their multifarious content, lend them all the more to marriage with mathematics.

The resulting ideas are not so much physical theories expressed in the language of mathematics as they are mathematical conceptions presented as physical theories. Under such a view, the distinction between laws and initial conditions disappears.

The extreme limit of this idea is the notion that natural realities are nothing but mathematical structures. Because such structures are timeless, so must the states of nature that they comprise be timeless. To each mathematical structure there corresponds a universe, instantiating that structure in all its particulars. Observational surprises reveal mathematical ignorance.

This second approach (whether or not in its extreme form) is an invention of the late twentieth and the early twenty-first centuries. It has been almost entirely foreign to the history of physics and cosmology until the last few decades. It arose by the circumstantial convergence of developments in particle physics, culminating in string theory, with the conjecture of a multiverse and the appeal to anthropic reasoning. It found inspiration and reinforcement in mathematics, given the central role that it assigned to mathematical ideas.

An initial objection to this approach is at once methodological and moral. It invents imaginary entities – all the other unobserved and unobservable universes (in cosmology) or states of affairs (in particle physics) – to save itself from having to confront, in either particle physics or cosmology, the failure of its theoretical conceptions to account for nature as we encounter it. In this way, it wastes the treasure of science, its enigmas.

A second, related objection is that, by using this stratagem, it inverts the relation of physical science to mathematics and elides the difference between them. If mathematics is a storehouse of ideas about the ways in which pieces of reality may connect with one another, then physics, in this account, becomes the identification of each of these mathematical connections with a physical reality. It is, we argue in this book, a practice resting on a misguided view of the relation of mathematics to science and nature.

A third objection – and the one that will be most telling to a scientist – is that at the end of the day this approach evades the work of explanation. It subsumes the unexplained laws and initial conditions under a vast framework of possible variations of nature, all but a tiny number attributed to unobservable universes and unknown states of affairs.

We develop a third approach. Its working assumption is that the more promising way to explain the regularities as well as the structure of nature, and so too the initial conditions of the one real universe, is to explain them historically. This approach proposes that cosmology complete its transformation into a historical science. It seeks empirical support in the most important findings of the cosmology of the last hundred years: those that have to do with the history of the universe and that have been codified incompletely in the now standard cosmological model. Structure results from history more than history derives from structure.

This third approach has many counterparts in the life and earth sciences as well as in the historical study of human society. However, unlike the other two approaches, it counts on few representatives in the history of modern physics and cosmology.

It fails to explain away the factitious character of the universe – that the universe just happens to be one way rather than another. However, it vastly enlarges the field of causal inquiry. As a result, it suggests an agenda of empirical research that communicates with the major discoveries that cosmology has made over the last hundred years and continues to make now.

The historical approach, as we here understand and develop it, makes use of each of the three chief claims of this book. It discards the fabrication of imaginary universes in favor of a focus on the one real universe and its history. It takes the reality of time so seriously that it refuses to exempt either the basic structure or the fundamental regularities of nature from susceptibility to change. It wants to put mathematics in its place, as an instrument of physical theory rather than as a substitute for it.

Such an approach raises daunting problems. I have already touched on two of them in this early stage of our argument.

The first problem is that if there is only one universe at a time, we must conceive the seemingly paradoxical endeavor of developing the science of a singular entity. The traditional way of avoiding this problem in cosmology is to scale up: to extend explanations developed to address pieces of the universe into ideas about the whole universe. In the cooled-down universe, with its discriminate structure, exhibiting laws and symmetries, such pieces of the universe – for example, patches of space-time – come in multiple instances conforming to the same regularities. Cosmology relies on the amalgamation of theories about local phenomena.

However, scaling up from piecemeal theories of nature to cosmological conceptions, at least insofar as it relies on a distinction between stipulated initial conditions and unchanging laws, deserves to be resisted. It is just what the argument against the first cosmological fallacy and its Newtonian paradigm – an argument developed later in this chapter – forbids. We must face, without the relief that this procedure offers, the difficulty that the universe, as a reality both unique and historical, presents to science: Aristotle's conundrum about the ineffable character of individual being.

The second problem is that if everything in this one universe, including its regularities and its structure, changes sooner or later, we cannot accept a move that has helped define the path taken by physics and cosmology at least since the formulation of special and general relativity and of quantum mechanics in the early twentieth century. The move is to combine denial of an absolute background of space and time, distinguishable from physical events, with the reaffirmation of belief in a permanent structure of ultimate constituents of nature and in an immutable framework of laws and symmetries. The rejection of these twin ideas requires us to change our view of causality and of the relation of causal connections to the laws and symmetries of nature.

THE SELECTIVE REALISM OF MATHEMATICS

The third idea central to our argument is a conception of mathematics and of its relation to nature and to science. Mathematics, according to this idea, represents a world eviscerated of time and phenomenal particularity. It is a visionary exploration of a simulacrum of the world, from which both time and phenomenal distinction have been sucked out.

Our causal explanations are steeped in time: the cause precedes the effect. If time were illusory, so would any causal nexus be an illusion. On the other hand, however, if time were real and inclusive to the point of resulting in the mutability of the laws of nature, our causal judgments would lack a stable warrant. Our conventional ideas about causation are confused; they assume that time is real, but not too real.

The relations between mathematical and logical propositions are, however, timeless: the conclusion of a syllogism is simultaneous with its premise. They are timeless, even though we reason them through in time, and use them in the analysis of events in time.

In the philosophical and scientific tradition within which the ideas of the singular existence of the universe and of the inclusive reality of time have remained decisive, mathematics has gained a power that none of the well-known positions in the philosophy of mathematics seem adequately to explain or to justify. The laws of nature appear to be written in the language of mathematics. But why and with what significance? The "unreasonable effectiveness of mathematics" remains a riddle without a convincing solution.

In the history of philosophical ideas about mathematics, two sets of conceptions have come close to exhausting approaches to the solution of this riddle. According to the first set, mathematics is discovery of an independent realm of mathematical entities and relations. According to the second set, mathematics is invention: made-up conceptual entities, manipulated according to made-up rules of inference. The problem is that neither of these approaches to mathematics seems to help explain the applicability of mathematics to the world.

We propose a different view, one that begins from the acknowledgment of the contrast between the temporal character of every causal nexus and the timeless quality of mathematical and logical relations. Mathematics is about the world, viewed under the aspect of structured wholes and bundles of relations, disembodied from the time-bound particulars that make up the actual world: effacement of particularity goes together with denial of time.

The world studied by mathematics is not quite our world, the one real world, soaked through and through in time. Neither, however, is it another world, of eternal mathematical objects, separated from ours by an unbridgeable gulf. It is a proxy for our world, a counterfeit version of it, a simulacrum, distinguished from it because in it everything is denuded of placement in time and of phenomenal particularity.

It is as if our mathematical and logical reasoning represented a Trojan Horse, placed in the mind against the recognition of the ultimate reality of time and difference. However, its selectivity – its disregard for time and particularity – is the source of its usefulness.

Instead of regarding our faculty of mathematical and logical reasoning as a way of overcoming the limits of our natural constitution, we should understand it as a part of that constitution. By enabling us to expand and recombine our ideas of how pieces of the world can connect with other pieces, independently of the particulars of any given time-bound circumstance, this faculty vastly enhances the scope of our problem-solving capabilities. It conferred an evolutionary advantage when we were simpler than we now are, and continues to confer one now that we have become more complicated.

The foundation of this advantage lies in its simplifying approach to the one real natural world, rather than in a direct access to another world of timeless and therefore unnatural objects or to an array of possible worlds that never wore the garment of reality. It is a natural faculty that has nature as its subject. However, it increases its power by virtue of its distinct approach to the particulars of this one time-bound universe of ours.

In its early stages, the relation of mathematics to the world of temporal change and of phenomenal particularity is direct: less by induction than by what Pierce called abduction – an imaginative jumping off from an open-ended series of particulars. Soon, however, the predominant relation of mathematics to nature becomes indirect. We begin to expand the range of mathematical ideas by analogy, without license or even provocation from natural experience. We go, for example, from the three-dimensional space of Euclidean geometry, with its simplification of our sensual experience, to geometries that have no counterpart in our perception. We move from the natural integers by which we count things in the world to numbers useless in counting anything we will ever directly encounter and experience with our senses.

The mathematics that we develop on the basis of this indirect relation to nature, driven by an agenda internal to mathematics itself, may or may not apply to the elucidation of natural phenomena. It may or may not be useful in the work of natural science. There is no assurance that it will be serviceable, although it often is. The ultimate source of its power is that it combines connection to nature with distance from nature.

This power perennially tempts us to succumb to two connected illusions. The first illusion is that we have in mathematics a shortcut to indubitable and eternal truth, somehow superior to the rest of our fallible knowledge. The second illusion is that, as the relations among mathematical propositions are timeless, the world itself must somehow participate in the timelessness of mathematics.

Here we offer an account of mathematics that has no truck with either of these illusions. It is a realistic and deflationary view. It claims that we cannot make adequate sense of the effectiveness of mathematics in natural science by treating mathematics either as the exploration of a separate world of timeless mathematical objects or as the free invention of ideas about number and space that turn out, mysteriously, to be applicable to nature.

We enjoy our mathematical powers for natural reasons. We develop them at first inspired by nature, eviscerated of time and particularity, and then at a distance from the original sources of our inspiration. Mathematics, however, is smaller, not greater, than nature. It achieves its force through a simplification that we can easily persuade ourselves to mistake for a revelation and a liberation.

The view of mathematics as the imagination of a counterfeit version of the world, robbed of time and phenomenal particularity, acquires its full force and meaning only when combined with the other two ideas central to our argument: that there is only one real world and that everything in this world changes sooner or later. One world. Real time. Mathematics is about the one world in real time, not about something else. Instead of trying to find what else mathematics could be about other than the world (there is nothing else), we should be concerned to understand in just what sense it can be about a world to the manifest qualities of which it is so strikingly and willfully blind.

THE FIRST COSMOLOGICAL FALLACY

There is one real universe. Time is real, and nothing lies beyond its reach. Mathematics has the one real, time-soaked world as its subject matter and inspiration. It is useful to the understanding of this world precisely because it explores the most general features of relations among pieces of the world abstracted from both time and phenomenal particularity.

These three propositions form the axis of the argument of this book. To recognize and to develop the truth that they express, we must reject two fallacies. Each of these fallacies enjoys widespread influence within and outside physics and cosmology. They are closely connected.

Taken together, they summarize much of what is misguided in our received understanding of the discoveries of science.

Call them the two cosmological fallacies. Both of them mistake a part for the whole. They make different but connected mistakes. The first fallacy applies to the whole of the universe methods and ideas that can be successful only when applied to part of it. It is a fallacy of false universality: it treats the whole universe as if the whole were one more part. The second fallacy embraces a view of nature and of its laws that is inspired by the forms that nature takes during part of the history of the universe. It is a fallacy of universal anachronism: it applies to the whole history of the universe ideas that are pertinent only to part of that history. Its view of the workings of the natural world is too parochial to do justice to the metamorphoses of nature.

* * *

The first cosmological fallacy – a fallacy of false universality – applies to the whole of the universe, and therefore to the central problems of cosmology, what we here call the Newtonian paradigm. The Newtonian paradigm is the chief method of explanation that physics and cosmology have deployed since the time of Galileo and Newton. Relativity and quantum mechanisms have not disturbed its ascendancy however much they may have modified its application.

Under the Newtonian paradigm, we construct a configuration space within which the movements and changes of a certain range of phenomena can be explained by unchanging laws. The range of experience defined by the configuration space and explained by the laws can in principle be reproduced, either by being found in another part of the universe or by being deliberately copied by the scientist. The recurrence of the same movements and the same changes under the same conditions, or the same provocations, confirms the validity of the laws.

The configuration space within which changeless laws apply to changing phenomena is marked out by initial conditions. These conditions are the factual stipulations defining the background to the phenomena explained by the laws. The stipulations mark out the configuration space: the space within which laws apply to the explained phenomena. By definition, they are not themselves explained by the laws that explain movements and changes within the configuration space. They are assumed rather than explained.

However, that they perform in a particular part of science the role of unexplained stipulations rather than of explained phenomena does not mean that they cannot reappear in another chapter of scientific inquiry as subjects for explanation. In the practice of the Newtonian paradigm what is stipulation for some purpose becomes the subject matter to be explained for another. That the roles of what is to be explained and what does the explaining can in this way be reversed ensures that we can hope to explain all of the universe, part by part.

The observer stands, both in principle and in fact, outside the configuration space. Conceptually, his relation to it resembles the relation of God to the world, in the Semitic monotheisms – Judaism, Christianity, and Islam: not as creator but as observer. He looks upon it, to use an astronomical metaphor, from the vantage point of the stars. The laws go together with this ideal observer. They govern what happens inside the configuration space. They have, however, no history of their own within that space – or anywhere else.

The laws determine changes or movements within the configuration space. Thus, they can be used to explain events in time. To explain changes of the phenomena, it is first necessary to represent them. The most familiar way in which to do so is to plot them as movements along an axis. Time is converted into space.

The laws are timeless. They have no history. They underlie and justify our causal explanations. They are, however, themselves without explanation. To ask why they are what they are is to pose a question that lies in principle beyond the limits of a natural science conforming to the Newtonian paradigm.

Those whose ideas about the practice of science have been formed in this mold may hope to find in mathematics the beginnings of insight into why the laws are what they are. This conjecture remains, however, no more than a metaphysical speculation with limited practical significance for the conduct of science under the guidance of the Newtonian paradigm.

The first cosmological fallacy consists in the application of this way of doing science to the universe as a whole, which is to say to the problems that are distinctive to cosmology. When the topic is the whole of the universe and its history, rather than a part of the universe, the distinction between law-governed phenomena within a configuration space and the stipulated factual conditions defining that space ceases to make sense. There is no place outside the configuration space for anything else to be; that space has become the entire universe. It is no longer thinkable, even in principle, to prepare or even to discover copies for what we are to explain, now the entire universe, so that we can test the constant validity of the laws.

Deaf to Newton's warning not to feign hypotheses, we may appeal to the idea of multiple, parallel universes in an effort to rescue the cosmological uses of the Newtonian paradigm. If, however, these other universes are, as they must be, causally unconnected with our own, and no light-borne information can travel from them to us, this conjecture will amount to no more than a vain metaphysical fantasy, disguised as science.

The process by which what is the factitious stipulation of an initial condition in one local explanation becomes an explained phenomenon in another is now interrupted. In an account regarding the whole universe and its history, no occasion arises for such a reversal of roles. Thus no hope can be well founded that by accumulating local explanations we slowly approach an explanation of the whole.

The observer can no longer stand outside the configuration space, and claim to adopt the godlike view from the stars; all the stars, and everything around them, are dragged down into the field of explanation. If the laws of nature are somehow exempt from the violent changes that nature undergoes, they must exist on some other plane of reality, in the company of mathematics, as it is understood by mathematical Platonists.

Thus, every feature of the Newtonian paradigm fails when its subject matter ceases to be a region of the universe and becomes the entire universe. The denial of this failure is the substance of the first cosmological fallacy. It results in a series of equivocations that corrupt the practice of scientific inquiry and prevent cosmology from remaining faithful to its vocation to be a master science rather than a sideshow.

A major stratagem by which to dismiss or diminish the implications of seeing the first cosmological fallacy for what it is consists in treating the problems of cosmology as peripheral to the agenda of physics in particular and of natural science in general. The parts of the universe, however, are parts of *the* universe. Our view of the universe and its history has implications for our understanding of all of its parts. If, for example, there is a succession rather than a plurality of universes and if the causal connection between successive universes, although stressed, is never broken, many features of our world may have their origin and explanation in the traits of the very early universe or of universes that preceded them.

An influential variation on the strategy of marginalizing cosmology the better to suppress the embarrassments it creates for established scientific ideas and practices is to represent the earliest moments of the universe as characterized by infinite degrees of temperature and energy. That is precisely what marks a singularity in the strict and conventional sense. Once we cross the threshold of the infinite in the representation of nature (rather than just in the exercise of the mathematical imagination), we can no longer make use of any of the explanatory practices, including the Newtonian paradigm, that we are accustomed to apply to the world of nature that we know. Thus, under the view that the present universe began in a singularity the parameters of which are infinite, rather than in a violent event of extreme but finite parameters, we can attribute to the enigmas of the infinite what are in fact confusions and contradictions resulting from the illegitimate universalization of local explanations. It is as if the jump from the finite to the infinite provided a generic license for ideas that, in the absence of such license, would readily be dismissed as untenable.

THE SECOND COSMOLOGICAL FALLACY

The second cosmological fallacy – a fallacy of universal anachronism – sees the entire history of the universe from the standpoint of ideas that may be pertinent to only part of that history. It incorporates into our practices of scientific explanation a view of the workings of nature that accounts for those workings only in certain states of nature but not in others. The substance of the second cosmological fallacy is to treat the form that nature takes in the differentiated, cooled-down universe as its sole and permanent form. This model of the workings of nature can then be read backward as well as forward, to earlier and later moments of the history of the present universe, as its one and only mode; hence the mark of universal anachronism.

It is a cosmological fallacy because it can arise only within cosmology and as a result of its most significant discovery: the discovery that the universe has a history. The second cosmological fallacy is thus no mere methodological misstep. It amounts to a misreading of the facts of the matter. It concerns the most important contribution that cosmology has made to our understanding of the world.

The import of the second cosmological fallacy is that our received image of both nature and natural science is modeled on a historically parochial view of how nature works. Cosmology has long since denied us any entitlement to such parochialism. We nevertheless remain reluctant to give it up.

There is no invariant or quintessential scientific method. Our views of the practice of science develop together with the content of our scientific ideas. The discovery that the universe has a history, and so therefore must everything within it be historical, has implications for the practice of science. We have so far failed to acknowledge them.

It is not just any history. It is a particular history. We already know enough about it to begin to form the idea that nature can exist in different states or wear different masks. Our prevailing conception and practice of scientific explanation take only one of these states for granted, and identify that state with the necessary and universal workings of nature. In so doing, however, they fail to take adequate account of what cosmology has already discovered to be the facts of the matter – at least if we interpret its findings undistracted by metaphysical prejudice.

Yet on any of the accounts of the origins of the present universe that now command authority, we have reason (although we have no direct evidence) to think that the workings and characteristics of nature were once very different from what they have since become. These views can be broadly grouped into two main families of ideas.

One family of ideas, predominant to this day, traces the origins of the observed universe to a singularity in the now conventional sense: an original state in which the energy density of nature reached infinite value. Another family of ideas, which the argument of this book accepts, follows the history of the present universe to an original state in which the energy density of nature was extreme but nevertheless finite. In this second family of ideas, the conjecture of an original state of extreme energy density is readily married to the further conjecture of a succession of universes.

What is striking is that on either of these two sets of conceptions, we have reason to suppose that the familiar divisions within the mature and evolving universe – the structural distinctions and rudimentary components of nature described at one level by the standard model of contemporary particle physics and at another level by the periodic table – may once not have existed. (Of course, the chemical description of nature, as summarized in Mendeleev's periodic table, is not fundamental. It is nevertheless connected through many intermediate links, such as the Dirac equation and the Pauli exclusion principle, to the fundamental description offered by particle physics. The idea of a permanent differentiated structure of natural phenomena is central to the dominant tradition of modern science. Darwinism and, more broadly, the earth and life sciences have barely made a dent in the ascendancy of this vision. The idea of an ahistorical differentiated structure does not live exclusively in the forms of science that explore fundamental levels of reality – particle physics first among them – but also in the sciences – chemistry, for example – that address nature at less fundamental levels. Unless we are to subscribe to a radical reductionism, incompatible with the way in which the physical sciences have developed, there is no reason to disregard the less fundamental descriptions and to focus solely on the more fundamental ones. The notion of a timeless structure must be contested and overthrown at all levels: the less fundamental ones as well as the more fundamental ones. In the meantime, chemistry, like particle physics, continues to be a structural science rather than a historical one.)

It is not simply, on this line of reasoning, that other structural distinctions and rudimentary components marked nature in the very early universe. It is that the presentation of nature as a differentiated structure, and its working as an interaction of clearly distinct forces or fields, may then have failed to obtain. Such distinctions and interactions could not have existed under the conditions of the very early universe. If they existed at all in the circumstance of the original extremes, they would have had to have been radically different, and to have worked in a radically different way, from how they later came to be and to work. A premise of much established thinking in cosmology and physics is, nevertheless, that nature works always and everywhere as a structure of distinct parts (particles, fields, forces) interacting with one another in conformity to unchanging laws.

The criticism of the second cosmological fallacy has as its aim to explore this contradiction within our present beliefs about the history of the universe and to consider its implications for the practice of science as well as for the content of some of our most comprehensive scientific theories. The whole argument represents a natural-philosophical reflection on what it would mean to take altogether seriously the idea that has been central to cosmology ever since Lemaître's conjecture about the origins of the universe gained widespread acceptance.

In this reflection, I resort to a heuristic device. I imagine two states of nature and say nothing about the transition from one to the other. This contrast, in the terms in which I sketch it, far exceeds the authority of the evidence. Moreover, it is couched in terms that could not figure among the formulations of a developed scientific theory. Nevertheless, it serves a legitimate analytic purpose: the aim of exposing the logic of the idea that nothing in nature lasts forever. In particular, it makes this logic explicit in the context of the second family of beliefs to which I have just referred: those that presuppose extreme but not infinite values of the earliest states of affairs in the history of the universe. The strategy of the heuristic argument is to contrast only two states of nature and to suggest nothing about the transition from one to the other.

The stark simplifications and the metaphorical language to which I here appeal in no way undermine the usefulness of the argumentative device. The core point is that the research agenda and the way of thinking inspired by Lemaître's conjecture fail to be fully achieved if we content ourselves with the idea that the early universe had a different structure. The implication of the conception of the original state is that it had no structure at all, in the familiar sense of the concept of structure to which the scientific study of the mature universe has accustomed us. Because it had no such structure, it must also be supposed to have worked in a different way.

Moreover, the significance of the device is not limited to finitistic views of the original state: accounts of the original state of the universe that restrict all parameters to finite values. It is pertinent as well, albeit in a different way, to views that invoke a finitude-defying singularity. For, according to such views, there must also have been a moment when the distinctions and interactions of the mature universe did not yet exist. There must have been a transition and a transformation leading from the universe then to the universe later. Indeed, the transition and the transformation must have been all the more far-reaching if they accommodated, as they must have for such conceptions to make sense, a passage from the infinite to the finite.

In one state, nature appears and works as it does in the formed, cooled-down universe: the universe that we observe. Nature is divided up into discontinuous elementary components, the most basic of which are the particles, fields, and interactions studied by particle physics. More generally, nature is constituted in this state by kinds of things or natural kinds: a fact that inspires the projects of classical ontology as well as of natural science. It is in this way that nature was seen in the tradition of Aristotle. It is likewise in this way that nature continues to be represented in the tradition that began with Galileo and Newton and continues to today.

Natural phenomena present themselves, according to this conception, within a limited range of parameters of energy and temperature. They display only modest degrees of freedom. The penumbra of the adjacent possible around each phenomenon – what it can become next, given what it is then – remains restricted or thin. The laws of nature – both the effective laws operating in particular domains, and the fundamental laws or principles cutting across domains – are clearly distinct from the phenomena that they govern. It is only a short step from these conceptions to the idea that changing states of affairs are governed by unchanging laws.

Nature, however, to follow the logic of this heuristic device, admittedly beyond the boundaries of the evidence before us but not contradictory to any of it, may also appear in another mode. It may have existed in this other way in the very early history of the present universe as well as at the beginnings or at the ends of other universes, if our universe was preceded by earlier ones. Nature may so appear again in its very late history. It may also from time to time present thus in particular regions, subject to extreme conditions. These local realities would then depart from the model of the workings of nature established in the cooled-down universe.

In this second state of the universe (the first, however, in the order of time), the structural distinctions among elementary constituents of nature have broken down or not yet taken shape. The parameters of temperature and energy are extreme but they are not infinite (as they are under the standard concept of a cosmological singularity). Consequently, no insuperable obstacle of principle exists to investigating and explaining them; it is not true that nature is open to our understanding only in its first state but not in its second.

Much higher degrees of freedom are excited than we observe in the cooled-down universe, and the penumbra of the adjacent possible around each phenomenon now becomes thick and rich. It does so whether we account for this wealth of transformative opportunity in the language of either causal or statistical determination. The laws – at least the effective laws applicable to particular domains – cease to be readily distinguishable from the states of affairs that they govern. If the phenomena change, the laws change coevally with them. This last characteristic of the second state of nature is intimately related to all the other traits: to the absence of clear and stable structural divisions (and thus of distinct domains to which different sets of effective laws would apply); to the extreme though finite physical parameters; and to the enhanced degrees of freedom enjoyed by the phenomena – the range of other phenomena that the existing phenomena can become and the facility with which they can turn into them.

The second cosmological fallacy is the disposition to take account of only the first state of nature while disregarding the second, and to do so in our methods as well as in our theories. When we succumb to this fallacy, our conception of how to practice science, as well as our view of the workings of nature, allows itself to be shaped by an intellectual engagement with only one set of the variations of nature. It becomes in a sense the science of a special case. It consequently remains limited in the reach of its insight even into that special case. The deepest enigmas of nature escape it.

It is not just the Newtonian paradigm that takes this path. It is an entire approach to science that has been shaped by the assumption that the first of these two states of nature (the second in the order of time) represents the ultimate and constant character of reality. In developing and supporting the idea that the universe has a history, cosmology, however, has already given us grounds to reject this assumption as false. On one interpretation of its findings (for which we argue in this book), everything is emergent – everything comes and goes – except time.

The emergence of everything except time is one of the ways in which the first state of nature ceases to represent the essential and enduring character of reality. It is not the only way. Every version of the now standard account of the origins of the present universe suggests that nature at the earliest moments in the formation of the present, observed universe may have displayed traits very different from those that it later came to exhibit as it cooled down and assumed the structured form in which we now observe it.

It is simply that under many of the most influential cosmological theories – those that appeal to the idea of an initial singularity – the alternative traits of nature remain hidden under the veil of the infinite. The state that these theories purport to describe is one in which the parameters of the phenomena had infinite values. To ascribe infinite values to them is to place them effectively beyond the reach of inquiry and understanding: the ultimate secrets of universal history would remain sealed behind a door that we could never open. The result would be – indeed, it has been – to allow us to treat the variations and workings of nature as we encounter them this side of that door as if they were its permanent traits. It would also be to regard the practice of science that relies on this assumption as what science must always be.

If nature wears multiple disguises – the states through which it passes – a science that presupposes a stable structure of ultimate constituents of nature – the structure represented at one scale by the standard model of particle physics and at another scale by the periodic table – and a framework of immutable laws or symmetries clearly distinct from the phenomena that they govern cannot be more than the science of a special case, even if a special case of broad and enduring application. Such a science – the science that we in fact have – will be bereft of the cosmological equivalent of the physics of phase transitions: an account of the transitions from one state of nature to another.

Unlike the physics of phase transitions, such an account is universal rather than local. Unlike the physics of phase transitions, it requires a style of scientific explanation that dispenses with both the idea of a framework of immutable laws of nature and the picture of nature as a differentiated structure, made up of distinct elementary constituents – forces, fields, and particles – interacting with one another in conformity to such laws.

* * *

The two cosmological fallacies are closely connected. They reinforce each other. They make each other seem to be unavoidable conceptions – indispensable to the practice of scientific inquiry – rather than the contestable options that they in fact are.

The second cosmological fallacy limits our understanding of the variations of nature. In so doing, it makes the cosmological use of the Newtonian paradigm seem less troubling than it would otherwise be. It fails to solve the problem of the breakdown, in a cosmological context, of any distinction between initial conditions and a local configuration space of law-governed phenomena. Similarly, it does nothing to show how we can be justified in using the Newtonian paradigm in a setting in which we have no hope of observing or preparing copies of the explained phenomena. Nevertheless, the second cosmological fallacy represents nature as working always and everywhere in the way in which the Newtonian paradigm supposes it to work: by the conformity of distinct elements or phenomena, within a differentiated structure, to changeless laws.

The first cosmological fallacy presupposes a view of the workings of nature that makes any other conception of how nature works seem to be incompatible with the requirements of science. All the better then if nature can provide us with an excuse for the limitations of our insight by taking refuge in an exceptional condition that, because it has infinite parameters, is forever barred to investigation and understanding. It is for this reason that the conventional idea of the cosmological singularity helps makes the universalization of the Newtonian paradigm seem legitimate. By associating the finite with the workings of nature in the cooled-down state of the universe and any other variant of nature with the impenetrable infinite, it lends appeal to the second cosmological fallacy.

Despite their reciprocal connections, the two cosmological fallacies have different characters and consequences. The second is more fundamental, and more far-reaching in its implications, than the first.

The first cosmological fallacy commits a mistake of method, with empirical assumptions and implications. The second cosmological fallacy amounts to a mistake about the facts of the matter, with wide consequence for the practice of science. The matter that it mistakes is the most important in science: the nature and history of the universe.

The argument against the first cosmological fallacy ends in a negative claim: the claim that we are not entitled to apply to the whole world the methods and habits of mind that modern science has applied to parts of the world. This negative claim in turn evokes the need for a way of thinking different from the one that the Newtonian practice expresses.

The argument against the second cosmological fallacy results in a positive claim: the claim that there is already more – implied if not shown – in what science has discovered about the universe than our established natural philosophy – the lens under which we read these discoveries – is willing to countenance. It suggests that this something more is baffling but in principle not inscrutable and that our understandings have not yet caught up to our findings. It inspires the need for a practice of science that can persevere in the endeavor of scientific inquiry even when the two features of nature that have seemed most indispensable to science are missing: the presence of distinct and constant elements or types and their interaction according to law-like regularities.

The arguments against the two cosmological fallacies require us to think historically about nature and its laws. As a result, they force us to confront what we here call the conundrum of the meta-laws. If the laws of nature have a history inseparable from the history of nature, it seems unacceptable to say either that their history is itself law-governed or that it is not. If the history of the laws of nature is law-governed, we seem to have rescued part of the standard view of science only by equivocating about the reality of time and by separating the content of the laws from the vicissitudes of the phenomena. If their history is not law-governed, it appears to lack an explanation, in violation of the principle of sufficient reason. Moreover, our causal explanations, relying as they do on the picture of a law-governed world, are rendered insecure. They will remain insecure until we change our understanding of the relation between causal connections and laws of nature.

The meta-laws conundrum is central to the agenda of cosmology. The solution to this conundrum bears on the meaning of every proposition within natural science. Cosmology is not an afterthought to physics. It is the part of natural science that has the most general implications for all the other parts.

CAUSALITY WITHOUT LAWS

The three central claims of this book (about the world, time, and mathematics) and the argument against the two cosmological fallacies cannot be advanced without revising our view of causality and of its relation to laws of nature.

The approach to causation that has been predominant for several centuries rests on two pillars. The first pillar is the notion of causal links as mental constructs rather than as real connections in nature. The second pillar is the principle that causal explanations presuppose laws of nature: the laws serve as the warrants justifying causal explanations. We cannot have the latter without invoking the former.

That we should understand causation as a device of the mind – a requirement of the way in which we cope with the world and seek to understand it – rather than as a description of the workings of nature has been the prevailing view in philosophy since Hume and Kant. According to this view, causality is an indispensable habit of the mind, a requirement of our efforts to make sense of reality, an unavoidable simplification, a proxy for ultimate truths about nature that are forever denied us. So long as the inquiries and actions that we undertake under the aegis of the idea of causality produce acceptable results, either as theory or as practice, we have no reason to rebel.

One of the many benefits that this view of causation as mental construct renders to the ruling ideas about nature and science is to disguise or muffle the disharmony between causal connections among parts of nature and relationships among mathematical propositions. Nature, it is believed, works according to laws that are written in the language of mathematics. But how can there be such a comprehensive consonance between nature and mathematics if causal relations imply time (as effects succeed their causes) whereas mathematical relations are timeless (as the conclusions of a mathematical inference are conceptually simultaneous with its premises or points of departure or, rather, have nothing to do with the passage of time)? By treating causality as a necessary projection of the mind onto the workings of nature, which we would otherwise be unable to decipher and which we can grasp only under the constraints of human understanding, we make the paradox of the application of the timeless to the time-bound seem less troubling.

That causal explanations depend on an appeal, however tacit, to laws as well as to symmetries and constants of nature is a proposition that may seem all but self-evident. If causality has a clear and constant meaning, its proper usage appears to imply an appeal to regularities of nature. These regularities are laws, symmetries, and constants. However, it is laws, rather than symmetries and constants, that are easier to enlist, and have been most commonly enlisted, in the effort to explain why or how the same effects follow in similar circumstances from the same causes. Under this view, the laws of nature not only account for recurrent causal connections, they also establish which circumstances count as similar.

Causality without laws would seem to be a senseless notion: what would make the effect follow the cause? Without laws, relations of cause and effect would, according to this widespread conception, be arbitrary – mere coincidences – or express something different from what they seem to reveal. For example, they might describe relationships of reciprocal implication, better represented in the language of mathematics than in the vocabulary of cause and effect.

The view of causality as mental construct is not, strictly speaking, inseparable from the thesis that causal explanations presuppose laws of nature. Nevertheless the two ideas reinforce each other. Each makes the other look yet more natural. If causality represents an enabling condition of our ability to reason about reality, then we can easily extend this supposedly indispensable syntax of concepts to include the partnership between causal accounts and law-like explanations. If causal explanations rely, implicitly or explicitly, on an appeal to regularities, especially laws of nature, then we can have more confidence that whatever the limits on our power to grasp "things in themselves" may be, we can at least bring order and clarity to our practices of inquiry, and hope to distinguish justified from unjustified beliefs about the workings of nature.

We do better to destroy these twin bases of the modern view of causation and of laws, and to think in another way. A different conception fits better with the ideas of the singular existence of the universe, of the inclusive reality of time, and of the selective realism of mathematics that we here develop and defend. It also conforms more closely to the empirical and experimental spirit of science. Causal connections, according to this alternative view, form a real feature of nature. They are not just an indispensable invention or projection of the mind.

Because they are real features of nature, they can take as many different forms as nature takes in the course of the history of the universe. Whether causal connections are always law-like is not a matter that we can determine by investigating the logic of our conceptual categories or the implications of our habits of mind. It is something that we can clarify only by finding out how nature in fact works, not universally and once and for all but rather variably, over time. It depends on facts of the matter about nature, not just on facts of the matter about human understanding.

If change changes, if the forms of connection and transformation evolve in the course of the history of the universe together with the states of affairs, then the real causal connections that bind nature together and that we describe in our theories may also undergo transformation.

* * *

In the most rudimentary sense, a causal connection is the influence that a state of affairs exercises over what follows it. The key presupposition of causality is therefore not the recurrence of the same connections: their law-like form. It is time. If time is not real, causality, understood in this way, cannot be real. It must be assimilated, or reduced, to something else: for example, to relationships of reciprocal implication, such as those that mathematics and logic represent.

Causal relations usually connect recurrent phenomena. Such will ordinarily be the case in what I earlier called the first state of the universe (which is the second state in order of time): the state in which a fixed structure of distinct elements of nature (as described by particle physics and by the periodic table) has taken shape, the laws or regularities of nature can be clearly distinguished from the phenomena that they govern, and there are tight constraints on the change from one state of affairs into another.

However, it may also happen that phenomena have not yet become, or no longer are, recurrent, if only because no structure of distinct elements or parts of nature has been established or maintained, the laws of nature are not yet, or have ceased to be, distinguishable from states of affairs, and the range of transformative opportunity – for the change of some states of affairs into others – remains ample. In such a circumstance – what I called earlier the second state of nature (but the first in the order of time, as in the early history of the present universe) – there can be causality without laws.

Causality will then continue to describe real relations in the one real, time-haunted world. However, there will not then be the element of recurrence or repetition enabling us, in similar circumstances, to attribute the same effects to the same causes. In this sense, the world will then be lawless.

The character of causation in each of these variants of nature is not a subject separate from scientific inquiry; it forms part of that inquiry. A theory in physics, cosmology, or any other branch of science is, among other things, an account of the real workings and changes of nature. We may, however, seek to develop a view of the similarities and differences among such causal connections: of what they are and of how they change.

Such a view will belong as much to natural science as to natural philosophy, and serve as an example of the porousness of the boundaries between them. Recognition of the real, rather than ideal, character of causal connections makes it possible to affirm their mutability and variety.

Under this conception, the character of causality cannot be uniform for the reason that everything in nature changes over time, including the forms of connection and of change. However, although change changes, it changes on the basis of what it was before. One state of affairs influences the next one. One way in which a state of affairs shapes its sequel influences a subsequent way in which it exerts this power over its aftermath. We should thus expect that despite the absence of a single form and meaning of causation there will be a substantial overlap among the forms and meanings of causal connection over time, the time of the history of an evolving universe. The common thread will be influence upon succession: causation is always about how every state of affairs in nature influences the states of affairs that succeed it in time.

* * *

"In time" is the decisive qualification: a universe in which causal connections form no part of nature (because they are mere constructions of the human intellect) is one in which time plays a secondary or epiphenomenal role. In such a universe there may be time-reversible laws of nature, as in Newton's mechanics. Reversibility of the laws diminishes the reality of time. Or there may be a timeless relational grid, as in Leibniz. The existence of a grid of that kind solves the contradiction between time-bound causality and time-denying logical or mathematical implication by reducing the former to the latter. Or the view may be offered that an appearance of causal succession merely disguises the workings of some other providential force coordinating events in nature and producing the false impression of causal connection. Such was the doctrine of the occasionalists, like Malebranche. These positions in the natural philosophy of the seventeenth and eighteenth centuries may seem more or less quaint until we realize that they remain alive in other less evident and all the more dangerous contemporary counterparts.

The preceding contrasts show that the reality of causal connection is closely or internally related to the reality of time. This relation has at least three aspects. First and most fundamentally, causation takes place in time and implies the reality of time. Second, time would not be inclusively real if causal connections simply enacted timeless laws of nature. From the idea of causal connection as such an enactment, it is only one step to the notion of time-reversible laws of nature (as in Newtonian mechanics). Third, the variety and mutability of causal connections – properties that they can meaningfully possess only if they are realities of nature rather than simply constructions of the mind – help us better understand what is implied in the claim that time is real.

That everything in nature can change – the kinds of things that there are as well as the ways in which they change – means that nothing stands outside time. It also modifies our understanding of what time is: part of what is at stake in the thesis of the reality and inclusiveness of time is that no absolute framework, whether of space or of laws or of mathematical truths and relations, envelops time. It is time, on this view, that envelops everything else. It is the only feature of nature that enjoys absolutely the attribute of non-emergence.

On this account, the long-held conventional view of the relation between causal connections and laws of nature is turned upside down. It is the causal connections, not the laws of nature, that are primitive and fundamental, though also time-bound, diverse, and mutable. By the laws of nature, we designate a feature that causal connections sometimes fail to possess: that they recur because they bind together recurrent phenomena.

In the mature, cooled-down universe, most natural phenomena possess this feature. Suppose, however, that we take the long, cosmological view, especially when we prefer the idea of a succession of universes, or of states or phases of the universe, to the idea of a plurality of universes, and reject the notion that the universe began in an infinite initial singularity. The way is then open to think that causal connections may at times have failed to work as recurrent connections among recurrent phenomena. They may have failed to exhibit the feature of recurrence in the early universe: the universe before (or after) a discriminate structure emerged and laws became distinguishable from states of affairs. They may again fail to exhibit that feature later on, in extreme states of nature during the evolution of the cooled-down universe.

In this conception, the laws, like the bonds of causality, represent real features of the workings of nature. They are no mere heuristic devices. Theirs, however, is a derivative reality by contrast to the primitive and fundamental reality of causal connections. The invocation of laws describes a special case – the standard case in the mature universe. By using the vocabulary of laws we allude, as if by shorthand, to defining features of this standard case: regularity in the ties among repetitious phenomena. It is, in more senses than one, the inverse of the now conventional account of the relation between causal connections and laws of nature. In that account, it is the causal connections that are derivative from the laws of nature, and affirmed only for the convenience of human understanding. If my argument is correct, we should invert this line of reasoning.

This inverted view has implications for the conundrum of the meta-laws: the problem of how to think about change of the laws of nature, given that either of two apparent solutions to this problem seems unacceptable. One of these solutions appeals to the idea of higher-order laws governing change of the laws. It triggers an infinite regress and circumscribes, unjustifiably, the inclusive reality of time. The other solution dispenses with the idea of higher-order laws. It makes the change of the laws seem to be uncaused, if indeed causation presupposes the operation of laws.

In discussing the conundrum of the meta-laws I suggested a response to the conundrum: the co-evolution of the laws and of the phenomena, an idea familiar in the life sciences as well as in social and historical study. This idea, however, remains incomplete and unnecessarily baffling if not complemented by the idea of the primitive reality of causal connections.

The idea of the co-evolution of laws and phenomena makes sense if, and only if, causal connections are real in nature. Because they are real, and imply time, indeed in a sense embody time, they can change over time. If causal connections were only mental constructs, to say that they change would be indistinguishable from saying that our ideas about them change. We would have no basis on which, and even no vocabulary with which, to distinguish change in theories about causal connections from change in such connections.

The idea of the reality of causal connections remains unfinished and enigmatic so long as it fails to be developed into a view of how, in the course of the history of the universe, causation acquires a law-like form. Such a view leads into an account of how the laws may change as the phenomena and their connections change.

Thus, it is a mistake to regard the idea of the co-evolution of laws and phenomena and the idea of the real and primitive character of causal connections as two separate conceptions, much less as rival ones. Rather they represent two aspects of the same approach. Together, they suggest the beginning of a solution to the conundrum of the meta-laws. They bring greater clarity and support to the central theses of this book: that there is only one real universe, that time is real and inclusive, and that mathematics gains its power by exploring a counterfeit version of the world, bereft of time and particularity. These ideas do their work at the cost of attacking the foundations on which much of our thinking about causes and laws has wrongly come to rest.

That the laws of nature supervene on causal connections, which are primitive in nature, is a view diametrically opposed to the conventional conception, according to which causal connections are mere instances of the laws of nature.* Causal connections regularly assume law-like form in the observed, cooled-down universe. There may, however, be states of natural reality in which they exhibit no such form. Such states (by inference from current standard cosmological ideas) may have played a central role in the formative moments of the present universe as well as in extreme conditions (such as those that prevail in the interior of black holes) occurring in its subsequent history.

That causality can exist without laws is a proposition that may seem paradoxical to the point of absurdity when entertained in the context of physics. Yet it has become a commonplace, though an inadequately explained one, in the life and earth sciences as well as in the study of society and history.

In Chapter 2, I discuss how this problem has been expressed in the history of social theory and of social science. Those who insist on the vital influence of formative institutional and ideological structures in society and of structural discontinuity in history are, for the most part, no longer able to believe in laws of historical change, driving forward the succession of such structures. They have, for example, largely abandoned explanatory practices, like the one Karl Marx embraced, that represent history as a law-like progression of indivisible

* It is also to be distinguished from views holding that the empirical discoveries of science are best understood without any reliance whatsoever on the idea of laws of nature in any state of the universe. See, notably, Bas C. van Fraassen, *Laws and Symmetry*, 1989, proposing that symmetries rather than laws deserve to be placed at the center of our understanding of scientific inquiry. In this argument I take invariant symmetries, just as I take laws of nature, to be a mode of causation rather than its basis. They characterize the workings of nature over much of the history of the universe. They need not characterize these workings always and everywhere. I focus on the relation of causes to laws rather than to symmetries because of the central role that the idea of timeless laws of nature has played in the development of the traditions that we here oppose. Regularities in the workings of nature are laws, symmetries, or constants. A symmetry may be defined informally as a transformation that leaves all relevant structure intact. Relevance is determined with respect to a theoretically chosen and interpreted context. The concept of symmetry is intimately related to the idea of invariance. Thus, symmetries, if invariance constitutes part of their nature, impose a restraint on the inclusive reality of time, as do immutable laws.

institutional regimes: the modes of production in his social theory. The task then becomes to do justice to causal influence and constraint in the succession of such regimes without appealing to unbelievable laws of history. This problem is analogous, in some ways but not in others, to the conundrum of causality without laws in cosmology and physics.

Change changes. That it changes is much of what the thesis of the inclusive reality of time means. The transformation of transformation implies that the laws of nature are in principle mutable. It also implies that the way in which a prior state of affairs can influence a later state of affairs, when causality exists without laws, can also change.

Causation works with what exists at any given time, including the established forms of change. It does not work by selecting from a range of states of affairs marked as possible according to the criterion of some abstraction from nature, such as the criterion of the varieties of phenomenal connection that we are able to represent mathematically. Nor does it operate by returning to some no longer existing form of connection, unless that prior form of connection retains a vestigial presence in the universe that now exists; otherwise, the recurrence would represent the temporal equivalent to action at a distance.

Wherever, as in most of the observed universe, there exists a differentiated structure, a clear distinction between states of affairs and laws of nature, and tight constraints on what can happen next, the change of change will be rare. It will take the form of the appearance of emergent phenomena, with new properties, displaying new regularities, or governed by new laws. Such is the case with the phenomena studied by the earth and life sciences, and then again with those realities that we address when we try, through the study of mind, society, and history, to understand ourselves. Complexity may expand the range of the adjacent possible – of the *theres* that nature can reach from any given *here*. In so doing, it creates a basis for emergent phenomena, exhibiting novel regularities.

Suppose nature can also exist in another form, the second state evoked in my discussion of the second cosmological fallacy, in which there is no differentiated structure and no clear contrast between laws and phenomena, in which many degrees of freedom are excited, and in which there persists ample transformative opportunity. In such a state, the restraints on the change of change will be weakened. Degrees of freedom, the adjacent possible, and emergent phenomena and properties will no longer be concepts that can be clearly distinguished when applied to such a presentation of nature.

We are accustomed, by the dominant tradition of physics, established as the supreme model of successful science, to regard historical explanation as ancillary to structural explanation. On the view that we here defend, this hierarchy must be reversed: structure results from history. Historical explanation is, thus, more fundamental than structural explanation. Cosmology affirms its ambition to be the most comprehensive natural science when it understands itself as a historical science first, and as a structural science only second.

The primacy of historical over structural explanation should give no offense to science, so long as we qualify the demand for causal explanation of everything in two ways (neither of which would be acceptable to those who espouse the metaphysical rationalism of Leibniz's principle of sufficient reason). The first qualification is that we allow a historical explanation to count as a causal account in cosmology and physics as in other branches of sciences, indeed as the most characteristic form of causal explanation when the subject matter of science becomes the whole universe. Under a historical view, a state of affairs is the way it is because of the influence of an earlier state of affairs, not because it conforms to timeless and invariant regularities. We shall not always be able to account for the influence of the earlier on the later by invoking such regularities. The second qualification is that we be willing to pay the price of a practice of historical explanation that is not subordinate to structural explanation.

This price has, in turn, two parts. The first part is that there is no absolute beginning. Time, we argue in this book, is not emergent. At any given moment in the history of science, our ability to draw inferences, supported by observation, is limited. Moreover, even if it were unlimited, we could not peer into the beginning of time; on this account, time has no beginning. Thus, historical explanation is by its nature incomplete.

The second part of the price is that change in how change occurs, as described by a historical science, has an ineradicable matter-of-fact-ness or facticity. We can increase the extent to which we are able to make sense of the transformation of transformation. At the end of the day, however, nature will always be found to have an irreducible factitious element: it is what it is. If it were not what it is, but rather the consequence of some mode of rational necessity, history would once again be subordinate to structure.

We can attenuate such just-so-ness. We cannot abolish it. Examples of how we can attenuate it are the proposals that we make later in this book for the resolution of the dilemma of the meta-laws in cosmology, conceived as a historical science: we have reason to resist accepting either that change of laws of nature is governed by higher-order laws or that it is not.

* * *

These propositions require us to believe that the workings of nature are not necessary, even though they are causally determined. There is no univocal, unambiguous notion of necessity in science. Necessity designates the limit of the least mutable realities that are represented in a given set of ideas: what is necessary is whatever, according to that way of thinking, could least be other than it is.

In the tradition of physics that began with Galileo and Newton, the content of this limiting ideal of necessity is given by the convergence of three commitments.

The first commitment is to what we call the Newtonian paradigm: the extrapolation to the whole universe of an explanatory strategy, distinguished by the contrast between initial conditions and timeless laws applying within a configuration space demarcated by stipulated initial conditions. This procedure is legitimate only when applied to parts of the universe. To repudiate its cosmological application was the aim of my argument against the first cosmological fallacy.

The second commitment is to the premise that the characteristic form of the observed, cooled-down universe, with its stable, differentiated structure, its apparent contrast between laws and phenomena, and its severe constraint on degrees of freedom, on the range of the adjacent possible, and on the facility for the appearance of emergent phenomena and properties, that is to say, of the new, is the only form of nature. To reject this temporal generalization of the form that nature takes in the cooled-down universe was the purpose of my criticism of the second cosmological fallacy.

The third commitment is to the sovereignty of mathematics over physics. On the view presupposed by that commitment, what is physically realized is what can be mathematically represented and justified. Mathematics stands to physics as both oracle and prophet, divining the ultimate nature of reality. Given the non-temporal and ahistorical character of the relations among mathematical propositions, this commitment is intimately related to the assumption of the immutability of the laws of nature and to the invariance of its symmetries, expressed as mathematical equations. One form of this ambition is to conceive the universe as isomorphic to a mathematical construction or even as a mathematical structure. Another form is to infer the laws and symmetries of nature from the most consistent and comprehensive mathematical ideas. To contest the third commitment is the goal of my discussion of mathematics in Chapter 6.

Neither any law or symmetry of nature, however fundamental it may appear to be, nor any working of causality, in the absence of such laws and symmetries, is necessary if by necessity we mean an idea of necessary realities and relations that is defined by the coexistence of these three commitments.

It does not follow, however, that the meaning of the thesis of causality without laws is to affirm the radical contingency of the way in which nature, at any given time, works. Radical contingency is a metaphysical, not a scientific, idea. Its function is to express a disappointment: that we cannot infer the way things are from the imperatives of reason (in the spirit of Leibniz's principle of sufficient reason). Its invocation betrays bad faith or confusion: a surreptitious genuflection to rationalist metaphysics by those who pride themselves on having cast off its shackles. It is an homage that has often had an ulterior religious, moral, or political motive.

The way things are is, for science, just what they are. The subordination of structure to history ensures the defeat of the rationalizing metaphysical project that the dominant tradition in physics has patiently served. It has served this project in the conviction that in so doing it would be able to wed mathematics, and serve itself. As dowry, it received from mathematics a poisoned gift: the means with which to explain temporal events by timeless laws.

Structure results from history. The combination of fundamental historical explanation with derivative structural explanation is the basis of science.

It falls to science to make sense of how and why the workings of nature are what they are. To guard against illusion, it must do so, however, without taking the why part of this endeavor as an invitation to infer natural reality from rational necessity. The universe is more neutrally described as factitious than as radically contingent: its most important attribute is that it is what it is rather than something else. It is what it is because it was what it was.

2 The context and consequences of the argument

THE ARGUMENT AND RECENT PHYSICS AND COSMOLOGY

The historical context of the argument helps clarify its intentions. Consider four such contexts: the physics and cosmology of the last few decades, the physics of the first half of the twentieth century, the rise of the life sciences and their relation to physics, and the study of human history and society. To place the argument in these multiple contexts is to understand how much is at stake in these disputes. It is also to undermine the fake authority that clings to widespread ideas about the plurality of worlds, the restricted reality of time, and the power of mathematics to serve as a privileged window on reality.

* * *

In its relentless quest for a definitive unification – a view that would bring gravity under the same theories that account for the electromagnetic, the strong, and the weak forces – much of contemporary physics and cosmology has despaired of explanations that meet the traditional and exacting standards of either deterministic or probabilistic causality. It has settled for explanations that admit a vast array of states of affairs, of which the observed states of affairs represent no more than particular variations. Rather than acknowledging such underdetermination as a limit or a failure of insight, it has tried to turn a detriment into a benefit by describing the former as the latter.

In particle physics, the chief vehicle of this operation has been string theory. The results have yet to meet the explanatory standards of classical physics. Only a few of the resulting equations, or of the specifications of their parameters, describe states of affairs that we find in the observed universe. The large preponderance of the equations, or of their admissible parameters, refer to circumstances that we have never found and may never, even in principle, find. The temptation is strong to convert the explanatory embarrassment into explanatory triumph by fiat. Such a conversion of failure into success relies on the thesis that the unobserved states of affairs allowed by the equations exist – somewhere else.

In cosmology, the move has been from the conception of an inflating universe, even an eternally inflating universe, to the idea of a multitude of unobservable universes – a multiverse. The reciprocal adjustment that I earlier mentioned applies here with a vengeance: the actual, observed world – the only one we do or ever could observe – lends some of its reality to the many other, unobserved and unobservable worlds. It becomes less real so that they can become more real. The distinctions between the mathematically conceivable and the physically possible, and then between the physically possible and the physically actual, are attenuated or even effaced.

In both particle physics and cosmology, the explanatory failures of the doctrine of multiple universes find partial relief in the appeal to the so-called anthropic principle. In its strong version, this principle seeks to explain the observed states of affairs backwards, as the uniquely selected variation on the possible or actual states of affairs from which we humans – who now observe and theorize – could emerge. One instance of special pleading is pressed into the service of another.

The outcome is a watering down of the rules and standards by which the practice of natural science has been conducted over the last few centuries. On one side, it is a dilution of the task of causal explanation: to explain why things are what they are, not just to show how they are possible, or susceptible to mathematical representation, among many other things that no one has seen or even could see. Instead of trying to show how the possible becomes actual, such a way of thinking rests content with the discovery that the actual can be brought under the aegis of ideas that are also compatible with a vast array of states of nature that no one has ever or could ever observe. Together with this inversion of the standard task of explanation goes further movement in a direction that much of physics has been taking for a longer time: the substitution of structural analysis for causal explanation: views of how things are put together instead of ideas about why they change into other things.

On the other side, weakening of the standards of natural science takes the form of a distancing from the discipline of empirical verification or falsification. It is one thing to develop a theoretical apparatus that becomes verifiable or falsifiable only at the periphery of its implications. The core ideas and presuppositions are subject to empirical challenge, albeit indirectly, through the testing of such implications. It is another thing to propose theories lacking in the power so to be challenged, through critical experiments or observations, even at such a periphery.

The twofold dilution, of determination and of empiricism, made more visible by the appeal to the strong anthropic principle, sounds as the thirteenth chime of a clock, which not only disturbs us but also makes us wonder about the previous twelve chimes. The tendencies in recent physics and cosmology that have produced this outcome demonstrate the consequences of relying on assumptions that it is the purpose of our argument to oppose: that our universe is best understood as simply one of many; that time is less real than it seems to be; that, in any event, it does not threaten the permanence either of the basic structures of nature or of its fundamental laws, which supposedly govern the crowd of universes, all but ours inaccessible to observation; and that mathematics, as the language in which such timeless laws about the multitude of worlds are written, offers privileged insight into ultimate reality.

The scientific practices and theories that have proceeded on these assumptions break with the requirements – of explanatory power and of vulnerability to empirical test – that, together with theoretical imagination, have enabled natural science to progress on its revolutionary course over the last three hundred years. Any assumptions that threaten to derail science from this course, by weakening the disciplines that chasten and guide it, deserve to be reconsidered.

THE ARGUMENT AND THE PHYSICS OF THE FIRST HALF OF THE TWENTIETH CENTURY

The physics of the first half of the twentieth century undermined the idea, characteristic of Newtonian mechanics, that we can draw a clear distinction between the physical events of the world and their setting in space and time. In the two-step movement of special relativity and general relativity, Einstein destroyed the basis for this distinction. Spacetime became part of the unfolding drama rather than just a piece of the unchanging background.

In this respect, the new physics reinstated what had always been a tradition, although a suppressed or recessive one, within the old physics: the relational view, most famously associated with Leibniz. According to this view, each event is the sum total of its relations to other events. Spacetime forms part of that relational grid; it is not its changeless seat.

The argument of this book bears on this momentous shift in the following way. In tearing down the contrast between the events and their background in space and time, the new physics nevertheless reaffirmed two other distinctions of many-sided importance, which it should now be our aim to overturn.

The first was the difference between the phenomenal world and an invariant background, not of space and time, but of universal laws and symmetries and elementary constituents of nature. (In the methodological disputes provoked by the rise of quantum mechanics, the focus fell on the respective claims of deterministic and statistical explanation; all parties to the dispute continued to accept the idea of unchanging and generally applicable laws.) The second was the distinction between initial conditions and the law-like workings of nature in a configuration space bounded by such conditions. A straightforward way to state the core thesis of our argument is that these two distinctions, reaffirmed by the revolutionaries of the early twentieth century, now deserve to be cast aside.

Consider the relation between the two contrasts that we reject. The difference, so central to established physics, between stipulated initial conditions and the law-like workings of nature within a certain region of reality – a configuration space – breaks down when we try to apply this distinction to the universe as a whole. The boundaries of the configuration space then become those of the universe.

There is no place outside this one real world from which to deliver the specified initial conditions. Nothing remains other than the universe and its history. We can no longer say, as we do when dealing with a segment of the universe or a part of reality, that what is given as brute fact from the standpoint of the laws that account for the workings of local phenomena may become a subject of explanation by other laws. Such other laws might apply when we redraw the boundaries of the configuration space. When, however, we deal with the whole universe we cannot redraw boundaries in this way. We have reached the top; we have nowhere else to go.

No sooner do we begin to subvert the distinction between initial conditions and laws applicable to particular configuration spaces, by generalizing the terrain of its application to the whole of the universe, than we are forced to question the idea of timeless laws governing a world the elementary structure of which is also timeless. Particular sciences report to us that the way in which things change also changes. For example, life on Earth introduces a new set of mechanisms of change, which in turn keep changing (as in the appearance of Mendelian mechanisms in the context of sexual reproduction). The idea of timeless laws begins to seem a convenience or an approximation that remains plausible and useful only so long as we focus on one configuration space at a time.

Cosmology is the part of physics in which we are brought up short in trying to preserve these two connected contrasts: between stipulated initial conditions and the law-like workings of nature as well as between spacetime and the timeless and universal laws of nature. The issue is whether we should think of the breakdown, within cosmology, of these two related distinctions as an exception, or a limiting situation, peculiar to that science or, on the contrary, as a revelation of truths relevant to all science and all nature. We here offer reasons to take the cosmological perspective – the view from the stars as it were – as a paradigm rather than as a peculiarity.

If our argument is correct, the overturning of the distinction between the events and their backdrop in space and time, which marked the physics of the first half of the twentieth century, should now be followed, in the first half of the twenty-first century, by the overcoming of the two related contrasts that were reaffirmed when that one was demolished. The method of distinguishing between initial conditions and law-like explanation within a configuration space should be recognized to be the less legitimate the more it is universalized. It is an expedient that depends for its legitimacy on the localism of its applications. The contrast between the physical manifold of events, space, and time and a set of unchanging laws, symmetries, and constants should be undermined. The laws of nature as well as its symmetries and apparent constants should be included in the history of the universe, rather than placed outside it, just as space and time ceased to be represented by the new physics of the early twentieth century as an independent and absolute background to natural phenomena.

This is an agenda for the future of science. It suggests, however, a reinterpretation of its past, which in turn helps shed light on what we can and should do next.

A standard view of the history of physics over the last hundred and fifty years distinguishes the main line of intellectual advance from what has been largely seen as a side line. At the source of both the putative main line and the supposed side line lies the combination of Newton's mechanics with Clerk Maxwell's electrodynamics, consistent with Newton, as well as with Maxwell's non-Newtonian conception of fields. The unification of Newton's and Maxwell's equations in the early twentieth century shaped the course on which physics has remained ever since: the quest for a view capable of unifying the theoretical treatment of the basic forces at work in nature.

According to the dominant interpretation of the history of physics, the main line is the one that goes from Newton and Maxwell to contemporary string theory by two distinct routes. One route is special relativity, followed by general relativity. The other route is quantum theory, followed by the "standard model" of particle physics as well as by the theoretical foundations of this model, including quantum chromodynamics and electro-weak theory.

On the same interpretation, the side line is the theoretical succession that goes from thermodynamics, before it was given atomic foundations, to Maxwell and to an atomically founded thermodynamics and from there to Boltzmann's kinetic theory and statistical mechanics. Remarkably, in this narrative of the history of physics over the last century and a half, otherwise so obsessed with the unification of theory, no clear link exists between the progress of the main line and the advance of the side line. The difference between the two histories is often trivialized as a distinction between the microscopic study of the ultimate constituents of nature (the main line) and the macroscopic study of aggregate phenomena (the side line).

The divergence between the main line and the side line was foreshadowed in the founding pair of Newton and Maxwell. Newton and Maxwell were reconciled. Nevertheless, what distinguished the side line from the main line sprung directly from Maxwell's discoveries and ideas.

In fact, the two lines have harbored contrasting approaches to some of the problems central to the argument of this book. In the main line, time began by being treated as part of the absolute backdrop to physical events. When it was later promoted from the scenery to a performing part, it was still assigned a role accessory to the role granted to space. General relativity, under its most influential interpretations, was more inclined to spatialize time than to temporalize space, as the geometrical metaphor of time as the "fourth dimension" suggests. The willingness to see mathematics, with its core focus on number and on space, as a vehicle of privileged access to fundamental and hidden truths about nature only reinforced the anti-temporal bias.

In the side line, however, the opposite took place. The understanding of time, real time, going all the way down and including everything, was given a real basis in the directional force of entropy and in the account of the workings of nature of which the concept of entropy formed part.

That the side line has a claim to be taken as at least as fundamental in its significance and as general in its scope of application as the theories generated in the main line was presaged by an incident in the history of physics whose significance has gone largely unrecognized. Einstein's demonstration in the arguments for special relativity of the primacy of the so-called Lorentz transformation showed that Maxwell did not deserve to be treated as the junior partner to Newton. Einstein taught that the coordinate transformation that held the Maxwell equations constant, rather than the one that preserved the Newtonian equations, was the most general and reliable transformation. Newtonian mechanics, however powerful, had to be reinterpreted as the theory of a limiting case.

A revisionist reading of the history of physics would seek inspiration for a historical way of thinking about the universe in the line that begins in thermodynamics before Maxwell, continues in thermodynamics after him, and leads to the contemporary study of cosmological difficulties such as the so-called horizon and flatness problems. It is a recessive strand in the past of physics that could become dominant in its future. In that strand, time is not accessory to space. Events are not time symmetrical. The historical character of natural reality is not an accidental or peculiar feature of certain sets of aggregate phenomena; it is an attribute of the one real universe. The analysis of microscopic structure is no substitute for the explanation of macroscopic history; on the contrary, the former can be understood only in the light of the latter. And the elusive final unification of theory is a fool's errand if we advance it only by putting ideas that analyze how the forces and phenomena of nature work in place of theories that explain how they came to be what they are.

Such is the view of the past and prospects of physics that the argument of this book suggests and from which it draws encouragement.

THE ARGUMENT AND NATURAL HISTORY

The history of our thinking about these issues has always suffered the influence of a prejudice about the hierarchy of the sciences and the exemplary practice of scientific method. According to this prejudice, physics, especially as represented by Newton's mechanics, is the supreme practice of science. Biology is a kind of weak physics: weak in the relative generality and simplicity of its law-like propositional claims. Historical and social analysis is, by the same token and in a similar sense, a kind of weak biology.

The prejudice lives in a form independent of any strict ontological reductionism: it need not require us to believe that all significant explanations at one level can be readily translated into explanations at the supposedly deeper or more fundamental level. All it demands is a view of what a scientific explanation should aspire to be at the height of its ambition.

The experience of the life and earth sciences – or, more broadly, of natural history – shows that the abandonment of the ideas that the argument of this book opposes need not compromise the practice of scientific inquiry.

Poincaré believed the idea of immutable laws of nature to be an indispensable presupposition of natural science. The working assumptions of many physicists and cosmologists go much further: they embrace the two cosmological fallacies, as well as the presuppositions about nature and mathematics underlying those fallacies. They embrace the fallacies and their presuppositions not just as contestable scientific theories or philosophical doctrines, but as necessary requirements of science. The development of the life and earth sciences since the eighteenth century shows the opposite: science can survive the overthrow of the ideas against which we here rebel. If it can survive their overthrow in natural history, it can also survive it in cosmology and physics.

The claim that biology must use explanations different from those deployed by physics has often been associated, in the history of ideas, with vitalism: the idea that life is not only an emergent phenomenon but also a radical novelty. According to this view, living beings conform to regularities entirely distinct from those that operate in lifeless nature. Thus, the life sciences would be safe in a clearly circumscribed domain of their own, neither subordinate to physics nor threatening its entrenched practices and established self-conception.

Nothing in the argument of the next few pages relies on the thesis of vitalism or on the acceptance of this strategy of peaceful, unthreatening coexistence with the styles of explanation that prevail in physics. The history of the universe witnesses the occasional emergence of new structures, new phenomena, and new forms of change. Such novelties do not begin with the origins of life; they begin before life. The forms of explanation deployed, however crudely and inchoately, by natural history extend backward beyond the life world to lifeless nature, on Earth and in the cosmos. The question of how far back into the history of the universe and how far up into cosmology the value of the way of thinking that we associate with natural history may go remains open and unanswered.

* * *

Before advancing, it is important to dispose of the confusions engendered by the longstanding controversy about reductionism. In all its versions, strong or weak, reductionism serves the idea that there is a hierarchy of forms of explanation. Physics towers at the top of the rank order. Relative place in this hierarchy conforms to a many-sided standard: how general and fundamental an explanation is; how fully it embodies the explanatory strategies and assumptions of the Newtonian paradigm; and how qualified it is to wed mathematics. At the summit of the hierarchy stands the tradition of physics that Galileo and Newton inaugurated and that Maxwell, Einstein, and Bohr continued. It is the same science whose matchless accomplishments are marred by the two cosmological fallacies.

One of the most important standards distinguishing the supposedly more perfect from the seemingly less perfect in this methodological hierarchy is the place of time, of history, and of historical contingency. There are three distinct and connected elements in this alleged descent from the high ground of the most exacting standard of scientific explanation. The first element is the extent to which the subject matter of the discipline is a unique and irreversible process. The second element is the looseness of the connections among the independent causal sequences that make up any real transformation in history. (To take a simple example: in natural history the effects of natural selection on speciation are shaped by, among many other factors, the connections and separations of the land masses of the planet.) The third element is the diminished measure in which subject matter and explanations lend themselves to mathematical representation.

The less powerful the explanation and the less complete the approach to the ideal of science, the more the events subject to explanation assume the form of unrepeated and even unrepeatable processes, mired in the accidents of causal sequences bereft of close causal connection, and recalcitrant to mathematical depiction and analysis. The hierarchical prejudice survives whether or not one accepts the strongest, ontological variants of reductionism.

A major ambition of our argument is to interpret what physics and cosmology have already discovered and to suggest what they might discover once unburdened from the incubus of these connected methodological biases. In the achievement of this goal, it is crucial that the idea of biology as weak physics and of social and historical analysis as weak biology not be replaced by the opposite superstition: the view of physics as weak biology, and of biology as weak history. It is also important that the introduction into cosmology and physics of a historical style of explanation not appear to represent a lowering of sights, a retreat from exacting explanatory ambition.

If we could only free ourselves from the established superstitions without surrendering to their mirror image we might suddenly see in a new light the discoveries of physics and cosmology. We might change our understanding of their agenda. We would learn how to seek in one domain inspiration for insight in another.

Strong reductionism claims that all truths about nature, including the truths of natural history, can ultimately be stated in the language of the laws of physics: that is to say, of this tradition of physics.

Strong reductionism is not a scientific theory: it is a metaphysical program. This program has never even begun to be implemented. Its professed aim is to unify science on the basis of the established model of mathematical physics. Its real role is to insulate this model from attack by presenting it as the gold standard of scientific explanation.

The most effective response to strong reductionism is the demonstration of the failure of the scientific practice that it regards as exemplary to make sense of what cosmology has found out about the universe: that it has a particular history. To provide such a response to strong reductionism is one of the aims of this book.

Weak reductionism recognizes that we are entitled to explanations different from those of basic physics. Alternative styles of explanation allow us to disregard some characteristics of certain phenomena and to focus on others. Thus, for example, with regard to living beings, we may want explanations that address their distinctive attributes, such as the reproduction of genetic invariance by result-sensitive or goal-directed structures, formed through independent ontogeny. A more complete and fundamental physical explanation may not be useful because it may fail to be adequately selective.

The mistake made by weak reductionism is to suppose, in conformity to the same idea of a stable hierarchy of scientific explanations, that there is correspondence between domains of nature and the established methods of the different sciences. Under such a view, historical explanations, because they enjoy the right selectivity, may have a larger place in biology and geology but can have only a smaller role in physics and cosmology. For the physicist and the cosmologist, historical reasoning must, according to this point of view, be subordinate to structural analysis.

The most effective response to this diluted dogmatism is to show that no such reliable correspondence exists. Biology is not weak physics, and physics is not weak biology. The hierarchical dogma and its inverse are preconceptions inhibiting the progress of science. There is as much reason to move forms of explanation established in biology back into geology, and from there into physics and cosmology, as there is to proceed in the opposite direction.

Suppose that the universe has a history, as cosmology has taught us, and that nature exists in widely different states, in some of which it fails to present in the form of discrete elements interacting according to fixed laws distinct from the states of affairs that they govern, as our fragmentary knowledge of the history of the universe suggests. Under these assumptions, the question of which styles of explanation are good for which domains is open to an extent much greater than even the weak reductionist is willing to allow. Such is the circumstance in which we find ourselves. It creates the possibility that the styles of explanation that are now characteristic of the life and earth sciences have a role to play in cosmology. They may even suggest ways to solve what we call the conundrum of the meta-laws.

* * *

Consider three styles of explanation that have wide-ranging use in natural history. Their use is by no means confined to evolutionary biology or even to biology as a whole. It includes, for example, geology as well. If their application is not limited to life but extends as well to the study of lifeless objects, we have no good reason to reject out of hand their application in thinking about the universe and its history.

Whether and how they apply in a cosmological context is an issue about how nature in fact works. It cannot be settled by a methodological dogma. If we redescribe these styles of explanation as principles, we can call them the principle of path dependence, the principle of the mutability of natural kinds, and the principle of the co-evolution of laws and phenomena.

The principle of path dependence affirms that in natural history (whether it is the history of lifeless phenomena or of living beings) a present state of affairs is decisively shaped by a history made of chains of events that may be only loosely connected. (Path dependence might also be called hysteresis were it not for the conventional and restricted usage of this term in contemporary cosmology.) To say that they are loosely connected is to claim that the regularities or effective laws underlying each of them fail to mesh together into a unified system. They converge only by their common reliance on more fundamental laws or principles. These fundamental laws or principles may be insufficient to explain the particulars that interest us.

The causal chains may be more or less independent of each other, even within the same science (e.g. geology) and with respect to similar phenomena (e.g. different kinds of rock formation). The more independent they are of each other, the more does their outcome appear to us to be marked by chance or contingency. No higher-order laws explain why a causal sequence interacted with a certain other succession of causally connected events rather than with a different one. A consequence of such particularity is that any given outcome depends on a particular history: a history that without violation of fundamental laws, formulated at a deeper and more general level, might have been different from what it in fact was.

The most important source of path dependence in natural history is not the relatively irreversible character of entropy. It is something at once more superficial and more basic: the particularity of nature, its division into distinct types or kinds, mired in different sequences of change, in what, in an earlier argument about the two cosmological fallacies, I called the first state of nature.

Path dependence is pervasive in the evolutionary history of living beings. If, for example, marsupial mammals are caught in an isolated part of the world, and there become subject to non-competitive extinctions, the main axis of mammalian evolution may become the placental line for reasons that have nothing to do with the competitive advantages of marsupials and placentals. The forces influencing the movement of the land masses described by plate tectonics may have no connection (except at the vanishing horizon of fundamental laws) with the structural or functional constraints on the evolution of mammalian body types. A different disposition of the land masses might, in this simplified hypothetical, have made the marsupials the predominant mammals.

The importance of path dependence becomes unmistakable in the origins and building blocks of the biosphere as well as in the details of Darwinian evolution. As it happens, the nucleic acids serve as the vehicle of genetic invariance, especially as it is inscribed in DNA, while the proteins perform the crucial role in the operation and development of the regulatory mechanisms on which even a unicellular organism depends. If the roles of these two classes of macromolecules capable of replication had been reversed, an emergent reality might have resulted that resembled life, as it has come to be, in some ways but differed from it in other ways.

Thus, it is not only the particular forms of life but also its basic structure and attributes that appear to be relatively accidental by the light of natural history. This history may not violate any of the effective or fundamental laws exhibited by the world before the emergence of life. However, its course cannot be inferred from such laws.

The same principle applies in the absence of life. Consider an elementary geological example. Igneous rocks, crystallizing from a molten liquid, can be explained directly by reference to the composition, temperature, and cooling rate of the parent magma. The historical element in the explanation remains limited.

Composition, temperature, and cooling determine the formation of metamorphic rocks only much more imperfectly. There are multiple pathways to a similar piece of gneiss; its formation is, to a greater extent, sunk in historical particularity, with the result that even the classification of metamorphic rocks into foliated and unfoliated is much looser than the classification of igneous rocks into phaneritic and aphanitic.

Sedimentary rocks, produced by the settling of particles through aqueous mediums, by organic secretion or by direct precipitation out of water or brine, result from a wide range of combinations and sequences. Such series cannot be shown to have simple or close connections to general physical laws, nor can their rich detail be explained by a small number of laws of any other sort. The classification of sedimentary rocks is complex and tentative; it is the subject matter of a special branch of geology, stratigraphy. Each sedimentary rock has a multifarious individuality of its own.

The modification of law-like explanation by irreversible path dependence thus extends beyond the biosphere to the lifeless natural world. If path dependence can operate at the sublunary scale of earth science, it can in principle also work at the cosmic scale of the history of the universe. The foundation of its applicability is a basic attribute of nature in the relatively formed and cooled-down universe that we observe: that it is a discriminate structure consisting of many parts. Once distinct, these parts can have histories of their own.

* * *

The world investigated by natural history is not a repository of permanent types or kinds: types of livings beings (the distribution of life into species) or types of lifeless things (the macroscopic distribution of matter into kinds of objects with distinct attributes and origins). Although often stable for long periods of time, all these types are mutable. They have a history. Despite their stability, they always remain susceptible to transformation. This fact is expressed by a principle of the mutability of types.

The mutability of types expresses itself most clearly at the scale of macroscopic objects: of species and of kinds of things. It reaches as well into the microscopic world. DNA, once created, is tenaciously stable and strikingly similar throughout the biosphere. However, it did not exist even relatively late in the history of the planet as well as of the universe. That it is subject to change both by mutation and by our intentional intervention we already know for a fact.

The periodic table and the elementary particles described by the standard model of particle physics have a much older history, going back to the very early moments of the present universe. Nevertheless, on our present cosmological views, they too have not existed forever. They have a genealogy that we are not yet able fully to describe. That they are susceptible to change, on a much larger time scale, as well as on a short time scale through our forceful intervention, is therefore the only reasonable conjecture. The alternative is to suppose that a natural kind with a definite historical origin has no future history other than to remain forever stable and identical to itself.

The principle of the mutability of the types, at least as applied to species of living things and to macroscopic objects, may seem trivial to the point of self-evidence. In fact, it is astonishing in the reach of its implications. Consider its significance first in a particular domain – the realm of living beings – and then for our thinking about the universe.

In the biosphere, the natural kinds are the species. They are often remarkably stable. Many have barely changed for hundreds of millions of years. The stability of species has a threefold basis: in the constancy of DNA and the modesty of its variations; in the unmatched power of the function that this constancy performs – the preservation of genetic invariance; and in the very restricted repertory of structural forms and materials with which autonomous morphogenesis works.

Nevertheless, despite all these forces, there is a prodigious history of speciation, marked by both bursts of innovation and long periods of relative stability. We cannot infer this history, in its significant details, from laws of any sort. There is no permanent repertory of forms of life. No species has a ticket to last as long as the planet.

What holds for speciation holds more generally for the emergence and transformation of natural kinds. In its familiar, cooled-down state, the universe exists in clearly differentiated form (which physics is tempted to understand by analogy to the mathematical idea of a differentiable manifold), composed of distinct structures or of parts interacting in certain ways. All of these parts, however elementary, are historical entities subject to transformation. At any moment, something can happen that is absolutely new – not only unprecedented but also impossible to predict on the basis of the regularities exhibited by the previous history of nature. One such novelty is the emergence on the planet Earth of the kinds of beings that we describe as living.

The principle of the mutability of types is thus not confined to life and to the life sciences. It is a general feature of what I earlier called the first state of nature (the second in the order of time). In this state, nature is differentiated but no aspect of its differentiation, expressed in a set of types or natural kinds, is essential or eternal. The principle of the mutability of types is only derivatively a biological principle. It is in the first instance a cosmological principle. It requires us to import into cosmology some of the ways of thinking that we associate with natural history.

It contradicts the project of classical ontology, which sought to provide an account of the abiding varieties of being. It conflicts as well with any practice of science that treats a permanent structure of being as one of its presuppositions.

An important and surprising aspect of the mutability of types is the transformation of the sense or the way in which they are types, that is to say of the nature of the distinctions among them. One species of animals does not differ from another in the same way in which one type of rock differs from another. Indeed, one igneous rock does not differ from another in the way in which one metamorphic rock differs from another. The processes of formation and of change impart a distinct character to the separateness of the type. This fact connects the mutability of types to a third principle of natural history: the co-evolution of phenomena and of the laws governing them.

It can be generalized in the following form in its broadest, cosmological application. Not only does the universe lack a stable and permanent repertory of natural kinds but the way in which the natural kinds differ from one another is also subject to change. If nature in its first and normal state presents itself as a structured and differentiable manifold, the character of its divisions is as impermanent as their content.

* * *

The laws and symmetries by which we explain events in natural history manifest themselves together with the phenomena whose workings we use them to explain. (In this context, I often use the term regularities and laws as synonyms, give that symmetries and constants do not play in natural history the role that they may play in physics and cosmology.) Contrary to the claims of strong reductionism, such laws could not, and cannot, be inferred from the laws governing the rest of nature, or of nature as it existed before these phenomena appeared. All we can say is that they are compatible with those prior laws. Nothing in what I called weak reductionism, much less in any way of thinking about science and nature that has rid itself of the illusions of weak reductionism, prevents us from regarding the law-like regularities of any part of natural history as coeval with the phenomena that they help us explain. To represent these regularities as part of the eternal and timeless framework of the universe is a philosophical move with no operational meaning or justification.

The phenomena change, and so, together with them, does the way in which they change: that is to say, their laws. That is the principle of the co-evolution of phenomena and laws: a third aspect of the explanatory approach that natural history habitually uses. In the history that the naturalist studies, change changes discontinuously and repeatedly. Once again the range of application of the principle of the co-evolution of phenomena and laws is not coterminous with the limits of the biosphere. It too extends backward to lifeless nature and thus has an open frontier of application to the history of the universe.

The methods of change, which we express as explanatory laws, shift with the appearance of life. They change again with the emergence of multicellular organisms. And then again with sexual reproduction and the Mendelian mechanisms. They change with the emergence of consciousness and its equipment by language. These are not just changes in the kinds of beings – in this instance, living beings – that exist. They are also changes in the way in which phenomena change as well as in the distinctions between them, as the broader interpretation of the mutability of types suggests.

The co-evolution of phenomena and of laws outreaches the biosphere. It characterizes geology or earth science, as well as applied and hybrid disciplines such as plate tectonics and the science of tides. In all these applications, it is closely associated with both the influence of path dependence and the mutability of types (and of the distinction of natural kinds from one another). The formation of crystals, for example, represents a mechanism for the reproduction of invariance that is very different from genetic replication in the biosphere and that conforms to entirely different principles.

Where does the range of application of the principle of co-evolution of phenomena and laws stop? Nowhere, it seems, short of the entire universe and its history. The co-evolution of regularities and phenomena that we observe on the planet Earth must be susceptible to occurrence anywhere in the universe and at any time in its history. There is no reason to suppose that it is limited to either a particular scale of phenomena or to a certain period of universal history.

The cosmological application of the principle of co-evolution of phenomena and laws does not repeat the error of the first cosmological fallacy: the life and earth sciences have never conformed to the Newtonian paradigm, although many attempts have been made to obtain their surrender. What the redefinition of cosmology as a historical science does force us to confront is the problem that we call in this book the conundrum of the meta-laws.

* * *

It follows from these considerations that nothing in this argument about the relevance of natural history to cosmology has turned on the idea that the principles of path dependence, of the mutability of types, and of the co-evolution of the phenomena and their laws are unique to the domain of living beings. They are not. Their range of application extends unmistakably beyond the boundaries of the life world to lifeless nature. As a result, we can mobilize for cosmological use a way of thinking that remains untainted by the illusions of vitalism.

Contemporary biology has opened the way to this conclusion by its development, in the context of life, of ideas that enjoy varied and demonstrated non-biological applications: complexity, self-organization, punctuated equilibrium, critical thresholds, and their endogenous but catastrophic undoing. By developing biology as a science of structure as well as of function and by reinterpreting and revising the neo-Darwinian synthesis in evolutionary biology in the light of this intellectual program, it has torn down the false walls between life and the lifeless.

This intellectual program makes it possible to represent the principles of natural history, as I have here discussed them, in a fashion overriding the difference between life and its absence. Such an approach does not imply that the biosphere lacks distinctive features. On the contrary, it has so many of them that its emergence is unpredictable and unaccountable on the basis of the laws of nature prior to the beginnings of life. However, the specificity of life on Earth is only a special case of a pervasive phenomenon in the history of the universe: the appearance of the new, manifest in the mutability of types and in the co-evolution of phenomena and of laws.

Recall what are conventionally described as the distinctive attributes of life: the reproduction of genetic invariance through the medium of an enduring biochemical structure – DNA; the development of an apparatus – the organism – that can literally have no purpose but that acts, through regulatory mechanisms, as if it were purpose driven (teleonomy, as Monod called it, without teleology); and the formation of this apparatus through independent morphogenesis (the more complex forms of which come to be the subject matter of embryology). Of these three attributes, the first two have rough counterparts in other natural phenomena.

It is, surprisingly, the third attribute, the independent self-construction of the body, which at first seems to be only the lowly instrument or embodiment of the other two attributes, that is most distinctive. The ontogenetic development of the organism represents a striking instance both of the mutability of types and of the co-evolution of phenomena and laws. A new kind of being appears, forming itself in a novel way and exhibiting unexpected regularities.

* * *

The significance of this argument about the analogical application to cosmology of ways of thinking characteristic of natural history is not that the life and earth sciences serve as a model for cosmology, much less that they hold the secret of solutions to the enigmas resulting from rejection of the two cosmological fallacies. The point of the argument is rather that the problems cosmology confronts once it rejects those fallacies and pursues unflinchingly the implications of its discovery that the universe has a history are not unique to its domain of inquiry. They reappear in other areas of science, in some of which their importance has long been recognized.

It is not enough to say that cosmology is a historical science that can resort with benefit to some of the ideas and methods of natural history. The deep question presented is how the study of the universe can be both historical and a science. The conundrum of the meta-laws is simply the sharpest expression of this more general problem.

Natural history provides only an imperfect model of the path toward a solution to this problem. It is an imperfect model because the scope of its inquiries is merely local. It is also imperfect because the generality and simplicity of its explanations remain limited for the reasons that the principle of path dependence makes clear. The question of how cosmology can be both historical and a science remains unanswered. To answer it is one of the chief goals of this book.

THE ARGUMENT AND SOCIAL AND HISTORICAL STUDY

The problems addressed in this book have striking analogies across the whole field of social and historical study. Once we free ourselves from the superstitions that prompt us to see the study of society and history as weak biology and biology as weak physics, we are free to recognize these analogies and to learn from them.

The ideas that everything changes – laws and even symmetries and supposed constants; that the stable and recurrent relations that our causal explanations ordinarily invoke cannot be constant or eternal but rather must co-evolve together with the states of affairs; that there may be causality without laws; and that, more generally, no particular organization of reality lasts forever may seem strange to those whose minds have been schooled in the traditions of physics. These ideas occur, however, inescapably to whoever engages the study of society and history. Puzzles related to those that are central to this cosmological argument have there long been evident.

The point is not that we can find in the theoretical study of society and history the solutions to our enigmas. It is that, as a theme and variations, the same riddles reappear, each time with a distinct character, in each domain of inquiry. Because, however, these problems appeared in social and historical study earlier and more clearly than in physics and cosmology, they provoked the development of habits of mind and stratagems of thought that may prove useful to natural philosophy. Students of society and history have not found the solutions to the problems that concern us here. Nevertheless, the discoveries that they have made, and the setbacks that they have suffered in their more longstanding search, can help light our way.

In this pursuit, the mind can stock itself with intellectual resources, richer than those that the traditions of physical science make available, with which to confront the tasks of natural philosophy. They are resources with which to reimagine the relation of laws, or other regularities, to states of affairs, of history to structure, and of the repetitious to the new.

It is futile to look, as natural scientists are accustomed to do, to mathematics for inspiration in the solution of these problems. What we find in mathematics is a peerless body of conceptions of the most general relations among features of the world, robbed, however, of all phenomenal particularity and temporal depth: a lifeless and faceless terracotta army. Mathematics is powerless to suggest how nature can escape any one established order without falling into anarchy, how the rules of nature can change together with the ruled phenomena, and how there could ever be something new in the universe that is not just a ghostlike possible – a pre-reality – waiting to be made actual.

Once we form views of these matters, we may find in mathematics instruments with which to represent them, or invent new mathematical or non-mathematical representations if the analytical instruments that we need are not yet ready to hand. Guidance can come from our reckoning with all of reality – social as well as natural. Through such confrontation, we can broaden our sense of how parts of reality may connect and of how something that exists may give way to something that never existed before.

Consider the status of structural change and of law-like regularities in both classical European social theory and in contemporary positive social science. The origins of modern social thought lie in the work of thinkers, like Montesquieu and Vico, who developed, against the background of doctrines as old as those of Aristotle's *Politics*, the view that social order can take radically different directions. Each of these directions draws, in its own way, on our pre-existing predispositions. Each makes possible certain forms of life, encouraging the development of particular powers and varieties of experience while discouraging others. Each relies for the integrity of its characteristic institutional arrangements on the cultivation of distinctive virtues or forms of consciousness.

It was the revolutionary accomplishment of the social theory of the nineteenth century to carry this conception into a more far-reaching claim: that the structures of social life are made and imagined. They are not to be treated as natural phenomena, as part of the furniture of the universe. (Many currents of nineteenth- and early twentieth-century thought, such as the sociology of Durkheim, nevertheless worked in the opposite direction, presaging the posture of contemporary positive social science.)

Vico remarked that we can understand the arrangements of society because we made them. If their mutability imposes a constraint on the understanding, it also opens an opportunity that we are denied in the study of nature: the opportunity to know the structure of society from within, in the manner in which a creator may know his creation.

The thesis that the structures of society are made and imagined has as one of its many corollaries the appreciation of structural discontinuity and structural alternatives in history. The most accomplished and influential expression of this insight in classical social theory was Karl Marx's critique of English political economy. What the economists took to be the universal laws of economic life were, by the terms of this criticism, only the laws of one particular "mode of production": capitalism. They were, in the conventional language of today's philosophy of science, effective rather than fundamental laws. The false universality claimed on their behalf rendered them misleading even for the historically specific domain to which they properly applied.

The idea that society is made and imagined can then be deepened into a view that has yet another range of implications: all the arrangements of society – its institutions and practices as well as the conceptions that represent the established order as an intelligible and defensible plan of social life – amount to a frozen politics. They result from a temporary containment or interruption of our struggle over the terms of social life: politics understood more broadly than contest over the mastery and uses of governmental power.

A corollary of this thesis is that the structures of society and of culture can exist in different ways. The harder they are to challenge and to change, the more they assume the false appearance of natural phenomena. The easier they are to reconstruct in the midst of the ordinary business of life, the less can they wear the semblance of natural objects. According to this view, we can change the quality as well as the content of our arrangements. We can so organize them that they enable us to shorten the distance between the ordinary moves that we make within an institutional and ideological framework that we take for granted and the extraordinary moves by which we revise pieces of that framework. By taking the arrangements of social life in this direction, we make change less dependent on crisis, weaken the power of the dead over the living, and strengthen our mastery over the otherwise entrenched regimes of society and culture.

The idea that the structures of society represent artifacts of our own creation, so powerfully evoked in the work of Karl Marx as well as in many other currents of classical social theory, failed to develop into such a broader account of social structures as frozen politics. It was stopped from such an evolution by its juxtaposition, in the work of Marx and others, with ideas that limited its reach and compromised its force. These compromises were the illusions of false necessity. Three such illusions have exercised paramount influence.

The first illusion has been the idea of a closed list of alternative institutional and ideological systems, such as feudalism, capitalism, and socialism, available in the entire course of human history for the organization of society. Every society must belong to one of these types. In fact, there is no such closed list of types of social, political, and economic organization. The impression that there is one becomes plausible only to the extent that each type is defined with so little institutional specificity that the definition can apply elastically and loses explanatory power. The most important example of this misunderstanding lies in the equivocal uses of the concept of capitalism. When defined at the level of institutional detail required to give it the power to explain the economic, political, and discursive practices of a particular social world, an institutional and ideological settlement, like the array of such settlements that we traditionally label capitalism, ceases to exemplify a type that we can plausibly take to recur across a wide range of social and historical circumstance.

The second illusion has been the idea that each such type is an indivisible system, all the parts of which stand or fall together. Politics must therefore be either the reformist management of one of these systems or its revolutionary substitution by another system: for example, feudalism by capitalism. In fact, the formative institutional and ideological contexts of social life change step by step and piece by piece. Change that is fragmentary and gradual in its method can nevertheless be radical in its outcome if it persists in a certain direction and comes to be informed by a certain conception. Such revolutionary reform is the standard mode of structural change in history. The wholesale replacement of one institutional and ideological order for another amounts to no more than the exceptional, limiting case.

The third illusion has been the idea that higher-order laws of historical change drive forward the succession of indivisible institutional systems in history. As there are effective laws governing particular institutional systems, so there are fundamental or meta-laws guiding the movement from one system to the next. In Marx's social theory, they are the laws of historical materialism, as summarized in *The Communist Manifesto*: the interaction between the forces and the relations of production that anoint a particular social class as the bearer of the universal interests of humanity in overturning the established relations of production for the sake of the fullest development of the forces of production.

If such meta-laws existed, they would endow history with a pre-written script. The script may be susceptible to discovery only in retrospect or at least late in its enactment. That we come belatedly to understand it only confirms and increases its power. There is consequently no legitimate role for programmatic thinking: the imaginative construction of the adjacent possible. History supplies the program.

In fact, the fundamental laws of history do not exist. History has no script. There is nevertheless a path-dependent trajectory of constraints and causal connections that are no less real because we are unable to infer them from laws of historical change. We can build the next steps in historical experience only with the materials – physical, institutional, and conceptual – made available by what came before. However, the force and character of this legacy of constraint is itself up for grabs in history. By creating institutional and ideological structures that facilitate their own revision and diminish the dependence of change on crisis, we can lighten the burden of the past.

In the subsequent history of social theory, these three necessitarian illusions have ceased, increasingly, to be believable. Yet students of society continue to use a vocabulary that relies on them and to display habits of mind formed through their use. For example, those who profess not to believe in any of them resort to a concept like capitalism as if they did.

The illusion of the higher-order laws of historical transformation has been the first to fall. The illusions of the closed list of alternative institutional systems and of their indivisibility have sometimes survived, in a climate of half-belief. When they persist, they imply a conception that, although it may seem plausible to many social theorists and historians, remains undeveloped and unsupported: that there are laws specific to different institutional and ideological formations in history. Such effective laws, however, emerge and evolve together with the formations themselves. No fundamental laws stand behind them guiding their co-evolution. It is a view reminiscent of ways of thinking long established, although also unexplained, in the life sciences, but, to this day, foreign to physics.

A major reason why the idea of the co-evolution of laws and of states of affairs has failed to be more developed in our thinking about society and history is that contemporary social science has for the most part taken an entirely different direction. Social science has repudiated the necessitarian assumptions only because it has rejected the central insight of classical social theory: the insight into the made and imagined character of social life and therefore as well into structural discontinuity and structural alternatives in history. Its dominant tendency is to naturalize the established institutions and practices by representing them as the outcome of a progressive, functional evolution.

According to this view, the established arrangements of contemporary societies result from cumulative trial by experience. What works better survives. What works less well, relative to the competing solutions on offer, fails. We may therefore expect to see in history a halting but cumulative convergence of societies to the same set of best practices and institutions. Nowhere is this view more fully developed than in the most influential social science, economics, at least as soon as economics abandons the refuge that it has taken, ever since late nineteenth-century Marginalism, in analytical purity and deploys its methods in the design of policy and in the explanation of behavior. According to this view, there is no special problem about the structure. A market economy works best, and it has a largely predetermined legal and institutional content: a content exemplified by the regimes of private property and of free contract that have come to prevail in the North-Atlantic societies.

The result of this way of thinking is to conceal under a veneer of naturalness and necessity what is most decisive and enigmatic in historical experience: the ways in which the institutional and conceptual presuppositions of social life get established and remade. In the absence of insight into this most fundamental problem of social and historical study, the vital link between insight into the actual and imagination of the adjacent possible is severed. Social science then degenerates into rightwing Hegelianism: the retrospective rationalization of a world whose historical vicissitudes and transformative opportunities it is powerless to grasp.

The task presented to social thought by this history of ideas is to salvage and radicalize the central insight of classical social theory into the made and variable character of the structures of social life: the institutional arrangements and ideological assumptions shaping the routine activities and conflicts of a society. These institutional and ideological regimes are frozen politics. We must rescue this insight from the necessitarian assumptions that eviscerated its meaning and reach in that theoretical tradition. We must recognize our stake in the creation of structures that are so arranged that they empower us to defy and revise them without needing crisis as the condition of change. We must acknowledge the reality of constraint and the power of sequence that help explain the prevailing arrangements and assumptions. We must acknowledge it, however, without conferring on such influences a mendacious semblance of necessity and authority. We must reestablish the indispensable link, in social and historical study, between insight into the actual and exploration of the adjacent possible. On this basis, we must exercise the prerogative of the programmatic imagination: the vision of alternatives, connected by intermediate steps to the here and now, especially alternative institutional forms of democracy, markets, and free civil societies.

Such a project provides no model for a cosmology that does justice to the singular existence of the universe as well as to the inclusive reality of time. It nevertheless has an affinity to such a cosmology. It is connected to it by its commitment to a practice of causal explanation that dispenses with the invocation of timeless laws governing events in time. It is bound to it as well by its insistence on seeing the basic constituents of the reality that it addresses – for social and history study, the formative institutional and ideological contexts of social life; for physics, the elementary constituents of nature – as evolving, discontinuously, in time. The institutional and ideological regimes melt down, periodically in those incandescent moments, of practical and visionary strife, and become, at such times, more available to reshaping. So, too, nature passes through times in which its arrangements break down and its regularities undergo accelerated change. A difference is that we can hope to change forever the character of the structures and their relation to our structure-defying freedom. Nature, so far as we know, enjoys no such escape.

The two kindred projects, of cosmology and social theory, cannot take for granted either an immutable catalog of types of being or a changeless framework of laws.

REINVENTING NATURAL PHILOSOPHY

This book is neither an exercise in the popularization of science nor an essay in the philosophy of science, as that discipline is now commonly understood. We seek here to recover, to reinterpret, and to revise a way of thinking and of writing that has long ceased to exist. It used to be called natural philosophy. Up to the middle of the nineteenth century, natural philosophy remained an accepted genre. It gained a brief afterlife in the work of Mach and Poincaré in the early twentieth century and continues today to be represented chiefly in the writings of philosophical biologists. Here are some of its enduring characteristics, all of them important features of the type of discourse most useful to the development of our argument.

Its first hallmark is to take nature as its topic: not science but the world itself. It engages in controversy about the direction and practice of part of science only as part of a larger argument about nature. The proximate subject matter of the philosophy of science, as now understood and practiced, is science. The proximate subject matter of natural philosophy is, and has always been, nature. Science and natural philosophy have the same subject matter, but not the same powers and methods.

A second characteristic of natural philosophy is to question the present agenda or the established methods in particular sciences. It does so from a distance rather than from within science. It makes no new empirical discoveries nor does it subject new conjectures to direct empirical test.

Natural philosophy tries to distinguish what scientists have discovered about nature from their interpretation of these discoveries. The interpretation is regularly influenced by metaphysical preconceptions, especially supra-empirical ontologies – views of the kinds of things that there are in the domain addressed by the science. Such views form an unavoidable part of scientific theorizing. The more ambitious the theory, the larger their role is likely to be. The cost for relying on them is an unacknowledged blindness: the progress of science requires that they be occasionally identified, resisted, overturned, and replaced.

Natural philosophy can be useful in the early stages of such an effort. It cannot accomplish, or even justify, a reorientation of the agenda of any science relying solely on its own limited resources. Yet from the outset, and unlike much of the now established philosophy of science, its intentions may be revisionist, not merely analytic or interpretive. On what basis and by what method it can hope to do this revisionist work is what I seek to elucidate in this section.

The argument of this book disputes widespread accounts of what cosmology and physics have discovered about the universe, including accounts that continue to exert influence within these sciences, not simply in philosophical or popularizing discourse about them. It contradicts, for example, leading interpretations of general relativity as well as cosmological conceptions such as the notion of a multiverse that have commanded a wide following.

A third trait of natural philosophy, as we exemplify it here, represents a break with much of the way in which natural philosophers used to view their own work when natural philosophy was an accepted genre. We deal with problems that are both basic and general. We do so, however, without depending on metaphysical ideas outside or above science. We do not think of the natural philosophy that would now be most useful, or indeed of philosophy in general, as a super-science in which untrammeled speculation can take the place of the dialectic of empirical inquiry and theoretical analysis that moves science forward. Our watchword is to take on foundational matters on terms that dispense with foundational doctrines.

A fourth characteristic of natural philosophy, as we here interpret and try to recover it, is that, as it intervenes in discussion of the agenda of natural science, it attenuates the clarity of the divide between a discourse within science and a discourse about science. It cannot claim the authority of a scientific inquiry: you will find here no hypotheses closely embedded in a context of empirical testing or falsification and equipped with any of the mathematical and technological tools with which natural science has armed itself.

Nevertheless, the issues that we address and the ideas that we present do not simply take the present direction of physics and cosmology for granted. They have implications for our beliefs about what should happen next in cosmology and physics. They even provide a perspective outside science from which to assess the path that contemporary science in these fields has taken. They have revisionist potential as well as revisionist intentions.

How can ideas manifestly lacking in any of the mathematical or technological instruments of science nevertheless claim to speak to the direction of a science like cosmology? The answer lies in an understanding of the proper relation between a first-order discourse and a higher-order or meta-discourse. It lies as well in the practice of three methods that make use of such an understanding in the advancement of its revisionist program.

A first-order discourse is a discourse within a particular science or discipline. It begins from where that science or discipline finds itself at a given moment and even at a particular place: its organizing controversies; its accepted methods of analysis, explanation, argument, and proof; and its guiding assumptions about nature and about thought.

For natural science, some of the most important presuppositions are those that have to do with laws and symmetries and their relation to structure and change in nature. Others deal with mathematics as well as with the relation of mathematical analysis to causal explanation. Even for a first-order discourse, however, the ruling ideas, the dominant theories, and the accepted methods need not to be the point of arrival although they are sure to be the point of departure. They can be revised piece by piece, under the pressure of discovery and imagination.

A higher-order discourse addresses such presuppositions directly and passes judgment on them in the name of considerations that may include those that would be acknowledged to carry weight within the particular science or discipline but that are not limited to such considerations. A defining move in a higher-order discourse may be to suggest a change in some of these presuppositions. Such a change may be motivated by the hope that it will throw surprising and revealing light on well-established facts and suggest a shift of direction: a new way of looking at the familiar, offering a path into the unfamiliar. To serve as a terrain for the development of such proposals has historically been the province of philosophy.

A common tendency in contemporary philosophy is to depreciate such higher-order discourse as an exorbitant attempt to claim for speculative reason an authority that belongs only to the specialized forms of inquiry and, in particular, to the distinct sciences. The only meta-discourse we need, according to this view, is a meta-discourse that discredits the pretensions of all meta-discourses. Natural philosophy has no place in such a view.

One of the most important justifications of natural philosophy is the relativity of the distinction between a first-order discourse and a meta-discourse. The more far-reaching a new conception at the first-order level is, the more likely it is to imply and to require a change in established presuppositions about method or content in a science. Conversely, any proposal in a higher-order discourse to revise substantive or methodological assumptions will and should be assessed by its effect on the insights to be gained down below: its consequences for the work of particular disciplines.

It is sheer dogmatism to stipulate from where forward movement will come. Normally, it will come from the internal conflicts of first-order discourses. As such controversies escalate, they soon begin to cross the frontiers that separate them from the higher-order conversation. Occasionally, however, breakthroughs of insight will begin in this second-order conversation and then gain interest as their implications for the first-order discussion become clear.

One criterion of intellectual ferment – and of the advance in insight that such ferment may encourage – is the frequency and the intensity with which this double movement, from higher-order to first-order discourse and back again, takes place. A byproduct of such double movement is to attenuate rigid divisions among sciences. Debates transcending the distinctions between higher-order and first-order discourse are likely to engage more than one field and to bring into question the orthodoxies of method and of vision around which each such field is organized.

Alongside the difference between higher-order and first-order discourse and the divisions among disciplines wedded to methods, a third distinction will be weakened by this intellectual turn: the contrast between normal and revolutionary science. We can best understand the significance of this third subversion by a political and historical comparison. After all, the contrast between revolutionary and normal science (as described by Thomas Kuhn) is itself the product of just such an analogy.

The dominant traditions of classical social theory, Marxism included, distinguished between the revolutionary substitution of one system (e.g. socialism) for another (e.g. capitalism) and the management of a system and its "contradictions." They imagined each such framework to be an indivisible whole, all the parts of which stand or fall together. Consequently they divided politics into reformist tinkering and revolutionary transformation, associating gradualism with the former and sudden, violent change with the latter.

These categorical contrasts are misguided. There are no indivisible and historically recurrent institutional systems such as capitalism, each with its built-in logic of reproduction and transformation. Change can be fragmentary and gradualist in its method and nevertheless radical in its outcome if it continues in a certain direction, especially if it is informed by an idea. Revolutionary reform represents the characteristic form of structural change; wholesale revolution supplies only the limiting case.

However, the relation between the reproduction of a certain institutional order and its transformation is far from being a constant in history. It is a variable. We can indeed distinguish between the normal moves we make within a framework of institutions and conceptions that we take for granted and the exceptional moves by which we change pieces of such a framework. Once again, the distance and the distinction between these two sets of activities vary. Our institutional arrangements and discursive practices can be arranged to either increase or diminish the distinction and the distance.

An institutional and ideological ordering of social life can have, in superior degree, the attribute of laying itself open to criticism and revision. As a result, it can allow the transformation of society and culture to arise more constantly out of the daily activities by which, as individuals and as groups, we pursue our interests and ideals within the established context. Our most powerful material and moral interests are engaged in the enhancement of this attribute of social and cultural regimes. Such an enhancement is causally connected to the conditions for the development of our productive capabilities (through economic growth and technological and organizational innovation). It is also causally related to the conditions for the disentanglement of our practical, emotional, and cognitive dealings with one another from entrenched social division and hierarchy.

Practical progress requires freedom to experiment and to recombine not just things but also people, practices, and ideas. Moral emancipation demands that we be able to relate to another as the context and role-transcending individuals that we now all hope to be, rather than as placeholders in some grinding scheme of hierarchical order and pre-established division in society. Neither of these two sets of requirements is likely to be satisfied unless we succeed in building societies and cultures that facilitate their own reconstruction, weakening the power of the past to define the future and diminishing the extent to which crisis must serve as midwife to change.

In addition to the service that it renders to these fundamental material and moral interests, this property of self-revision has independent value. It attenuates the contrast between being within an institutionalized or discursive context and being outside it. We can never establish the definitive context of life or of belief: the one that would accommodate everything that we have reason to prize. The next best thing to finding the definitive, all-inclusive context is to develop arrangements and assumptions that in satisfying our fundamental material and moral interests also best lend themselves to correction in the light of experience. Corrigibility supersedes finality.

We can engage in such an order, even single-mindedly and wholeheartedly, without surrendering to it. In the midst of our ordinary business, we can keep the last word to ourselves rather than giving it to the regime. In this way, the social world that we inhabit becomes less of a place of exile and torment; it no longer separates us from ourselves by exacting surrender as the price of engagement and isolation as the price of transcendence.

An institutional and ideological framework of social life that is endowed with this power to facilitate its own remaking enjoys an evolutionary advantage over the rivals. However, we do not select from a closed list of ways of organizing society and culture an institutional system readymade to seize this advantage; we choose from the messy materials of a relatively accidental history and turn them into something else.

Every word in these remarks about culture and society applies, by analogy and with adjustment, to the structure of our scientific beliefs and practices. The nature and the extent of the contrast between normal science and scientific revolution are at stake in the history of science. A stronger, deeper science would be one exhibiting in its normal practice some of the characteristics that we ascribe to its revolutionary interludes.

One of these characteristics is a more ample dialectic among theories, instruments, observations, and experiments than is ordinarily practiced. Another is the investigation of problems that require crossing boundaries among fields as well as among the methods around which each field is organized. Yet another is the deliberate and explicit mixing of higher-order and first-order discourse. Viewed in this light, natural philosophy works to overcome the contrast between normal and revolutionary science.

We aim simultaneously to recover and to reorient the eighteenth-century genre of natural philosophy. Not idle speculation, but engagement in the agenda of science and in our ideas about the relation of science to the rest of our world view, should be the ambition of a reconstructed natural philosophy. It must be, as I argue below, an engagement defined by concerns, limitations, and methods distinguishing natural philosophy from science.

Today, natural philosophy has not disappeared completely. It lives under disguise. Scientists write popular books, for the general educated public, professing to make their ideas about the science that they practice accessible to non-scientists. They use these books to speculate about the larger meaning of their discoveries for our understanding of the universe and of our place within it. They also have another audience, however: their colleagues in science, addressed under the disguise of popularization. The popularizing books have become a secret form of the vanished genre, a crypto natural philosophy.

Here we propose to cast the shield down, and to do natural philosophy in just the sense we have specified without disguise or apology. We reinterpret the meaning of some of what physics and cosmology have discovered about the world and argue for a revision of some of the attitudes, assumptions, and expectations with which we do science.

* * *

In the pursuit of these goals, natural philosophy can rely on three strategies among others. Each of these strategies plays a major role in the arguments of this book.

A first strategy is to identify and exploit the distinction that exists in any ambitious scientific theory between its hard core of empirically validated insight and of operational procedures and the supra-empirical ontology with which this hard core is ordinarily combined. Nowhere is this combination more evident and more important than in the most comprehensive systems of scientific ideas, such as those of Aristotle, Newton, and Einstein. The same combination marks as well theoretical systems in science that are much less far-reaching.

Viewed from one perspective, a physical theory is a guide to practical orientation and transformative action in part of the world. It teaches us how certain initiatives can produce certain effects. It shows us, as well, how we can and must adjust our limited and misleading perceptions to take account of what happens. Our observational and experimental equipment is decisive in extending the reach of unaided perception and of transformative intervention in the workings of nature. It serves the nearly blind as a walking stick.

In this respect, the arbiter of science is practical success: success at guiding intervention and at correcting perception. Science, in the performance of this role, has no message about how things really are, only about what we must assume them to be like for our limited purposes. Its assumptions about the workings of nature can be both parsimonious and accommodating because they are likely to be compatible with a range of different conceptions of how part of nature is organized.

In the history of science, however, there is always another element: a representation of how nature works and of how it is structured in a particular domain. Uniquely for cosmology that domain is the whole universe. Insofar as science plays this second role, the role of the revealer, it subscribes to an ontology, though often a fragmentary one: a conception of the kinds of things that there are in the part of nature that it investigates. This ontology is supra-empirical both in the sense that it can never be read off directly from observations and experiments and in the sense that what we learn from the experiments and observations is invariably compatible with more than one such view of the kinds of things that there are.

The ontological element in science has a twofold source in the aspiration to make sense of the world and in the conflicted relation of scientific discovery to perceptual experience. To guide our transformative interventions in nature, we must know to what extent we can rely on what we perceive. The need to organize, to extend, and to correct our perceptual experience, without abandoning it, provides a permanent incitement to ontological speculation. It does so even in those forms of scientific practice that are most determined to remain close to the ground of observation and experiment. For this reason, explanatory modesty fails to exempt science from ontological pre-commitments. If science cannot avoid such commitments, it becomes crucial to make them explicit, to weigh their advantages and disadvantages, and to understand their implications.

The broader the scope of a scientific theory and the greater its explanatory ambitions, the more significant the presence of this supra-empirical ontology is likely to be. It is most pronounced in systems, like those of Aristotle, Newton, and Einstein, that have defined epochs in the history of science.

Nevertheless, the distinction between the empirical and the supra-empirical aspects of a theoretical system is likely to be elusive or even invisible to the author as well as to its converts. The hard core of insight confirmed by observation and experiment and the metaphysical interpretation superimposed on it appear seamlessly joined as if they were indissoluble parts of the same discoveries and the same understanding. The aura of empirical confirmation falls, undeservedly, on the philosophical gloss as well as on the empirical subtext.

For this reason, a major scientific system represents, in part, a frozen natural philosophy, just as an established institutional and ideological regime amounts to a frozen politics, resulting from the temporary interruption and the relative containment of conflict over the terms of social life. Having been accepted by the adepts of the theories to which it belongs as a fact of the matter, it becomes relatively entrenched against challenge.

In the argument of this book, a major example of this phenomenon is the role of Lorentzian spacetime in general relativity and the spatialization of time – the treatment of time as accessory to the disposition of matter and motion in the universe – that the notion of a spacetime continuum has been used to promote. The empirical tests adduced in favor of general relativity bear, for the most part, only an oblique and questionable relation to that notion (a matter discussed in Chapter 4). Yet the conception of the spacetime continuum is almost universally viewed as validated by the classical and post-classical tests of general relativity.

In this way, the operational and empirical element in science is married to a supra-empirical ontology. From time to time, an advance in science requires that this marriage be dissolved. Here natural philosophy has a vital task. It can provide an antidote to metaphysical bias, when such bias is disguised as empirical truth. It can help open the way to an alternative interpretation of the observational and experimental insights of an established theory. In this practice, it can find a powerful instrument for the pursuit of its revisionist goals.

A second strategy available to natural philosophy is to confront the practices followed in one branch of science with those that are preferred in another. The aim is to undermine belief in a necessary relation between method and subject matter. A consequential change of direction in any science is likely to have methodological as well as substantive aspects: no way of practicing a science is likely to survive unchanged a major innovation in the content of the ideas in that science.

However, just as empirical discoveries may appear to be naturally and necessarily joined to an ontological program that they need not, in fact, imply, so too the relation of a particular science to a conventional repertory of methods may wear a false semblance of naturalness and necessity. A methodological bias, like an ontological one, may then prevent a science from seizing an opportunity to go forward. It may inhibit it from seeing its own empirical discoveries and experimental capabilities in the light afforded by the methods employed in another science.

Rather than inventing *ex nihilo* a new method for a new conception, the best prospect of advance may be to begin by jumbling up the relation of subject matter and method across a range of distinct scientific disciplines. By looking next door to the neighboring sciences and asking to what extent some of their practices can be imported, we begin to free up the connection between method and substance. We enlarge our sense of intellectual possibility. The point will rarely be to replace the procedures of one science with those of another; it will more often be to remove the impediments that a methodological prejudice imposes on a substantive reorientation. The comparative and analogical exercise may thus serve as an early step in an itinerary of theoretical reconstruction. In this way, it too comes to support the revisionist purposes of natural philosophy.

An example of this strategy in this book is the consideration of the extent to which moves characteristic of the life and earth sciences and even in social and historical study may have cosmological uses. They may help create a cosmology untainted by what I earlier described as the two cosmological fallacies. Thus, the principles of path dependence, of the mutability of types, and of the joint evolution of types and regularities, familiar in natural history and, more generally, in geology and biology, may all have counterparts in a cosmology that has completed its transformation into a historical science. The ideas of causation without laws and of an alternation between formative periods in which structures are rendered relatively inchoate and other, longer periods in which they take definite and stable shape have an important place in social theory. They may also prove useful to thinking about the history of the universe.

A third strategy on which natural philosophy can count is the attempt to establish a direct connection between speculative conceptions and opportunities for empirical and experimental discovery, and to do so, tentatively and suggestively, without passing, as science normally must, through an intermediate stage of systematic theory. Natural philosophy, as we here view and practice it, is not natural science. Neither, however, is it what the philosophy of science has largely become: a commentary on scientific ideas, delivered from the distance of analytic self-restraint and unencumbered by any intention to intervene in the agenda of a particular science.

If it is to play such a role, even its most speculative conceptions must be able to form part of a set of ideas that at least at its periphery of implication, if not in its core conceptions, lays itself open to empirical challenge and confirmation. Its proposals grow in interest if, despite their generality and abstraction, they express physical intuitions and anticipate pathways of empirical inquiry.

The bridge between the speculative conceptions and their empirical vindication is scientific theory. It is not within the power of natural philosophy to develop systematic theoretical ideas in science, much less to demonstrate how such ideas can be upheld by observation and experiment. What natural philosophy can and should do, in its role as scout of science and enemy of the metaphysical and methodological preconceptions that restrain its progress, is to foreshadow theory. It is to prefigure the contours of the theories that could connect its speculative proposals with an agenda of empirical research.

Having helped overturn the metaphysical and methodological obstacles to a reinterpretation of what science has already discovered, in the service of a solution to problems that otherwise remain unsolved, natural philosophy can go on to envisage next steps for scientific inquiry. It can suggest how speculative ideas that may at first seem paradoxical can in fact begin to take theoretical shape – in fact, alternative theoretical shapes. It can help draw around the canon of established science a larger penumbra of untapped intellectual opportunity.

Because it is not a science, but only a prophecy of science, or a prolegomenon to theory, natural philosophy cannot choose among these roads. Much less can it travel on them. It can point to that promised land, but not enter it.

The following chapters provide many examples of this strategy of seeking to connect the speculative with the empirical by means not of a single organized and tested theory but of a range of alternative theoretical possibilities, sketched rather than developed. Among these examples are the development of the conception of a non-cyclic succession of universes, as a deepening of the idea of the singular existence of the present universe; the defense of the existence of a preferred cosmic time as an aspect of the inclusive reality of time and the reconciliation of such time with the strictures of general and special relativity; the conjecture, by analogy to the local physics of phase transitions, that in the course of universal history nature may take forms different from those that generally prevail in the cooled universe; and, above all, the effort to address what we call the conundrum of the meta-laws – how we can make sense of a joint evolution of the regularities and of the structures of nature and lay this proposal of co-evolution open to empirical inquiry.

What is most prominently missing from this account of three strategies of natural philosophy is an idea both more controversial and more consequential than all of them. If I fail to list it as a fourth strategy, I do so because it represents one of the central claims of this book. This idea is the refusal to take mathematics as more than an indispensable tool of cosmology and physics: an ante-vision of the ultimate structure of nature and a supreme judge of right and wrong in physical science.

The rejection of this view of mathematics and of its role, argued in Chapter 6, results in an understanding of the prospects of basic science at odds with the one that is now in command. This understanding suggests intellectual problems and opportunities arising from the divergence between nature and mathematics. It prefers to make of mathematics a good servant rather than a bad master. It insists on correcting the biases of the mathematical imagination. Preeminent among these biases is the trouble that mathematics has with time.

If mathematics were everything that those who believe in its premonitory powers make it out to be, natural philosophy would be both less useful and less dangerous than it is.

WHAT IS AT STAKE

What is at stake in the argument of this book is the future of ideas that have shaped both how we do science and how we interpret the meaning of some of its major discoveries.

Is real novelty possible in the world, or is what seems to be new simply the working out of a program inscribed in the ultimate nature of reality, the actualization of possibilities that awaited their cue to come onto the stage of the real? Is there only one universe, or is this universe of ours simply one of many? Are we to think of what lies beyond the observable universe as the unobserved part of the same universe, as other universes, in the language of plurality, or as past and future universes, or past and future states of the universe, in the language of succession? Is time real, inclusively real, to the point of holding sway over everything? Or is part of ultimate reality, notably a framework of unchanging regularities of nature and a structure of ultimate constituents of nature, outside time? If time goes all the way down, must we admit that the laws, symmetries, and apparent constants of nature might change and have in fact changed in the course of the history of this one real world? How should we revise our conventional ideas about causation so that they accommodate change in the laws and other regularities on which causal explanations usually rely? And how should we think of the causation of the change of the laws of nature on which causal judgments are ordinarily thought to depend? If there are no higher-order or meta-laws governing how the laws of nature change, are there nevertheless principles or hypotheses to guide us? If so, can we assess and confirm or disconfirm them in the light of their empirical implications or predictions? Are we entitled to say in physics and cosmology, as we have learned to say in the life and earth sciences and in social and historical inquiry, that insofar as explanatory laws exist they may evolve coevally with the explained phenomena and even that there may be causality without laws? What light do the inclusive reality of time and the singular existence of the universe throw on the nature of mathematics and on its uses in natural science? How can we make sense of the "unreasonable effectiveness" of mathematics in science while both affirming the reality of time and recognizing the timeless character of the relations among mathematical propositions? How is it that we can come to understand mathematics as being an analysis of the one real, time-bound, and fragmented universe, but the universe seen from a vantage point that robs it of both time and phenomenal particularity? And how can all these questions, and the answers we give to them, come to form part of a conversation within science, decisive to its future course, rather than just of a conversation about science, conducted from a philosophical distance?

In formulating these questions and in proposing answers to them, we must contend with not one but two adversaries. To grasp the intention of our argument, it is useful to understand the relation between them.

The chief opponent is a distinct but immensely influential stand within the tradition of physics and cosmology from Galileo to relativity and quantum mechanics. It is a way of thinking to which Newton's science gave the most powerful impulse, but which has survived in physics ever since. This tradition devalues, diminishes, or denies the reality of time. It has done so in two main moments.

The first moment is that of classical mechanics. The decisive move is the unwarranted generalization of the Newtonian paradigm: the explanatory practice distinguishing between stipulated initial conditions and a configuration space of law-governed events. It is a distinction readily applied to part of the world. However, it is unsuited, for the reasons I have described, to deal with the universe as a whole: it has no legitimate cosmological use.

The observer stands outside the configuration space. For him, everything that happens in the configuration space is present at once to his mind. The end of each process is in the beginning. The relation of the observer to the events in the configuration space resembles the relation of God to the world.

The ideas informing this tradition deny or devalue the reality of time twice: first, because within the configuration space everything is governed by deterministic or statistical laws that, once adequately understood, foreordain the outcome; and second, because the observer, in his godlike position, is not himself within time, even in the highly qualified sense in which time can be said to exist within the configuration space.

The human experience of time has no significance for such an observer. In his scientific capacity, he frees himself from the dross of humanity, sunk in time. He sees the world through the lens of timeless laws of nature, expressed in mathematical propositions standing outside time.

A second moment in this way of thinking against time is the one most closely associated with Einstein's special and general relativity. By affirming, in special relativity, the relativity of simultaneity, it denies the existence of a global time, often the first step in the denial of the reality of time altogether. By representing, in an influential interpretation of general relativity, spacetime as an unchanging four-dimensional block and by representing time spatially, as an additional "dimension," it robs time of reality. The spatialization of time in this "block-universe" view goes further toward denying or circumscribing the reality of time than classical mechanics had ever done. In Newton's physics, despite the time-reversible or symmetrical character of the laws of motion, time is preserved as an absolute background, distinct from the phenomena of a three-dimensional world. (I later distinguish the hard empirical residue of what special and general relativity have discovered about the workings of nature, from the ontological pre-commitments that have shaped the interpretation of these discoveries.)

The denial or devaluing of time, in these two successive moments of the history of physics, has as one of its most revealing implications the privileged position accorded to mathematics. If the laws of nature are written in the language of mathematics, it must be, according to this view, that they share in the nature of mathematics. No feature of mathematics is more striking than the timelessness of the relation among its propositions. It is a feature that stands in stark contrast to the time-bound relation among causes and effects in the world, as exemplified both in our first-hand acquaintance with nature and in our conventional use of causal language.

The view that mathematics, with its timelessness, provides privileged insight into the ultimate language of reality fits, like hand and glove, with the "block-universe" picture of the world. It accords as well with the idea that the physical events and the entire manifold of spacetime happen within a framework of natural laws that is itself timeless. This framework is, according to the tradition unbroken at least since Newton, the embodiment of the godlike intellect, the mind looking in, from a place outside time, upon the world in which, for human beings, time seems all too real.

Yet this whole tradition, from Galileo and Newton to Einstein and Bohr, never severed its link with the idea that there is one universe, all of the parts of which are in causal communion with one another. The incomparable reality of the one real world, embraced by natural science in reinforcement of the testimony of experience, has coexisted, in the history of the tradition, with the demotion of time, in defiance of other aspects of the reality that the untutored human being perceives and undergoes.

We argue against this conception, the orthodoxy of natural science, at least in physics, from the mid-seventeenth century to today. Our thesis is that all things considered – all things understood as both what physics and cosmology have already discovered and as how we can best connect what they discover with what we also know about the world and about ourselves from other sources and by other means – we should drop each of the characteristic elements of this tradition. They are not science. They amount to a baseless metaphysical gloss on the hard core, or the empirical residue, of the discoveries of science. What science has discovered about the world, as distinguished from what scientists often say about these discoveries, gives us mounting reason to reject the philosophical prejudice.

We do better to put the Newtonian paradigm in its place, to drop the block-universe picture of the universe, to recognize the reality of time all the way down, to dispense with the notion of a framework of natural laws outside time, to admit that the laws of nature may change, and to deflate the claims of mathematics to represent a uniquely privileged channel of insight into reality. The philosophical assumptions needed to establish and develop these views, contrary to the tradition we resist, are less heroic than the ones we repudiate: they require much less of a break with how other sciences understand nature, as well as with our pre-scientific experience of the world. Most importantly, they accord better with what cosmology has discovered about the universe and its history as distinguished from the ways in which these discoveries have been interpreted under the lens of the theoretical traditions that we criticize.

None of these considerations imply that natural science should obey the lesson of the senses, unassisted by scientific instruments and theories. After all, part of the point of science is to loosen the restraints on insight resulting from our condition as fragile, ramshackle, and mortal organisms, situated in time and in space. It is no goal of this argument to defend our pre-scientific experience against science or to reconcile the latter with the former. The view developed here stands in contrast to many features of our pre-scientific experience, including our experience of time, as well as to now influential theories in science. Willingness to defy that experience should, however, be subject to two qualifications.

The first qualification is that radical denial of the reality of time is not comparable to a localized correction of our perceptual view such as the correction by virtue of which we come to understand that the Earth is round rather that flat and that it can be round without our falling off it. Uncompromising denial of the reality of time undermines the sense of causality as well as many other conceptions that deniers of time continue habitually to invoke. The temporal element in our experience is not a thread that we can pull out while leaving all the others in place. Pulling it out deranges every part of our experience. It does so to such an extent that it is no longer clear how we can then continue to rely on a corrected version of perception either to make our scientific discoveries or to interpret them. It is not a move to make lightly, without overwhelming reason to make it and understanding of its implications. (I consider these issues in Chapter 4.)

The second qualification is that the basis on which we defy untutored and unequipped perception matters. It is one thing to loosen the restraints of ordinary experience under the prompting of theory-guided observation and experiment. It is another to deny them under the influence of a metaphysical program. If we suspend belief in the extra-scientific program we attack, we will not be at a loss to move forward, even to advance in the direction of views that are counterintuitive, perplexing, and subversive of the present form of the marriage between empirical discovery and supra-empirical ontology in our cosmological ideas.

The picture of the history of the universe that became predominant in the cosmology of the twentieth century, expressed in the now standard cosmological model, should have been enough to make questionable the time-denying tradition against which we rebel. According to this picture, the universe has a beginning and a history. The standard cosmological model offered powerful reasons to believe that the present universe began in violent events, whether the values of these events are represented as finite or infinite. In a moment close in time to these occurrences, the application of the laws of nature, as we now understand them, seems to fail and the elementary constituents of nature, as they are described by the standard model of particle physics and, at another level, by chemistry, could not have existed or must have been very different. All the subsequent events that take place in the universe form part of a history that must include the evolution of those regularities and of that structure.

The blinkers imposed by the extra-scientific tradition that I have identified as the chief target of our argument diminish or deface the significance of these cosmological discoveries for all of physics and indeed for all of natural science. The denial or diminishment of the reality of time, and the related view of mathematics as a shortcut to the understanding of ultimate reality, have survived in the face of the discovery of the historical character of the universe only through a series of conceptual maneuvers. The combined and cumulative effect of these maneuvers is to disguise the contradiction between that tradition and what we have already found out about the universe and its history.

One such maneuver is the application of the Newtonian paradigm (of initial conditions and law-governed phenomena within a configuration space bounded by those conditions) to the whole of the universe, where it cannot work, rather than to a part of the universe, where it can. A second maneuver is the acceptance of the features of the present, differentiated universe as if they were traits that nature possesses at all times in its history. A third maneuver is the reification of an idea of scientific practice and of causal explanation that is wedded to the notion of an immutable framework of natural laws, deterministic or statistical, as well as to the idea of a permanent stock of ultimate components of nature, as if we could not continue to do science or explain natural events without embracing these assumptions. A fourth maneuver is the marginalization of cosmology. The premise of this marginalization is that we can understand the workings of nature through the study of its basic constituents without regard to their origin and future, which is to say without regard to time and history.

All these maneuvers, deployed to reconcile the new (and now not so new) cosmology with the tradition that we oppose rely on the very line of ideas that they are designed to protect. As a result, all of them are tainted, to a greater or lesser extent, by circularity.

We propose to cast the philosophical prejudices enshrined in that tradition aside and to consider, free from their restraints, the implications for physics as well as for science more generally, of our present view of the universe. We ask the reader to suspend disbelief and to consider what the cosmological discoveries of the last hundred years might be taken to mean once we relinquish the impulse to reconcile them with the tenets of the time-denying and mathematics-worshipping tradition that we dispute.

The substance of this part of our argument is contained in the second and the third of the three main theses of this book: that time is real and inclusive all the way down (everything changes, including the laws of nature) and that mathematics is useful to understanding the world precisely because it abstracts from certain features of the world (namely time and phenomenal distinction), not because it affords us privileged insight into timeless truth.

To carry forward this intellectual program, it is not enough to rebel against the tradition that we have described as our chief enemy. It is also necessary to contend with a second, lesser adversary, represented by more recent developments in physics. According to this secondary target of our argument, the theme that should command the agenda of physics is the final unification of our theories of the forces of nature, and in particular the unification of gravity with the electromagnetic, the strong, and the weak forces, as represented in the so-called standard model of particle physics.

Physics (in what I earlier called its main line, in contrast to its side line) has been engaged, at least since the mid-nineteenth century, in an effort to bring all the known and basic forces of nature under the aegis of a single, cohesive set of laws. Only one final, definitive battle supposedly remains to fight and win: the struggle to unify our ideas about gravity with our understanding of the other natural forces. The history of the universe, as presented by contemporary cosmology, is, by the lights of this intellectual project, no more than an interesting sidelight on the ultimate topic. The ultimate topic is the structure of the natural world and the content of the laws that explain it.

According to this view, the structure can be explained – indeed, according to this view, it must be explained – without reference to the history. The structural explanation is much more likely to help explain the history of the universe than the history of the universe is to explain the present structure. History would lead us toward narrative and away from theory. (Here again we see the power of the time-denying and mathematics-idolizing tradition.)

All efforts to achieve this unification, on the basis of such an approach to the relation between theory and history, have thus far failed. They have failed in a particular way. The unhistorical theories that would advance the unification project and explain the workings of nature in its ultimate constituents – in particular, contemporary string theory in particle physics – all turn out to be compatible with a vast number of other ways in which nature might also work but does not in fact work, so far as we can observe.

Moreover, within the observed universe there are a number of constants or parameters that have mysteriously precise but as yet unexplained values. Some of these constants we may treat as "dimensional": that is to say, as measures, like rulers, of the rest of the nature. Then, however, the seeming arbitrariness of the remaining, dimensionless parameters stares us all the more starkly in the face.

There is, however, this secondary opponent of ours assures us, a solution to all these problems, or at least a direction in which to look for a solution. The solution is to treat our universe as simply one of many universes. In the strongest formulation of this idea, and the one that has exercised the greatest influence on contemporary cosmology, these are not just many possible worlds; they are many actual worlds: the universe that we observe and the many universes that we could not even in principle observe.

The laws formulated, or discovered, in the course of the unification project address the full array of hypothetical universes. What seems to be a failure – the failure to explain what we observe in terms that do not also explain what we do not observe – is in fact, they insist, an achievement: the unified or grand laws account for the totality of universes, not just for the universe that we happen to inhabit.

The enigmatic constants or parameters, with their disturbingly precise and unexplained values, will eventually be accounted for by lower-order deterministic or statistical explanations addressing our segment of the vastly larger multiverse. We shall be guided toward these lower-order explanations by a reverse epistemological engineering: the selection of that subset of the larger set of laws that would be capable of producing and of accommodating our existence (the so-called anthropic principle, in one of its stronger or weaker versions).

Against this descent of science into allegory, circularity, special pleading, and factless speculation, we affirm the first of our three central theses: that there is one real universe, all of the parts of which are in tighter or looser causal communion with one another. There is better reason to believe today in a succession of causally connected universes than there is to believe in a plurality of causally unconnected universes.

The lesser enemy against which we direct our argument – the attempt to promote the unification of physics by multiplying imaginary universes – has as its ulterior motive the concealment of the vulnerabilities of our chief enemy – the tradition of thought that throughout the history of modern physics has denied or devalued the reality of time. The most important service, or disservice, rendered by the secondary opponent has been the postponement of a reckoning with the main antagonist.

The lesser enemy has taken care not to bring into question the tradition of thought that has entangled modern physics, from Galileo and Newton, to Einstein and quantum mechanics, in the diminishment of time and in the treatment of mathematics as a royal road to timeless truth. It has therefore failed as well to expose the limits that this tradition places on our ability to grasp the full range of the implications of the empirical discoveries of science. Instead it has built a wall of defense around these equivocations.

A working assumption of the argument of this book is that the two intellectual campaigns – the one against the denial of time and the worship of mathematics, the other against the plurality of universes – are connected. Given the history of these ideas, the first campaign is the decisive one. The second campaign is important but accessory.

At issue is how we should best approach the future agenda of physics and cosmology. The association of string theory with the idea of a plurality of universes (a multiverse) threatens to deepen the tradition of thought that has refused fully to recognize the reality of time and that has insisted on seeing mathematics as a shortcut to privileged insight into the ultimate nature of reality. Instead of deepening this tradition, we seek to rid ourselves of it. Cast this way of thinking aside the better to reinterpret past discoveries, and make future ones, without the restraints that it imposes.

3 The singular existence of the universe

THE CONCEPTION OF THE SINGULAR EXISTENCE OF THE UNIVERSE INTRODUCED

There is one real universe. This universe may extend indefinitely back in time, in a succession of earlier universes or of earlier states of the universe. We have no sufficient reason to believe in the simultaneous existence of other universes with which we have, and cannot have, now or forever, causal contact.

Causal communion is the decisive criterion for the joint membership of natural phenomena in the same universe. The parts of a universe are causally connected, directly or indirectly, to all the other parts over time. Over time is the first and most important qualification. Two parts of nature belong to the same universe if they share any event in their causal past, even if they have subsequently become causally disjoint. It is the network of causal relations viewed backward into the past that determines the scope of causal communion and thus the separate existence of a universe. The criterion is dynamic rather than static, historical rather than exclusively structural, and presupposes the reality of time.

It is then not the constancy of the laws and symmetries of nature that distinguishes a universe. It is causal connection over time. Causal connections, as the discussion of the second cosmological fallacy suggested, may not present themselves, in certain states of nature, as regularities: the laws, symmetries, and constants that we observe in the cooled-down universe, with its differentiated structure and its recurrent phenomena. The law-like workings of nature are best understood as a mode of causality rather than as the basis of causation. What matters to the discrimination of a single universe is that the constituents of such a universe display an uninterrupted causal history, whether or not in law-like and symmetrical fashion.

From the fundamental qualification relating the unity and identity of the one real universe to causal connection over time there follows another qualification. Regions of the observable universe may not now be in causal contact, and what we can observe may form only a small part of a much larger universe. Nevertheless, causally disjoint parts of the universe continue to count as parts of the same universe, the one real one, if they share a common history.

There is no clear distinction between the idea of causally disjoint parts of the one real universe and the notion of branching, bubbling, or domain universes that may arise in the course of the history of the universe. I shall later argue that such universes, if they exist, form part of a history of succession. The conception of branching universes describes incidents in a history of succession or of transformation. A view of such a history is in turn only a variant on the idea of the singular existence of a universe, formed in the course of events stretching back indefinitely in time.

Everything in the one real universe, we claim in this book, including both the fundamental structures and the most general regularities of nature, changes sooner or later, although both the regularities and the structures are remarkably stable in the cooled-down universe that we observe. That everything in the universe changes sooner or later cannot be inferred from the idea of the singular existence of the universe. The development and defense of this thesis is the concern of Chapters 4 and 5 of this book.

Nevertheless, an understanding of what the singular existence of the universe implies must anticipate some elements of the later argument about the mutability of both structures and laws of nature. No part of the identity of a universe requires that it conserve the same laws and structures, only that causation, albeit stressed, never be interrupted. Both the continuity of causation and the susceptibility of everything to change, including change itself, represent expressions of the inclusive reality of time.

The proposal of the singular existence of the universe leads immediately to what I shall call the antinomy of cosmogenesis. The history of the universe, under the thesis of singular existence, may extend indefinitely back into the past. It seems at first that of two things, one must be true. At some moment in this past, the universe and, with it, time may have emerged out of nothing. However, as Lear said to Cordelia, nothing will come of nothing. An unstable vacuum state, for example, is not nothing. It is something. It must have a history.

Alternatively, the universe may be eternal. Eternity is infinity in time. Nothing, however, in nature can be infinite, and time is part of nature: according to the argument of this book, the most fundamental part, but part nonetheless. The infinite is a mathematical contrivance. Only a way of thinking that sees mathematics as the oracle of nature and the prophet of science can make room for the infinite in its explanatory practices. One of the aims of this book is to combat that way of thinking and to offer an alternative to it.

There is an infinite difference between an indefinitely long history and eternity. The invocation of an eternal universe is no more defensible than the appeal to an infinite initial singularity at the beginning of our present universe. In both instances, a mathematical idea, with no counterpart in physical nature, is made to do service for missing insight.

We have reason to resist both sides of the antinomy of cosmogenesis. Neither science nor natural philosophy has any prospect of dissolving this antinomy, by justifying one of its sides or by finding a third, synthetic solution. Rather than pretending that we can find our way out of this antinomy by a conceptual maneuver, or by the subordination of natural science to some species of metaphysical rationalism, we should recognize this antinomy for what it is: a sign not only of the limits to the powers of science and of its ally in natural philosophy but also of our groundlessness – our inability to grasp the ground of being or of existence. Science, and natural philosophy along with it, can only be corrupted by claiming to have unlocked the secrets of being and of existence and by seeking to occupy the place of a lost religion.

It does not follow from our inability to resolve the antinomy of cosmogenesis that the idea of a history of the one real universe extending indefinitely back into the past, to earlier universes, or earlier periods of the universe in which we find ourselves, is either incoherent or vulnerable to the objections aroused by the two sides of the antinomy.

Something useful is gained by placing the confrontation with this antinomy in a remote and inaccessible past, far beyond what, by the lights of the now standard cosmological model, is represented to be the incandescent and explosive beginning of our present universe. The work of empirical inquiry and causal investigation can then retreat, step by step, to that horizon of past time, within the limits of our equipment and of our ingenuity. None of that work need make assumptions offensive to causal reasoning or to our natural experience, as either the making of something out of nothing or the eternity – that is to say, the temporal infinity – of the world would. Under this self-denying and skeptical response, the antinomy of cosmogenesis continues to hang over us as a reminder of our radical and insuperable limitations of insight. It need not, however, prevent us from going forward, without mistaking science for the source of a solution to the enigma of existence.

This antinomy of cosmogenesis recalls Kant's first antinomy of pure reason in his *Critique of Pure Reason*. In his formulation, the unacceptable thesis, following immediately upon his introductory discussion of the system of cosmological ideas, is that "the world has a beginning in time, and in space it is also enclosed in boundaries." The unacceptable antithesis is that "the world has no beginning and no bounds in space, but is infinite with regard to both time and space." In accordance with the spirit of his system, Kant presents both time and space, as he does causation, as presuppositions of the way in which we represent phenomena. By contrast, we regard time and space, like causation, as features of nature and believe that the power of science to correct our pre-scientific understanding of the world is open-ended although it is not unlimited.

Moreover, Kant supposes time and space to share a common fate: they must both and together be either bounded or infinite. He then argues that neither possibility is acceptable to reason.

In contemporary cosmology, the insistence on treating space and time as one, with regard to their boundedness or infinity, is preserved only in the conception of Lorentzian spacetime, assumed (mistakenly, I argue) by the predominant interpretations of general relativity and of its field equations to be an integral part of this theory and a beneficiary of its empirical validations. In rejecting that conception both as an account of nature and as an indispensable aspect of general relativity, we open the way to the distinct and unequal treatment of time and space. Space may be emergent, indeed repeatedly emergent in successive universes, or periods in the history of the one real universe. According to accepted topological principles, the universe may be finite in spatial extent, and yet have no boundaries. Time, however, we argue, is best regarded as non-emergent, in the sense that it derives from nothing else and thus, as the susceptibility of what is to change, represents the most fundamental aspect of natural reality.

We may be tempted to conclude that time can be non-emergent only if it is eternal, thus admitting the infinite into our account of nature. However, all that the non-emergence of time may require is that time have continued indefinitely back into the past. That the world is temporal, and justifies a temporal naturalism in our understanding of it, is an aspect of its being, factitiously, what it is, rather than something else. As with every other aspect of natural reality, the inclusive and non-derivative reality of time cannot be inferred from any higher-order rational necessity.

To resolve the antinomy of cosmogenesis exceeds the capabilities of science. We cannot look into the beginning and the end of time, if it has a beginning and an end. We cannot explain temporal reality, or the reality of time, as emergent out of timeless being or out of nothingness. We cannot justifiably circumscribe the reach of time so that some parts of nature – its basic structures and the regularities that they exhibit – remain, immutable, outside it. We cannot infer the temporal character of nature from supposed constraints on what can or must be, in obedience to the deliverances of metaphysics or of mathematics. All that we can do is to recognize the real as temporal, according to the lyrics of the country-music song: it is what it is, till it aint anymore. Yet the recession of the history of the universe to a remote past gives a cosmology that refuses to inherit the pretenses of metaphysics and that recognizes its inability to settle the antinomy of cosmogenesis a vast field in which to work.

In the remainder of this section, I define concepts useful to making sense of the debate about one or many universes. The next section ("Arguments for the singular existence of the universe") outlines the arguments for embracing the thesis of singular existence to the detriment of its rivals. The rest of the chapter develops the thesis by exploring its implications for the agenda of cosmology.

At the time of this writing, the most influential version of the idea of many universes was the conception labeled the multiverse, of a multitude of distinct universes, neither now nor ever in causal contact with one another (except for the conjecture of collisions among them), and each possessed of distinct structures and regularities. The most promising version of the idea of singular existence was the idea of a universe the history of which extends backward before the formation of the present cooled-down universes to earlier universes or to earlier periods in the history of our universe. With the qualifications that I earlier enumerated, there is, according to the idea of singular existence, only one universe at a time. The contest between the conjecture of divergent universes, distinctly ruled and organized, and the conjecture of one real universe going back in time is the present shape of the argument.

* * *

According to the idea of a plurality of universes, there are many, even indefinitely many, universes coexisting at the same time, or at least (given the difficulty of ascertaining their temporal relation to one another, if indeed time has any reality) existing in such a way that they cannot be said to have a history. For the reason stated earlier, branching, bubbling, or domain universes, emerging out of a common history, do not, for the purpose of this terminology, represent plural universes. Plural universes, in the strong sense in which I use the term, share no common history.

Such universes have no causal communion with one another, and have never enjoyed such contact. We are barred, in principle and forever, from access to them. A qualification to this property of plural universes is the conjecture of collisions among them. These collisions might leave traces of their occurrence, "ripples" in the cosmic microwave radiation background. Such ripples, however, have never been observed and, if observed, might have many other causes. Another qualification, which certain adepts of the multiverse idea have proposed, is that these inaccessible universes may leave in our universe traces of their existence unrelated to collisions. For example, in the early twenty-first century some cosmologists claimed to find evidence of a "dark flow" in the motion of galactic clusters. In such flow, they discerned a mark of the presence of other universes. Even if confirmed, however, such a phenomenon need not be a signature of another universe; it can be explained in other ways. As there is no prospect, even in principle, of accessing these other universes, regardless of the development of our observational and experimental equipment and capabilities, we cannot hope to demonstrate a mechanism of causal interaction between the other universes and our own.

There are two main variants of plurality in this strong sense. Parallel universes are universes sharing the same fundamental structures and regularities as our universe but not sharing causal contact or history. Divergent universes are universes likewise having no such shared causal contact or history but, unlike parallel universes, possessing distinct regularities (laws, symmetries, and constants) and structures. In contemporary discussion, the idea of divergent universes commonly goes under the label multiverse, although the label has sometimes also been used to express the idea of parallel universes, with much resulting confusion.

This simple classification of plurality into parallelism and divergence leaves open, conceptually, two intermediate possibilities: universes that have the same basic structures without the same regularities, and universes that have the same regularities without the same structures. However, the former possibility is absurd: the same structures would exhibit the same regularities. The latter possibility makes sense only if we suppose that the regularities underdetermine the states of affairs, countenancing a range of possible states of affairs, manifest in different universes. Such was the reasoning that in the mid-twentieth century led a few opponents of the Copenhagen interpretation of quantum mechanics to an early version of the multiverse idea and that in the late twentieth century prompted many more adepts of the string theory school of particle physics to find in the multiverse view a justification for the latitudinarianism of their doctrine. Better, then, to disregard the two intermediate possibilities.

Both the ideas of parallel and of divergent universes have long genealogies in the histories of science and of natural philosophy. I shall disregard this history except insofar as its more recent incidents bear on the issues at stake in my argument.

The notion of parallel universes, with no causal contact or shared history, but with the same fundamental structures and regularities – universes that are therefore mirrors of each other – performs no explanatory role in contemporary cosmology. If the structures and the regularities are the same, and are not credited with any potential for transformative divergence over time, the existence of such postulated mirror universes does no work other than to express the fecundity of a repeated cosmogenesis. It would be, however, unlike any similar fecundity in the evolution of life forms. In the evolution of life, even before the development of sexual selection and the corresponding Mendelian mechanisms, multiplication prefigures variation.

The closest that contemporary cosmology comes to exemplifying this idea has been in the loose allusion to multiple cosmogenesis in theories of eternal inflation, with no explicit account of variation of structures and regularities in the universes that would be generated by the multiple explosive events foreseen by eternal inflation. Moreover, if such universes existed, they would be better regarded as instances of succession. It is therefore unsurprising to find the idea of parallel universes to be a largely unoccupied position in contemporary cosmology.

The influential contemporary representative of the idea of plurality is that of divergent universes, the multiverse. Each universe, causally cut off, at all times, from all other universes, has its own laws and its own organization at the level of its most elementary constituents. Each is a world unto itself.

The conception of divergent universes has been proposed at least twice over the last sixty years. On each occasion, the alleged basis has been different. However, the theoretical motivation and logic in the two episodes have been strikingly and revealing similar. In the 1950s Hugh Everett intimated, and John Wheeler more explicitly proposed, the genesis of a multitude of universes out of quantum realities; each outcome of a possible quantum state would exist in a different world. Thus, the underdetermination of quantum theory, under the predominant Copenhagen interpretation, would be redressed by a proliferation of worlds, each of them enacting one of the otherwise unrealized quantum possibilities. What seemed to be underdetermination was reinterpreted by a theory that took every possible state of affairs to be real, albeit somewhere else.

In the closing decades of the twentieth century, the chief impulse to postulate a vast number of universes came from string theory, and more generally from the marriage of string theory, in a cosmological setting, to theories of eternal inflation and to anthropic thinking. To each of its mathematical possibilities, there was imagined to correspond a different vacuum state or universe, in fact 10^{500} or more. This view in particle physics radically underdetermined nature as we observe it in the cooled-down universe and failed to provide any criterion of selection among the states of affairs with which it was compatible. The problem was converted into a solution, and the failure into an achievement. The conversion relied on the simple device of imagining that each of the unobserved states of affairs was enacted in a different universe, replete with its own distinctive structures and regularities, in conformity to one of the countless (but not infinite) variations admitted by the theory.

The circle was closed with the appeal to anthropic reasoning: the universe in which we find ourselves would be one of this crowd of universes. Its extraordinarily improbable initial conditions (improbable by the standard of ideas used to account for the workings of nature in the cooled-down universe) and its fine-tuned properties and constants were to be explained retrospectively as the sole combination of features capable of having resulted in us, the human race, which discovers these truths.

It was, with a basis in this style of particle physics rather than in quantum mechanics, essentially the same line of reasoning that Everett had proposed a few decades before. However, it went further in radicalizing the attitude to mathematics as a prefiguring of natural reality. It went further as well in its deployment of retrospective anthropic rationalization as a proxy for causal accounts more canonical in the dominant tradition of physics.

Much of the argument of this chapter is devoted to a criticism of the multiverse idea and to a development and defense of the thesis of the singular existence of the universe, going back indefinitely in time: succession and transformation rather than plurality. There is, however, a sense in which the idea of plurality, in the form of divergent universes or a multiverse, resembles the thesis of singular existence and succession.

The multiverse idea in its contemporary form and in some of its more radical developments suggests a notion of regional laws of nature. Although there may be very general and fundamental laws, marking the perimeter of alternative universes, the effective laws represent the distinctive regularities of each universe. They are regional rather than universal. They are in a sense determined by the environment rather than determining it, as the standard way of thinking in the physics inaugurated by Galileo and Newton would require. Their regional character will be more salient, the more we discount the power of mathematics to reveal and explain, rather than simply to represent, the variations of nature.

The thesis of singular existence and of non-cyclic succession may be interpreted to apply to periods in the history of the universe, or to successive universes, a similar idea of domain-specific laws, similarly opposed to the idea of the universality of the regularities of nature. The idea of local laws, or of laws specific to different universes, has unwittingly prepared the ground for the rejection of the universality and constancy of the laws and other regularities of nature. There are, however, two differences of far-reaching consequence.

The first difference is that for the thesis of singularity and succession, natural variation works through time. The weakening of the absolute character of the laws of nature is temporal rather than spatial. The conception of the reality of time becomes inseparable from the thesis of the singular existence of the universe. The laws and other regularities of nature are mutable features of the one real universe.

The second difference is that if we regard time as inclusively real we cannot exempt either the regularities or the structures of nature from its reach. Neither the laws, symmetries, and supposed constants of nature nor its elementary constituents are permanent features of nature, moving and unmoved bystanders to its history, untouched by reciprocated action. On this view, there are no laws and structures, no matter how fundamental, that fail to change sooner or later, and that have not changed, or emerged as the outcome of change, in the past. Even the extent to which causality displays recurrent, law-like features may vary, marking some states of nature but not others. In such a universe, everything changes sooner or later, including change itself.

* * *

Consider now the chief variants of the idea of the singular existence of the universe. The universe may be solitary, without predecessor or prior history, before the fiery beginning that the standard cosmological model assigns to it, coming abruptly and out of nothing.

Alternatively, the universe may have a history extending before the "big bang" inferred by that model to have taken place in the earliest moments of the history of the universe. The reference to such a history is the thesis of succession, which appears as a development of the idea of singular existence rather than as an alternative to it. We may picture this earlier history either as a succession of universes or as a succession of periods in the history of the one real universe. A preference for one of these vocabularies over the other is of little consequence if we allow that causal continuity between successive universes, or between successive periods in the history of the one real universe, may be stressed, but never broken.

To say that causal continuity remains unbroken is to signify that causal succession – the *after* shaped by the *before* – persists without interruption even in such extreme circumstances. To say that it is stressed means that the distinction between laws and states of affairs may, in this extremity, break down and causal connection may cease to present in repetitious law-like form. That would be nature as a world of singular events: the possibility of which (according to the argument against the second cosmological fallacy) is predicated on the view that causal connections are a primitive feature of nature rather than instances or enactments of laws and symmetries. Consequently it is also predicated on the idea that laws and symmetries are a mode of causality – the mode prevailing in the cooled-down universe – rather than the basis of causality and the warrants of causal explanations.

There are in turn two main variants of the idea of succession: cyclic and non-cyclic. According to the cyclic view, the basic regularities of nature remain unchanged throughout the history of the universe or of successive universes. The structural forms of nature change, but in conformity to these unchanging regularities and in recurrent stages. These stages remain identical in each iteration of the cycle. Despite the differences among proposals of a cyclic cosmology made, over the course of the twentieth century, by Soddy, Tolman, Friedmann, Sakharov, Rozental, Rosen and Israelit, Penrose, Steinhardt and Turok, and others, the cyclic view has retained, across these variants, a discernible identity and a characteristic argumentative strategy.

According to the non-cyclic view, there is no unchanging feature of nature, other than its susceptibility to changing change, which we call time. The non-cyclic view of succession forms an integral part of the temporal naturalism developed and defended in this book.

The distinctive explanatory challenges faced by each of these three forms of the thesis of the singular existence of the universe – absolute beginning, cyclic succession, and non-cyclic succession – help elucidate the distinctions among them. The next section ("Arguments for the singular existence of the universe") takes up the argument in favor of non-cyclic succession against the other two variants of the thesis of singular existence (absolute beginning and cyclic succession) as well as against the thesis of plurality.

* * *

In contemporary cosmology, the idea of an absolute beginning is suggested by the initial infinite singularity that a long line of twentieth-century cosmologists argued to be implied by the field equations of general relativity. It was, however, widely recognized (even by Einstein himself) that this inference, rather than describing a physical state of affairs, revealed a breakdown of the theory when it was carried beyond its proper domain of application. More generally, as I argue throughout and as many have recognized in the history of both physics and mathematics, the infinite that is invoked in this view is a mathematical conception with no presence in nature. As with the multiverse idea, the notion of an absolute beginning attempts to convert a limitation of insight into a conception of nature and its history.

Moreover, to interpret the thesis of singular existence as if it required such an absolute beginning is to embrace the first horn of the antinomy of cosmogenesis: the emergence of something out of nothing. If we lift the screen of the mathematical idea of the infinite, illegitimately applied in cosmology, we are then faced with a choice between two accounts of the absolute beginning. On one account, nothing is nothing: we impose an arbitrary cutoff on causation and time, implying that they emerge not of something but out of nothing. On the alternative account: nothing is something – an unstable vacuum field, for example. Then, however, we are entitled, indeed required, to ask where this something comes from and what its earlier history, before the "big bang," may be. We have then effectively abandoned the notion of an absolute beginning in favor of the thesis of succession.

The challenge faced by the thesis of an absolute beginning is thus to escape these available but unacceptable choices. It does not seem that it can do so.

* * *

When we turn to the views of singularity that combine singularity with succession (described either as a sequence of universes or as a sequence of periods in the history of the one real universe), the first problem that both the cyclic and the non-cyclic variants of succession must confront is their relation to the agenda of empirical science. Unlike both the causally disjoint universes of the multiverse conception and the infinite initial singularity that may be used to represent the idea of an absolute beginning, both cyclic and non-cyclic variants of succession are in principle open to empirical research, if not directly, then indirectly by their signatures, vestiges, or effects. However, it is not good enough to say that they are open to such research if they fail in fact to be opened to it, and to help open it, by informing an agenda of investigation that cosmology can implement as its observational and experimental equipment becomes more powerful and its theoretical insight more acute. In the absence of such developments, the thesis of succession can be justly accused of being as speculative as the multiverse conception that it opposes. The presumption of causal continuity and temporal extension will not suffice to defend it against this accusation.

Another difficulty that the idea of succession must confront, in both its cyclic and non-cyclic variants, is the problem presented by the second horn of the antinomy of cosmogenesis. The cyclic or non-cyclic succession either has a beginning or it does not. If it has a beginning, further back in time than the explosive inception of our present universe that is pictured by the standard cosmological model, we face once again the problems of absolute beginning, having only pushed them into the past. If it has no beginning, the universe, as a succession of periods or of universes, is eternal. We shall then have reintroduced with regard to time the infinity that we rejected in other departments of our cosmological thinking. The reasons to regard time as, unlike space, non-emergent (which I explore in Chapter 4) may seem to provide grounds to exempt time, as eternity, from the rule against the banishment of the infinite from nature and from science.

The adoption of the thesis of the eternity of the world, however, is neither a necessary consequence of the view of time as non-emergent nor easy to reconcile with the rule against infinity. It is more appropriate to the spirit of a self-denying ordinance in science, relinquishing metaphysical and theological pretense the better to claim and to exercise other powers, to assume that the universe, or the succession of universes, extends indefinitely back into the past. Cosmology, at least in its present condition and with its present insights and instruments, is not entitled to describe the world as either eternal or as emergent from an absolute beginning. Time may be held to be non-emergent because there may be nothing more fundamental than it, and nothing from which it derives, without it being the case that we have any basis, other than rejection of the making of something out of nothing, to describe it as eternal.

A third challenge with which the idea of succession, in both its cyclic and non-cyclic versions, must deal is its apparent contradiction to the now predominant interpretations of general relativity. These interpretations, with their insistence on "many-fingered" time, their treatment of the spacetime continuum as a central and indispensable aspect of general relativity, and their approach to time as a derivative feature of the disposition of matter and motion in the universe, exclude the possibility of a cosmic or global time that is also preferred: the sense in which I use cosmic time here. They permit only the choice of spacetime coordinates that are cosmic in the sense that they cover the whole universe but that are not preferred. The choice of any such spacetime coordinate remains arbitrary from the standpoint of the theory.

A preferred cosmic time allows every event in the history of the universe, or of successive universes, to be placed, in principle, on a single time line. It is hard to see how we can make sense of the idea of succession, and thus of the singular existence of the universe, without appealing to a preferred cosmic time. A concern of Chapter 4 is to discuss the reasons for which, and the manner in which, we can reinterpret the empirical hard core of general relativity to allow for the existence of such time, required by the thesis of succession.

The cyclic version of this thesis faces, in addition to these general challenges to all variants of succession, a further difficulty. This version affirms the continuity of both the basic structures of nature, recurring in particular stages, and the regularities of nature, which account for such persistence and recurrence. In so doing, it allows part of nature – its fundamental constituents and regularities – to remain outside the reach of temporal change and reciprocated action. We have reason to think, I later argue, that nothing remains outside that reach.

Moreover, we know already that many aspects of our universe, for example its chemical constitution, could not have existed early in its history. Even the most fundamental constituents of nature and their interactions must have been different in the early, hot, and supercondensed universe. If everything was different structurally, how could the regularities of nature have been the same?

We would have to suppose that the same structural transitions recurred many times before, and did so under the governance of laws, symmetries, and constants that remained themselves immutable. This idea, however, is a metaphysical leap, unsupported by anything in our knowledge of nature other than the stability of the regularities that we observe in our cooled-down universe and that we infer to have been stable down to relatively early periods in its history. It is more economical, and more in accord with explanatory practice in sciences that are accustomed to deal with the mutability of their subject matter, such as the earth and life sciences, to think that the regularities evolved together with the structures.

The non-cyclic variant of the thesis of succession, with its insistence that everything, including the structures and the laws, changes sooner or later, must also overcome two major obstacles, in addition to the challenges presented to all variants of succession.

The first obstacle is the undeniable stability of the regularities and constituents of nature in the cooled-down universe that has been, until recently, the sole subject matter of cosmology and physics. If we come to think that these constituents and regularities have changed in the past and may change again, we must reconcile the idea that they are mutable with the fact that they have been stable.

The second obstacle is the riddle resulting from the idea of their mutability, which we call the conundrum of the meta-laws. If the laws and structures evolve jointly, it seems that, in accordance with this way of thinking and with the continuity of causation, their co-evolution must be neither law-governed nor uncaused. To address the conundrum of the meta-laws is one of our main goals in the subsequent arguments of this book.

ARGUMENTS FOR THE SINGULAR EXISTENCE OF THE UNIVERSE

I now outline arguments for the singular existence of the universe, as well as for the non-cyclic succession of universes, and against the plurality of universes. I refer to the notion of plurality with emphasis on the form of this notion that has recently acquired influence: the one that I described in the previous section ("The conception of the singular existence of the universe introduced") as divergent rather than parallel plurality and that is now commonly known as the multiverse.

These arguments have different characters. Some of them are negative, directed against the idea of plurality. Others are affirmative, in favor of the ideas of singular existence and of non-cyclic succession. Some invoke a conception of what the relation between theoretical speculation and empirical validation can and should be like in cosmology. Others appeal to claims about the most persuasive interpretation of past cosmological discoveries. Others yet emphasize the implications both of the views that I propose and of those that I criticize for the additional matters that we address in this book: the reality of time and the role of mathematics. Some of the arguments touch on foundational issues in science. Others are subsidiary to these basic claims.

Despite their heterogeneity, however, the arguments connect and overlap. They exhibit a point of view that should be judged by the fecundity of the agenda of theoretical reasoning and of empirical research that it may inform as well as by the merits of its particular propositions. The rest of this chapter explores the implications of this way of thinking about the solitary character of the universe for a number of problems that are important to the present and future of cosmology.

1. *The argument from the non-empirical character of the multiverse idea.* Plural universes are in principle beyond the reach of empirical inquiry. They resist empirical investigation because they are not, and have never been, in causal contact with our universe. They intersect no light cone crossing the history of our universe; they share no history with us. The impossibility of confirming or disconfirming their existence reduces them to a fabrication. Reliance on any such fabrication represents a major flaw in a scientific theory.

To establish this point, it is important to exclude from the idea of plurality branching, bubble, or domain universes that may have resulted, just as black holes do, from events in the history of our universe. Such universes should be considered instances of succession rather than of plurality. Our universe may not enjoy causal contact with them now. Nevertheless, there was once a time when it did. Causal communion must be defined historically rather than statically: another example of the historical character of basic cosmological concepts.

We may not be able, with our existing equipment and powers of observation and simulation, to take empirical advantage of these past episodes of historical intersection between universes, or among parts of the universe, that later fell out of causal contact with one another and with us. Nevertheless, they are in principle accessible – if not now, later, and if not directly, indirectly – to empirical study.

It is connection over time that provides the criterion by which to distinguish a universe, whether singular or one of many. Notice that nothing in this notion of causal connection assumes action at a distance, the suggestion of which led Einstein to reject "Mach's Principle," after having embraced it and named it. (Such action is nevertheless now admitted in thinking about fields and quantum entanglement.) The local inertial field is completely determined by the dynamical fields of the universe, but it is not determined by the matter content of the universe without regard to the shape of those fields.

The net of causal connection must remain uninterrupted even if at a distance. Everything in a universe must influence everything else through connecting links or mechanisms for the transmission of causal influence to occur. The influence, however, may be historical. It then becomes a major concern of cosmological or physical theory to supply an account of such links or mechanisms that can be put to observational or experimental test.

What this criterion of the separate existence of a universe does presuppose is the continuity of causal connections, whether or not they assume law-like form, together with the reality of time as a condition of causality. We cannot affirm the identity of a universe by viewing it without regard to its evolution; we can affirm it only by seeing it in evolutionary context, which is to say in time.

What was in causal communion may at some point cease to be in causal communion. Whether such a division does or will occur is not something that can be known a priori, or confirmed on the basis of theoretical considerations. It depends on how the universe in fact evolves and, in particular, on the nature and rapidity of its expansion.

That branching, bubble, or domain universes do not deserve to be considered instances of plurality is shown by the absence of any clear distinction between the idea of universes branching out of a single universe and of parts of a universe no longer in causal contact with one another. That parts of even the observed region of our universe cannot now be in causal contact is a widely accepted tenet of contemporary cosmology, helping motivate the theory of cosmological inflation. The unity of our universe is established by causal continuity over time, not by inclusive causal contact at all times.

Plurality, according to the nomenclature proposed in the earlier section ("The conception of the singular existence of the universe introduced"), can take the form of an idea of parallel universes, all exhibiting the same regularities (laws, symmetries, and constants) or of divergent universes, each of which displays different regularities and structures. The idea of divergent universes goes, in contemporary cosmology, under the name multiverse.

There is no reason to uphold the idea of parallel universes: it combines an absence of empirical validation, or of susceptibility to empirical challenge, with a lack of explanatory function. It is therefore unsurprising that it attracts little interest among present-day cosmologists. By contrast, the idea of divergent universes combines the same lack of empirical support or vulnerability with an explanatory role. This role, however, rather than providing a basis to accept it, supplies an additional reason to doubt it.

The driving motive to postulate many divergent universes is to convert an explanatory embarrassment into an explanatory triumph. The explanatory embarrassment is the failure of the most influential variant of contemporary particle physics – string theory – to apply narrowly to the universe that we see rather than applying as well to countless universes that we do not – and cannot, now or ever – observe. Even if we impose on perturbative string theories a long series of constraints (of which conformity to de Sitter spacetime is only one), no more than a tiny portion of them apply to the observed universe. We can tell ourselves, so long as we are willing to accept the idea of plurality, that each of the inapplicable theories is realized in one of those unobservable universes. By such special pleading, we pretend to turn a failure into a success, as if the setback to prevailing ideas were a taint to be disguised rather than an opportunity to be seized.

It has been argued in favor of the multiverse idea that the particle theories suggesting this idea have been successful at many other predictions and that therefore their prediction of a multiverse deserves deference. This defense, however, is doubly defective. First, it confuses the extraordinary predictive success of the standard model of particle physics, which bears no relation to the multiverse conception, with the claims of string theory, which has had a more questionable record. Second, and more fundamentally, it abuses the idea of prediction. To postulate the existence of universes on the ground that each such universe realizes one of a vast number of possible states of nature countenanced by the mathematics of a physical theory is a prediction only in a contrived sense. If the postulated entities cannot ever be observed, and no trace of them even indirectly found, the application of the idea of prediction has lost touch with what prediction has meant in science.

We go even further in redefining riddles as solutions and failures as triumphs when we combine the multiverse notion with the practice of anthropic reasoning. Only a few of the possibilities established by string theory in particle physics approach the realities registered in the observed universe. (In fact, no version of string theory has completely reproduced the standard model of particle physics, which has been amply confirmed by observation and experiment, or even its supersymmetric extension.) All the other equations and solutions generated by string theory must describe the inaccessible universes in which the realities to which they would apply are supposedly realized. In each of these universes, the regularities are the ones that conform to equations and solutions not realized in our universe and must therefore differ from the laws, symmetries, and constants observed in our cooled-down universe.

By this combination of moves, we seem to dispose of the difficulty presented by the massive underdetermination of reality by theory in some of the variants of particle physics now commanding the greatest support. We do so, however, only at the cost of devising a research agenda that weakens the vital connection between theoretical insight and empirical discovery.

The opposing idea, of the singular existence of the universe, conforms to a tradition of several centuries. It avoids resort to metaphysical fictions. However, its conservatism turns out to be revolutionary. It denies us a facile solution to connected and major conundrums. As a result, it helps open up another, little explored way of addressing those same puzzles: a way giving a central role to the ideas of the reality of time and of the mutability of the laws of nature. These ideas in turn compel us to reconsider the relation of mathematics to nature and to science.

Such a turn invokes no fantastical entities. Many of its aspects cannot yet be subject to observational or experimental test, while others can. None of its claims, however, is in principle and forever immune to empirical challenge.

2. *The argument from the preference for a view making it possible to begin answering the question: Where do the initial conditions of the universe and the laws of nature come from?* The idea of a multitude of universes having no causal contact with one another and no shared history makes it impossible to answer the question: Where do the laws and the initial conditions of our universe come from? Or, more precisely, the idea of plurality leaves open only one answer to this question: they are one of the countless possibilities envisaged by a fundamental physical theory that also accommodates many other possibilities. All the other possibilities – the ones not realized in our universe – are realized in the other universes that we can only postulate but never inspect, even indirectly.

Let us generalize the argument, without regard to the specific content of the multiverse idea.

There are in principle three approaches to the question of the basis of the laws (and other regularities) and the initial conditions of the universe.

a. The laws and initial conditions just are what they are. They are a primitive feature of nature. They cannot be inferred from anything else. Then the regularities of nature are like surprising singular events, except that what is not only surprising and singular but also beyond the reach of further explanation is the universe itself.

   Science then presents the apparent disorder of the universe under the semblance of regularities. This representation, however, only postpones the confrontation with brute factitiousness. The phenomena are to be explained by the laws, symmetries, and constants, together with the unexplained initial conditions of the universe. At the next step of reasoning, however, both the laws and the initial conditions remain unexplained.

b. The laws of nature and the initial conditions of the universe can be inferred from some higher-order set of abstractions: a set of laws of many possible universes (as in the multiverse idea) or an account of why only the features of this one universe of ours are possible because, for example, only they realize in the richest and fullest way the potential of being.

   This approach (expressed in the thought of philosophers as different as Leibniz and Hegel) always amounts to a mystification of one kind or another. It seeks to make the brute just-so-ness or factitiousness of nature appear to vanish under the spell of rational necessity. The relative ease with which the surprising features of the one real universe can be made to follow from metaphysical a prioris discredits such efforts at wholesale retrospective rationalization.

   A variant of this approach is the idea that every structure conceived by mathematics is expressed in some universe as its laws and symmetries. At the limit, each such universe *is* one of these mathematical structures. Because the realities that we observe in our universe embody only a tiny part of these mathematical structures, all the other structures must be embodied in other universes. The rational that is real is mathematics. Taken to the hilt, such a view dismisses the reality of time and dispenses with the concept of initial conditions.

c. The laws and initial conditions can be explained historically. Like everything else in nature, they are the outcomes of earlier states of affairs. To be sure, the appeal to historical explanation fails to exempt us from the problem of factitiousness ultimately: that the universe and its history, or the universe viewed historically, just happen to be one way rather than another.

This third approach nevertheless has several advantages over the other two. A first advantage is that it opens an agenda of empirical inquiry, closely connected with the most important discoveries that cosmology has made over the last century. All these discoveries have to do, in one way or another, with aspects of the history of the universe. They provide incitements to the continuing transformation of cosmology into a historical science. That transformation remains incomplete so long as the laws and initial conditions fail to be viewed historically. A pressing task is to find ways to render these questions amenable to observational and experimental research, by considering the empirical consequences in the present universe of alternative conjectures about its early history, or about the history of previous universes.

A second advantage of this approach over its rivals is that it enables us to delay confronting the factitiousness of the universe: that it is what it is rather than something else and that what it is cannot be inferred from any higher-order rational necessity. Science is powerless to determine the ground of being: why there is something rather than nothing, or why, to take a thesis that we develop in this book, space may be emergent whereas time may not be. It matters decisively, however, to science whether the confrontation with the brute just-so-ness of natural reality, undisguised by the pretense of rational necessity, takes place early or late in a course of scientific inquiry. From the perspective of the interests of natural science, the later it takes place, the better: the dialectic between theoretical imagination and empirical study can then advance over a broader field and expose itself on that field to the surprises of experience.

A third advantage of this approach, in relation to the alternatives, is that it enables and requires us to dispense with any appeal to the idea of the infinite. It replaces the infinite with history. At the beginning of the twentieth century, the most influential interpretation of the field equations of general relativity argued for an infinite initial singularity at the beginning of the universe. It was widely recognized that the invocation of the infinite revealed a breakdown in the application of the field equations to conditions in the earliest history of the universe rather than providing an account of a physical state of affairs. (See the later discussions of the infinite in Chapters 4 and 6.) Here as always the infinite is a mathematical idea rather than a reality present in nature. A cosmology insisting that causal continuity between this universe and its possible predecessors may be stressed rather than broken need make not invoke the *deus ex machina* of the mathematical infinite.

Both cyclic and non-cyclic views of succession may enjoy these three advantages. Non-cyclic views, however, enjoy them more fully than do their cyclic alternatives. The cyclic views of succession suppose that the regularities of successive universes remain the same. The initial conditions of each universe must therefore also either be the same, or represent a stochastic instance of the states of affairs that such laws of nature make possible. As a result, any opportunity to explain why these laws hold rather than others, as well as why these initial conditions occur rather than others, is drastically foreshortened.

An additional difficulty is that the known laws and symmetries of nature fail to account for the initial conditions of the universe. These conditions are in fact highly unlikely in a world described by those laws and symmetries. It is then hard to see how a cyclic view can adequately address the task of explaining the initial conditions in early universes or in earlier states of the present universe. The enigma rather than being dispelled would be multiplied many times over.

By contrast, a non-cyclic view of succession enables us to ask how the laws, symmetries, and constants of nature and the initial conditions of the universe came to be what they are and to seek answers in the history of succession. Under such a view, we are freed, as I next argue, from the need to distinguish between the history of laws and the history of initial conditions. We must reject any such distinction if we are to complete the transformation of cosmology into a historical science and avoid the first cosmological fallacy: resort to what we call the Newtonian paradigm.

The question remains whether we have more reason than not to think that there is such a history of the regularities of nature, despite their remarkable stability in the cooled-down universe that we observe. Several of the following arguments imply that we do.

3. *The argument from rejection of the first cosmological fallacy.* The ideas of the singular existence of the universe and of non-cyclic succession allow us to avoid the illegitimate cosmological application of the Newtonian paradigm.

All major physical theories in the history of modern science make use of the Newtonian paradigm: classical mechanics, statistical mechanics, quantum mechanics, and special and general relativity. The defining feature of the Newtonian paradigm is the distinction that it draws between stipulated initial conditions and changeless laws governing changing phenomena within a configuration space bounded by the initial conditions. Throughout this book, we argue that the Newtonian paradigm has no proper cosmological use: the distinction between laws of nature and stipulated initial conditions cannot be maintained when applied to the universe as a whole rather than to patches of nature or to regions of the universe.

A multiverse cosmology may appear to mimic the conditions for the application, across the multitude of universes that it postulates, of the Newtonian paradigm, properly applied only within a universe. The analogy, however, is superficial and flawed.

For one thing, in the proper use of the Newtonian paradigm, which is its application to parts of the universe, only the initial conditions change from one instance of the practice to another; the supposedly timeless laws remain always the same. However, in the multiverse version of plurality – the only variant to have any theoretical interest and influence today – the regularities as well as the initial conditions differ among universes, expressing states of affairs, allowed by the motivating theory, that are not manifest in our universe.

For another thing, in the proper use of the Newtonian paradigm, what is a stipulated initial condition for the purpose of one instance of the explanatory practice becomes an explained phenomenon for the purpose of the next instance. It is as if the moving searchlight that defines the configuration space, and thus the distinction between what is stipulated and what is explained, illuminated all of nature, piece by piece. No opportunity for such an iteration of the practice exists under the multiverse conception. In that conception, each imaginary universe is a complete and closed entity, bereft of causal contact with any other universe.

The replacement of plurality by succession and the subordination of structural to historical explanation open the way to dispensing with the cosmological application of the Newtonian paradigm. If there is only one universe (with the qualification of the existence of branching universes as well as of the existence of unobservable parts of our universe), and succession takes the place of plurality, we need no longer, with regard to the universe and its history, try to distinguish unexplained initial conditions from a configuration space of law-governed changes. The whole cosmological evolution, without any such distinction, becomes the topic of explanation. The realities to be explained, not all at once, but step by step, include what the Newtonian paradigm distinguishes as initial conditions, law-governed phenomena, and explanatory laws.

Once again, the non-cyclic variant of succession enjoys, in these respects, an advantage over the cyclic one. The regularities of nature cease to be eternal bystanders to the history of the universe or of successive universes and become, instead, protagonists in that history. I argue later, in this section and in this book (Chapters 4 and 5), that we have independent reasons to regard the regularities of nature as susceptible to historical explanation.

4. *The argument from the compatibility of the associated ideas of the singular existence of the universe and of non-cyclic succession with cosmological inflation and from their irreconcilability with eternal inflation.* The standard cosmological model faces a number of problems that have to do, in one way or another, with the need to explain how the observed universe can be flat, homogeneous, and isotropic, when it might be expected to be highly curved and inhomogeneous. Prominent among these riddles are the so-called horizon and flatness problems.*

These problems played a major role in motivating the conjecture of cosmological inflation: of a super-rapid expansion early in the history of the universe. Instead of a standard causal interaction resulting in a convergence of temperatures and of densities within the expanding

* The horizon problem has to do with the relation between features of the observed universe and the lapse of time required by the physical processes needed to produce them. At the time that it has become conventional to call decoupling, when the atoms became stable, and light could travel freely, the universe was already, and has since remained, remarkably homogeneous and isotropic; it everywhere exhibited and exhibits the same temperature to high accuracy and the same spectrum of small fluctuations in density. Much too little time had elapsed between the cosmological singularity, as conventionally viewed, for all of the regions of the universe at its then size to have been in causal contact. Nothing that we are able to infer about the initial conditions of the universe, without making arbitrary and *ad-hoc* stipulations, can account for such a surprising result. Restated, the horizon problem is that models of the early universe based on general relativity and assuming only the matter now known to us, fail to explain the uniformity of the cosmic microwave background. The parts of that background – vestiges of the earliest universe – could not have interacted by the time they are observed to be everywhere in equilibrium at the same temperature.

The flatness problem is that in our universe light beams neither converge as they would under general relativity if space were positively curved, or diverge as they would under the same theory if space were negatively curved. The flat universe must have begun close to the preternaturally improbable circumstance in which the expansion rate and the energy density of the universe compensate for each other's effects.

universe, there is supposed to have occurred, according to this doctrine, an explosive ballooning that scaled the universe up. The scaling up would then have produced what could not have been achieved by a causal interaction for which there was not enough time.

Proponents of cyclic succession have often claimed that their theories offer a solution to these problems without invoking cosmological inflation, for which, they have argued, there has been, at least until very recently, no direct evidence and only conjectural mechanisms. Theories of succession, whether cyclic or non-cyclic, would offer such a solution by extending back the time horizon for setting the pertinent features of the observed universe: its fundamental homogeneity and isotropy, its local and detailed inhomogeneities (indispensable to the formation of the observed celestial bodies), and its flatness.

The premise to such solutions is that the properties of the universe at decoupling should be understood historically, in the light not only of what happened between the earliest formative events and decoupling but also of what took place before those events, in the prior or pre-traumatic universe. Once we admit succession, and redefine the cosmological singularity as a moment at which temperature and density were extreme rather than infinite, and as an incident in a longer history rather than as an absolute beginning, there may be enough time for the parts of the universe to have become homogeneous and isotropic. Thus, if cosmological inflation fails ultimately to be confirmed, no difficulty would result for either cyclic or non-cyclic views of succession.

Suppose, however, that evidence for cosmological inflation mounts. A fundamental difference exists between the implications for cyclic and non-cyclic succession. The ekpyrotic version of cyclic succession (as proposed by Steinhardt and Turok) would then be unequivocally falsified if only because it supposes the present universe to have begun in a low-energy state irreconcilable with the inflationary view.

The many proposals of cyclic succession that have made no such assumption of an early low-energy state contradict the inflationary picture in less obvious ways. If we refuse to understand inflation as a recurrent event, determined by timeless laws (as it might be represented under theories of eternal inflation), we need to comprehend it as triggered by the special and exceedingly improbable initial conditions of our universe. It thus fits easily into a historical account of the universe: one that does not suppose that we can infer the history from the laws, not at least from the known laws. Such an approach to inflation would demand attention to the question: From where do the initial conditions of the universe come? If they are not randomly generated within a multitude of mathematical possibilities (as the defenders of the multiverse idea propose), they must be explained by what happened before. We must, that is to say, account for them historically.

If such evolution is to be explained by immutable laws, they should be laws that generate over and over again the same initial conditions. Such are the regularities that theories of cyclic succession (including the ekpyrotic theory and Penrose's conformal cyclic cosmology) presuppose: carried to their logical extreme, these theories dispense with the concept of initial conditions of the universe. Everything, for them, resides in the laws; they see no need to speak of initial conditions as distinct from the laws and their consequences.

However, the regularities that can explain the initial conditions of the universe resulting in cosmological inflation could not resemble the laws, symmetries, and supposed constants of nature with which we are now acquainted; otherwise, the initial conditions of the universe would not seem as extraordinarily improbable to us as they do. They seem improbable because they are not the expected outcome of the regularities with which we are familiar: those that apply to the cooled-down universe that we observe.

No such antipathy opposes the combination of the ideas of the singular existence of the universe and of non-cyclic succession to cosmological inflation, as distinguished from eternal inflation. (Assume, for the sake of this argument, that the conjecture of the super-rapid initial expansion of the universe may be vindicated by conclusive inference from observable remnants of the very early universe.) Cosmological inflation, in such a view, represents the immediate aftermath of a superdense and superhot moment. In the course of the events of which it formed part, causal continuity was never entirely broken; the values of the parameters, although extreme, never became infinite. If causal continuity remained uninterrupted, the presumption of science is that it continued indefinitely back into the past.

We subtract nothing from our present ignorance of that past by resorting to the speculation of the existence of a vast or infinite array of other universes. We merely provide ourselves a pretext to stop looking and to seek in mathematics what we have so far failed to find in nature.

Inflation fails to contradict non-cyclic succession, proposed against the background of the idea that there is one causally connected universe at a time. Eternal inflation does contradict it. It contradicts it insofar as theories of eternal inflation reenact part of the intellectual program of steady-state theories in cosmology: a changeless cosmological process going on forever and ceaselessly reestablishing the conditions for its own continuance. It contradicts it even more to the extent that eternal inflation is married to the multiverse idea. The product of this marriage is the conjecture of an infinity of unobservable pocket universes generated by eternal inflation. The contradiction is only mildly attenuated in variants of eternal-inflation theory that depict inflation as eternal into the future but not into the past, according to the view that, under reasonable assumptions, the inflating region must be incomplete in past directions.

Although it is common to suppose that the idea of eternal inflation represents a natural extension of the conjecture of cosmological inflation, it has a wholly different character, especially when associated with the conception of a multiverse. It evokes the picture of a recurrent process governed by immutable laws. It eludes empirical confirmation or falsification. It justifies its recalcitrance to empirical challenge by its appeal to what we here argue to be a misguided view of the relation of mathematics to nature and of the place of mathematics in science. In certain respects it shares the spirit of theories of cyclic succession. As they do, it teaches that the most important facts about nature – those that regard its basic structure and fundamental regularities – never change.

The thesis of the singular existence of the universe, as developed by the conjecture of non-cyclic succession, is compatible with cosmological inflation. It cannot be reconciled with eternal inflation. Rather than being a flaw, this incompatibility amounts to a virtue: it defines one of several ways in which that thesis can be put to the test. Any evidence for eternal inflation – if, as its proponents claim, there can be such evidence – amounts to evidence against the twin ideas of the singular existence of the universe and of the historical, non-cyclic succession of universes or of states of the universe.

5. *The argument from rejection of the second cosmological fallacy.* The ideas of the singular existence of the universe and of non-cyclic succession enable us to begin making sense of the emergence, in the course of the history of the universe, both of fundamental structures or constituents of nature and of law-like regularities in the interactions of these constituents. The now standard ("big bang") cosmological model has been immensely successful in providing a framework within which to understand what has already been discovered about the history of the universe.

This model suffers, however, from two basic weaknesses. One frailty is the continuing reliance of its proponents on the metaphysical element in the theory of general relativity: the representation of time as an aspect of a spacetime continuum, described as a four-dimensional semi-Riemannian manifold, that can be arbitrarily sliced by an infinite number of spacetime coordinates. I argue in Chapter 4 that Riemannian spacetime, with its exclusion of a preferred cosmic time, its denial of the fundamental, non-emergent character of time, and its consequent treatment of time as an accessory incident, together with space, to the disposition of matter and motion in the universe, represents a philosophical gloss on the empirical hard core of general relativity. I go on to argue that the metaphysical conception of such a spacetime continuum inhibits our understanding of what has already been discovered, as well as our openness to what may yet be discovered, about the history of the universe.

Another weakness of the way in which the standard cosmological model has developed concerns its picture of events in the very earliest universe. So long as these events remain hidden behind the screen of the infinite – an infinite initial singularity, mistaken for an account of physical events rather than recognized as a mathematical notion indicating limits to the applicability of the physical theory – we cannot hope to progress in our understanding of those events. It is only when we begin to represent the earliest history of the universe as a condition in which temperature and density had extreme but nevertheless finite values that we can hope to subject it to physical reasoning and to lay it open, at least in principle, to empirical inquiry.

In so doing, we satisfy the basic condition for causal continuity between our universe and any universe that may have preceded it, or between the expanding and contracting of the one real universe over time.

The idea of succession then arises as the alternative both to the stability and eternity of the universe (as in the steady-state cosmology of the early twentieth century) and to the conception of an absolute beginning of the universe, out of nothing. Succession requires much less of a break than do its rival conceptions not only with what we already know about the history of the cooled-down universe in which we find ourselves but also, more generally, with what we know about how nature works in the many domains studied by the specialized sciences.

Up to this point in the analysis, there is no reason to prefer either a cyclic or a non-cyclic view of succession. Both can accommodate causal continuity, make sense of the fiery beginnings of the present universe, and avoid the appeal to the mathematical conceit of the infinite or to the making of something out of nothing.

The different implications and unequal advantages of cyclic and non-cyclic views of succession begin to become clear when we consider the problem of the emergence of the basic structure of the universe. We already know enough to conclude that chemistry, as described by the periodic table, could not have existed at the earliest moments of the history of the universe. Research conducted under the guidance of the now standard cosmological model also gives increasing grounds to infer that the elementary constituents of nature, as described by the similarly successful standard model of particle physics, could also not have existed, at least not in their present form, very early in the history of the universe.

The fundamental structure of nature, as we observe it in the cooled-down universe, has a history. Each of its pieces emerged in the course of this history. The plasma existing at the beginning of the universe did not resemble the rudimentary structure portrayed, but not historically explained, by the standard model of particle physics. There is no evidence that the standard model of particle physics describes conditions prior to nucleosynthesis.

There are two distinct sets of reasons for the difficulty that we experience in developing a historical account of the step-by-step emergence of this structure. A first reason is that, even after we throw down the principled bar to empirical research represented by the notion of an infinite initial singularity, we continue to lack the experimental means with which to simulate conditions in that earliest history. Temperature was at that time higher than the energies generated in our present particle colliders. There is, however, no reason of principle why we cannot hope to develop means for investigating those conditions in the future.

A second reason is that a gap remains between our understanding of the laws, symmetries, and constants in the earliest history of the universe and the evolution resulting in the fundamental structure that we observe. The regularities of nature that we register, and the astonishing stability of which in the cooled-down universe we cannot fail to acknowledge, may be compatible with the changes producing this structure. They fail, however, to explain it. They are the regularities governing the regime forged in the early history of the universe. They do not, however, account for either the emergence of this regime or, consequently, for themselves.

This gap confronts us with a choice between two modes of thought. According to the way of thinking characteristic of the tradition of physics inaugurated by Galileo and Newton, the regularities must have been already and always there, as eternal and unmoved bystanders, rather than as temporal and shaped participants, in the history of the universe. It is a view that not only circumscribes the reach of history and of time but also violates the principles of reciprocated action: we imagine that part of nature – the laws, symmetries, and constants – acts without being acted upon. They fail to explain themselves: the distinctive and even "finely tuned" content of the laws, symmetries, and constants, which (the efforts of many philosophers notwithstanding) we are unable to deduce from any more general rational necessity, remain mysterious.

This approach gains its semblance of plausibility from the undisputed stability of the laws, symmetries, and constants in the cooled-down universe. However, it could not understand these stable regularities to be eternal if it failed to make, as well, two other moves. The first move is to treat the stable forms that nature takes in the consolidated though evolving universe as its permanent canon of forms (the second cosmological fallacy). The second move is to take the idea of immutable laws, symmetries, and constants to be an indispensable prerequisite of scientific explanation. Neither of these moves is justified.

The objection to the first move is that we may have reason to infer that nature in the early universe, as well as in extreme states of its later history, must have been, and must be, organized in a different way from the way in which we see it organized now. What remains open is whether any part of the fundamental order of nature – that it is to say, its elementary fields and particles as described by the standard model of particle physics – existed then as it came to exist later. The safest assumption – the most compatible with the direction of our discoveries about the history of the universe – is that none of it was then what it is now.

The objection to the second move is that unstinting recognition of the mutability of nature, in all of its elements, need not undermine our powers of causal explanation. That it need not do so is shown by the explanatory practice of the life and earth sciences as well as by the more remote lessons of social and historical study.

To recognize the force of these objections (developed in earlier and later parts of this book) is to open the way to another mode of thought. According to this alternative, the structure of nature and its regularities evolve jointly. Indeed, causal connections, rather than being instances of immutable laws of nature, constitute primitive features of nature. Although we are accustomed to see them display recurrent and general form in the observed universe, they may exhibit no such law-like character in extreme states of nature. Among such extreme states are those that seem likely to have prevailed in the earliest history of the present universe.

The conception of a plurality of divergent universes, the multiverse, each with its own distinctive regularities and structures, loosely subsumed under the ideas of a physical view such as string theory, cannot help make sense of these problems. On the contrary, it simply multiplies the enigma by as many universes as it postulates. For in each of these universes, the fundamental structure and the regularities must have come from somewhere through a physical mechanism that the conceit of their mathematical possibility fails to explain or even to describe. Moreover, in each of these postulated universes, if we understand them by analogy to our own (and how else can we understand them, given that they are forever inaccessible to us?), the distinct fundamental structure may have been different at the outset (unless we deny altogether the reality of time). Consequently, for each of them, the same question arises as to what we imagine the regularities were doing when the states of affairs that they supposedly govern did not yet exist.

Only the idea of a singular universe, placed in a historical context that makes room for a succession of universes or of states of the universe, can enable us to address these issues, and to do so in a fashion that remains, at least in principle, open to empirical study. In this respect, the non-cyclic view of succession enjoys a decisive advantage over the cyclic one.

Under the cyclic account of succession, both the fundamental structure of nature and its laws, symmetries, and constants remain the same. There is then no prospect of explaining what came later by what came before. If it is admitted that, in each formative moment of a new universe, the basic constituents of nature reemerge, they must, under this conception, always cluster into the same natural kinds and conform to the same regularities.

For such an account to be complete, the regularities of nature that it identifies must explain by what mechanisms the same structures repeatedly reemerged at each formative moment of each succeeding universe. It is true that we can mount computer simulations encoding our present understanding of the laws of nature that simulate the evolution of the universe back to relatively early stages. What we cannot do, however, is to reverse engineer such a formative process beyond or before the moment when the basic constituents described by the standard model of particle physics existed.

The cyclic views of succession are in no better position in this respect than is the application of the standard model of particle physics to the early universe; they simply extend the same riddle back into the eternity of a time without transformation. If then, as I argue in Chapter 4, time is the change of change, they equivocate about the reality of time. Moreover, they require, for these prior universes, an explanation of the recurrent genesis of structure that we do not possess even for our own universe, or for its present state.

In all these respects, the non-cyclic view of succession is in a better place. It conjectures that we will never be able to explain the genesis of the fundamental structure of the cooled-down universe on the foundation of the regularities that we observe in a universe marked by the stability of both this structure and of the regularities that it exhibits. Such a non-cyclic account of succession provides a basis for taking the less heroic and more economical position: that regularities do not antedate the structures manifesting them. It is a view that may seem paradoxical only to those who have embraced the dogma that the laws of nature are either eternal or non-existent and that only if they are eternal can science do its work.

We should be clear about what in this alternative way of thinking is speculative and what is not, and about the sense in which its speculative element may nonetheless connect with the empirical agenda of natural science. That the observed universe exists and that it must form part of a larger universe, which we cannot observe in its entirety, are matters of fact. All other universes, whether parallel or divergent under the aegis of the idea of plurality, or predecessors to our own, are, at the present time, speculations. The difference is that the parallel or divergent universes of the thesis of plurality are not only unobservable now but also forever beyond the reach of direct or indirect empirical investigation, whereas previous universes, or previous states of the present universe, are likely to have left marks or vestiges that we can observe. Moreover, we may be able to simulate in our universe some of the conditions that attended its beginnings and pre-history. Even then we cannot reenact the totality of an earlier state of affairs.

The difference is also that we come to the ideas of plurality and of succession in radically different ways, with different relations to the interests of natural science. We devise the conception of plurality out of an attempt to make up for the radical underdetermination of many of our present physical theories, notably the string theory variant of particle physics: their compatibility with a multitude of universes other than our own. By contrast, we arrive at the idea of succession as a direct consequence of banishing from the interpretation of the standard cosmological model the mathematical notion of an infinite initial singularity, which, like all versions of the infinite, can represent no real physical state of affairs. We reach it as well by fidelity to the practice of causal explanation and to its underlying assumption that every state of affairs must have causes and that the relation between cause and effect is by its very nature temporal. Each *after* must have a *before* until we can see and think no longer. We are not entitled to mistake the limits to our understanding of the universe for a moment when something came out of nothing.

We are then faced with a choice between the cyclic and the non-cyclic views of succession. We continue to lack any decisive empirical evidence or theoretical compulsion to choose one of these variants of succession over the other. Both views are speculative. There are nevertheless clues in what we already know about the history of our universe as well as about the study of the workings of the parts of nature to which we have direct access on our planet and in the life around us. These clues suggest the mutable and historical character of all types of being or natural kinds, including the basic constituents of nature. From the recognition of this mutability and historicity it is only a step to apply in cosmology the working assumption that we deploy elsewhere: where the structures emerge and change, so must the regularities that they exhibit and to which they appear to conform.

These clues, taken together with the other arguments that I have here outlined, favor the non-cyclic variant of succession over the cyclic one. The grounds for preference are neither robustly empirical nor merely speculative. They display the hybrid form characteristic of the work by which natural philosophy acts as the scout of science and instigates the dissolution of the marriage between empirical insight and metaphysical preconception in ruling scientific ideas.

6. *The argument from the reciprocal support of the ideas of the singular existence of the universe, of the inclusive reality of time, and of the selective realism of mathematics*. The thesis of the singular existence of the universe, extended by the idea of a succession of universes, should be judged not only on its own merits but also as part of a more comprehensive view. It is the conception that most fully coheres with the ideas about time and about mathematics representing the other chief proposals of this book. It supports those ideas and receives support from them. It shares in both their strengths and their weaknesses.

The claim that time is inclusively real can be established only if nothing, including the regularities of nature, remains outside the reach of time, invulnerable to change. The inclusive reality of time, I argue in Chapter 4, remains in jeopardy unless there exists a preferred cosmic time such that everything that has ever happened in the history of nature can in principle be placed on a single unbroken time chart. Only then does time cease to be an accessory to something else: in particular to the placement of matter and the occurrence of motion. There must be such a preferred cosmic time for time to be inclusively real and for us to be able to ask and to answer questions such as how old the universe is. There must be such a time notwithstanding the objections resulting from the standard interpretations of general relativity, with their adoption of Lorentzian spacetime, as well as from the relativity of simultaneity, established by special relativity.

These requirements are in turn readily satisfied only if there is one universe at a time, with the qualifications that the suggested existence of branching universes may suggest. A succession of universes provides a physical basis for the existence of a preferred cosmic time, and thus as well for the inclusive reality of time. Such a succession makes it possible for the unified history of our universes to extend backward into a history of successive universes, or of phases of contraction and expansion of the single universal reality.

However, as I acknowledge in Chapter 4, the singular existence of the universe, extending backward in time through succession, is not a sufficient condition for the existence of a preferred cosmic time, although it may be a necessary one. We need to know how such a time can even in principle be perceived and measured. If it is not perceptible and measurable, the notion of a preferred cosmic time would be no better than the untestable conjectures of the multiverse view; its theoretical advantages would be insufficient to justify its reception by cosmology.

For a preferred cosmic time to have a legitimate cosmological role, the universe must also be so arranged, by virtue of its relative isotropy and homogeneity, that it provides a clock of cosmic time, in the twin forms of its equal recession in all directions from preferred observers, situated in positions expressive of that homogeneity and isotropy, and of the equal temperature with which its cosmic microwave radiation background strikes these same preferred observers from all directions in the sky. The clock of cosmic time is the universe itself, viewed with regard to some of its features. We have reason to believe that we live in such a universe: one in which cosmic time can in principle be recognized and measured, not simply asserted as a theoretical pre-commitment or dissolved into the many-fingered time of the predominant interpretations of general relativity.

There is another way in which the idea of a succession of universes supports thinking of time as non-emergent, rather than as derivative from some more basic reality. If time stopped at an initial infinite singularity, or if it were an aspect of a spacetime continuum, as the leading interpretations of general relativity imply, it could not be fundamental. The dynamics of that singularity, or the geometry of that continuum, rather than time, would be fundamental. The ideas of singular existence and of succession offer an alternative to these views. In this alternative, time can be represented as non-emergent and inclusive.

An implication of the inclusive reality of time is the mutability of the laws, symmetries, and alleged constants of nature, despite their overwhelming stability in the universe that we see. Another implication is the metamorphosis of all natural kinds – of the types of being that there are – down to the most elementary particles and fields. In the non-cyclic variant of the idea of succession we find a way of thinking about the coeval history of the structure and of the regularities.

These ideas about the history of the universe are also reciprocally connected with a view of the relation of mathematics to nature and to science. The multiverse conception is intimately associated with a view of the prerogatives of mathematics in natural science. A theory that appears on its face to be physical, like the string theory variant of particle physics, turns out on closer inspection to be largely mathematical in its inspiration as well as in its expression. It countenances many more arrangements of nature at its most fundamental level than we can see realized. The theorist then conjectures that all the unrealized arrangements must be enacted in universes other than our own, forever closed to our inspection. In so doing, he reveals a prejudice about the privileged insight of mathematics into nature.

If, however, there is no such immense plurality of universes, ensuring the correspondence of a mathematical invention to a natural reality, if the universe is singular, if it is the outcome of a singular succession with features that cannot be inferred from mathematical abstractions (although mathematics may help represent the relations among its parts), and if the stable structure of the cooled-down universe amounts to the product of a unique history, rather than the other way around, a deep chasm opens up between nature and mathematics.

We must then rethink the relation of our mathematical ideas to the facts of nature and to the discoveries of science. The universe cannot be homologous to a mathematical object, much less can it be such an object. The applicability of mathematics to nature must be selective and conditional. We must reject the assumption that all constructions valid in mathematics – such as the notion of the infinite – have a guaranteed place in nature by virtue of their mathematical validity. Above all, we must guard against the anti-temporal biases of the mathematical imagination, which threaten to mislead us into discounting or even into denying the reality of time.

The idea of the singular existence of the universe is not an isolated proposition. It forms part of a wider contest in natural philosophy and cosmology. The more complete our understanding of what is at stake in this quarrel, the more likely we are to do justice to each of its aspects.

IMPLICATIONS FOR THE AGENDA OF COSMOLOGY

The remaining sections of this chapter address a number of debates that force upon us a choice between the conceptions of many universes and of a solitary universe, or, more precisely, between the idea of a plurality of universes and the idea of a succession of universes or periods in the history of a singular changing universe. Before addressing these debates and their implications, it is useful to say something further about succession and its relation to plurality.

If the universe has a history and began in a moment of extreme concentration during which – and maybe before which – its present laws, symmetries, and regularities failed to hold, it developed either out of nothing or out of something. The idea that it developed out of nothing amounts to a way of acknowledging our inability to advance further in the work of scientific explanation.

Alternatively, the universe developed out of something. Consequently, it has a pre-history as well as a history: something preceded the traumatic events that cosmologists call the cosmological singularity (the "big bang"). In these very early moments of the history of the universe, as the earlier discussion of the second cosmological fallacy suggested, causal connections may not yet have assumed law-like form and the division of nature into enduring natural kinds may not yet have taken shape. To describe the situation in this way is to conjecture that some form of causal connection or continuity exists between successive universes, or between different states of the universe, before and after the earliest, formative moments of the cooled-down universe that we observe.

To be sure, such causal connection or continuity between the present universe and a universe that may have preceded it cannot, on this view, be a connection or continuity that has for its basis immutable laws, symmetries, and constants. Change that is so radical poses a conundrum that we have already mentioned and to which we shall repeatedly return. Either the change of laws is itself law-governed or it is not. If it is, the higher-level laws must themselves be subject to change. Otherwise, we would have to stipulate, without reason to do so, that something in the world is exempt from time and change. If, however, the change of the laws is not itself law-governed, even in the sense of statistical determination, we would find ourselves face to face with arbitrariness in nature and with impotence in our explanatory efforts.

These are real, not fanciful difficulties, requiring real answers. They highlight some of the riddles generated by the hypothesis of succession. However, they do not apply solely to our thinking about the origins of the universe. They shadow all our causal judgments and the whole of science, if we have reason to believe that the framework of natural laws, symmetries, and constants is within time (and therefore susceptible to change) rather than outside and if we must therefore reconsider the relation of mathematics to nature and to science. Such change of the regularities of nature, however, may be discontinuous, as change in the part of the world of which we have direct experience generally is: sometimes fast and dramatic, at other times slow and imperceptible.

Given these refinements, the idea of a succession of universes should be understood as shorthand for the conception of a history of the universe, passing through distinct phases or periods. Causal continuity and connection may be stressed and shaken rather than broken.

On such a view, the substitution of succession for plurality in our ideas about the universe and about what lies beyond our ken does not amount to trading one enigma for another, of the same order. The puzzles attending the idea of succession are those of the relation of the history of nature to the basis of our causal judgments. States of the world behind the veil of a traumatic transition, such as the formative moments of the observed universe, present obstacles to direct observation and experiment. However, they do not contradict the assumption of causal connection and continuity. Nor do they appeal to metaphysical entities, such as parallel worlds, inaccessible from our own, that could not even in principle be open to scientific investigation, however remote or indirect. They arise from a willingness to take the reality of time seriously and to treat the history of the universe as exhibiting a discontinuity more radical than the discontinuity that we see displayed in the history of each of its parts.

Succession is not adequately described as plurality in the realm of time rather than of space. It is radical transformation, in time, of the one real universe.

THE FINITE AND THE INFINITE AT THE BEGINNING OF THE UNIVERSE

The characterization of the cosmological singularity (in the contemporary technical sense of this term) is the first of several problems in contemporary cosmology that bring into focus the need to choose between plurality and succession, as well as succession with and without continuity, across successive universes, or periods of universal history, of the same regularities of nature.

Penrose and Hawking argued that given a cosmological spacetime without spatial boundary, satisfying a small number of conditions, there is a temporal boundary in the past of finite proper time (for any observer) beyond which the application of the field equations of general relativity cannot continue. This result is commonly interpreted to show that given the equations of general relativity, together with reasonable assumptions about the distribution of matter, there is a cosmological singularity in the past of our universe. By such a singularity (used in the familiar cosmological sense rather than in the special sense of the thesis of the singular existence of the universe) is meant a moment of simultaneity at which all physical quantities such as temperature, density, and strength of the gravitational field are infinite. Upon reaching the circumstance of the cosmological singularity, the field equations break down. It is impossible to continue solving them further into the past.

The theory seems to give more precise expression and foundation to the dominant view of the origins of our universe: the "big bang" cosmology. It appears to imply that the universe began a finite time ago and out of nothing. In the moment of the infinite natural quantities and of the undivided unity of all matter and all forces, the laws, regularities, and symmetries of nature did not yet hold. Time itself, according to this view, did not exist; it emerged once the threshold of the singularity was past. Once out of the singularity, the universe began to assume the form that we are able, directly or indirectly, to observe: a manifold of distinct structures and forces, deeply united, and subject to a timeless set of laws governing events situated in emergent time.

Although this conception does not necessarily imply a plurality of universes, it characterizes the initial formative moment as so arbitrary – so far beyond the reach of observation and explanation alike – that it removes the initial bar in our thinking against the idea of other universes, also causally closed and causally unconnected with our own, that might have begun or might begin in similarly mysterious circumstances. If the miracle of making something of nothing can be performed once, why cannot it not be performed often?

There are four major objections to the inference of an infinite initial singularity.

A first objection, signaled earlier and discussed further in Chapter 4, on the inclusive reality of time, is that the inference of an infinite initial singularity is best understood as an indication of the breakdown of the underlying theory (general relativity) when its field equations are carried beyond their proper domain of application rather than as the description of an actual state of affairs. The mathematical idea of the infinite is not realized in nature. Its introduction represents a salient example of the dangers of succumbing to the view that mathematics offers a reliable shortcut to insight into the workings of nature.

A second objection is to the very notion of something arising out of nothing: no material, no agent, no circumstance, no time. A plausible interpretation of such a view is that it is less, in this respect, an account of the beginnings of the universe than it is a confession of our inability to provide such an account. It seems better simply to acknowledge that inability than to conceal it under a genealogy that we cannot translate into any discourse of causal explanation familiar to us in science.

A third objection is that the passage from the lawless world before time and distinction to the lawful world of distinction and time is not only unknown but impossible to represent persuasively in either the verbal or the mathematical language of science. At some point, we are asked to believe, infinite quantities became finite, a distinct structure of nature emerged, the laws governing it began to apply, and time started to flow. That we cannot explain such a transformation may be conceded. What is harder to accept is that we cannot even describe it. The *how* remains as impenetrable as the *why*.

A fourth objection is that these suppositions about the infinite initial singularity are not only hard to square with the rest of the way in which we think about nature; they are also difficult to reconcile with some of the empirical implications of contemporary physics. The picture of the singularity as the emergence of something out of nothing cannot easily be reconciled, if it can be reconciled at all, with the view from quantum mechanics. When that view is taken into account, the cosmological singularity appears to be an extraordinarily but not infinitely violent event, in which very large but not infinite temperatures and densities prevailed. There is an infinite difference between the finite and the infinite.

The view of the earliest, formative events in the history of the universe as capable of breaking and changing laws, symmetries, and constants and of confounding forces and phenomena that we observe divided in the subsequent universe does not require any suspension of causality and time. On the contrary, it points toward an extension of the universe back into time: back into a time prior to this violent event, when other laws may have held. It points us toward succession, in the qualified sense in which I have described it, rather than to plurality and, for the reasons that I earlier adduced, to non-cyclic rather than to cyclic succession.

It is a matter of indifference or convenience whether we call the world before this formative moment an earlier state of the universe or an earlier universe. Given the postulate of causal connection and continuity, this previous universe, or this earlier state of the universe, is in principle open to empirical inquiry.

Between it and ourselves stand the fiery beginnings of the present universe, reinterpreted in a fashion that brings them into the realm of time and change. If we suppose that the composition and structure of the world changed more rapidly and radically around the time of those events than they have changed since then, we can expect that our ability to inquire into this early history of the universe will be limited. However, it will not vanish.

For one thing, there may, on this view, be only a difference of degree, although a substantial one, between the mutability of the laws of nature, as well as of the causal explanations that rely on them, during that formative trauma and their mutability before and after it. I later consider the reasons to believe that time is more real than the dominant tradition of natural science has generally acknowledged it to be and that an implication of its reality, together with its inclusiveness, is that nothing in nature, not even its most basic structures and regularities, remains exempt from time and change. It is nevertheless important to reconcile the mutability (in principle) of the laws, symmetries, and supposed constants of nature with their stability in the cooled-down universe.

The mutability of the laws (and other regularities) of nature introduces a series of intractable difficulties: whether we should regard these problems as insuperable antinomies or merely as enigmas that will only slowly yield their secrets forms part of what there is to discuss. The laws of nature may develop coevally with the phenomena they govern. Causality may exist without laws. These ideas may shock in physics. They have, however, long been familiar in the life sciences as well as in our ideas about society and its history.

For another thing, a previous state of the world must have left some trace of itself in a later state. If the difference between the cooling universe, organized as a differentiated structure, and its superhot and conflating initial state is finite rather than infinite, no absolute barrier exists to the survival of such traces. Among them may be, for example, the seemingly arbitrary but precise values of some of the constants of nature, unexplained by the laws of the post-traumatic universe.

THE INITIAL CONDITIONS OF THE HISTORY OF THE UNIVERSE

A second controversy revealing the implications of the choice between the thesis of plurality and the thesis of uniqueness and of non-cyclic succession is the way in which we account for the initial conditions of the universe.

Because the universe is believed to have begun with a non-zero temperature, it requires an indefinitely large or even infinite amount of information to describe its initial microscopic state. Another way of saying this is that the universe sprang into existence complete with an infinite number of degrees of freedom excited to particular states. General relativity together with quantum field theory – our most credible present comprehensive account of the universe – has an infinite number of solutions. Indeed there are an infinite number of solutions that agree with observations and have a cosmological singularity in their past.

Such alternative solutions – and the alternative states of affairs that they represent – differ by the exact states of the particles and of radiation; the universe started very hot and hence full of photons and other particles. The alternative solutions and states of affairs also differ by the disposition of the gravitational waves: small deviations from symmetry. On these facts, the initial condition of the universe, just after the singularity, seems to defy explanation.

An appeal to stochastic causation to account for such a circumstance will confront the obstacle that the conditions are lacking for a well-formed probabilistic determinism. Ordinary probabilistic analysis conforms to the condition that the sum over all the probabilities for exclusive outcomes is equal to one. We have no basis under these conditions to define that sum. We do not know the range of the alternative states to which the probabilities apply. We cannot treat as stable and well-defined the elementary constituents whose combinations the statistical accounts would be designed to explain, for we are witnessing the process of their emergence and differentiation. We are not able to trace the boundaries between the phenomena subject to statistical explanation and those subject to deterministic causation.

At the root of these problems lies confusion between probability within the universe and probability about the universe. Stochastic reasoning applies within the universe. It cannot apply to the making or to the history of the universe if the universe is one of a kind. The multiverse conception tries to reestablish conditions for the application of probabilistic thinking to the universe. It does so by fabricating imaginary universes. The array of universes under the multiverse idea is limited only by the vast number of conjectural universes accommodated by the mathematics of an underdetermining physical theory: the string-theory variant of particle physics. We cannot confidently use probabilistic reasoning in such a circumstance, even under cover of a view that was designed, in part, to facilitate its use.

We will have no easier a time in deploying deterministic causation. The canonical model of deterministic explanation in science for the last several hundred years has been the Newtonian paradigm: the distinction between the initial conditions of a set of phenomena and the laws governing their movement or change within a given configuration space bounded by those conditions. What counts as stipulated initial conditions in one explanation of one part of the universe can figure as the subject matter to be explained – the law-governed phenomena within the configuration space – for another explanation of another part of the universe. However, the initial conditions of the universe can never become the *explanandum* in such an iterative process. They are not a piece of nature; they are the whole of it at an earlier time. In the end, the initial conditions of the universe as well as the supposedly timeless laws, symmetries, and constants of nature remain always unexplained by successive applications of the Newtonian paradigm.

It seems that the circumstances conventionally labeled the initial conditions of the universe cannot be explained by the standard explanatory moves of cosmology and physics. They are just factitiously there. The only way in which we could explain them would be to explain them historically, by the states of affairs that preceded them. Such an appeal to historical explanation is prohibited if, as the conventional, technical view of the cosmological singularity supposes, the universe began in a limiting event of infinite density and temperature. Moreover, the frontier between the factitious stipulations and the law-governed phenomena will be elastic; whatever we are unable to explain, we can simply incorporate into the definition of the mysterious initial conditions.

We may be tempted to escape our difficulties in applying stochastic reasoning to the explanation of the initial conditions of the universe by seeking refuge in deterministic explanation. We may try to redress our difficulties in deploying deterministic reasoning by finding succor in stochastic reasoning. However, the juxtaposition of two inadequate solutions to our problem will not provide us with an adequate one.

If we now compare the deep reasons for the failure of the two forms of explanation, we can see that they are in fact two species of the same quandary. The difficulty results, in both instances, in the realm of probabilistic causation as well as in the domain of deterministic causality, from a misguided attempt to apply to the whole of the world forms of reasoning that work only when applied to part of it. Their partiality is a condition of their success.

To these difficulties, arising from the attempt to explain the whole by the methods with which we explain the parts, we must now add a second order of complication, resulting from the special and surprising content of the initial conditions. The very hot initial universe might have been expected to turn out more irregular and entropic than it in fact has. From present observations, combined with the laws of the standard model and general relativity, we can infer that the whole universe formed with a remarkably symmetrical geometry. The asymmetric movement of particles within the early hot gas barely perturbed this geometry. There is, for example, evidence that few black holes existed very early in the history of the universe. One way to state this point is to say that matter started very hot, but that the gravitational field began at absolute zero.

Such a situation is very unlikely, by the standard of the variations of nature within the cooled-down universe, had the initial conditions been simply random. The significance of this unlikelihood is to strengthen further the reason to seek a larger context of explanation in which to make sense of the development of initial conditions with such characteristics. The simple application of the familiar forms of statistical and deterministic reasoning is powerless to solve this problem, for the reasons just stated.

There are then two large classes of ways in which we can go about solving it: plurality and succession. In the spirit of plurality, we take the universe formed in a cosmological singularity (in which the parameters of the formative events had infinite values) as one of an indefinitely large or even infinite number of universes shaped by similar traumatic events. To the extent that we explain the initial conditions stochastically, the large set of universes can in principle define the scope of the alternative states of affairs over which the calculus of probabilities will operate. The strangely symmetrical and isotropic features of our universe, from its initial conditions on, can be dismissed as the consequence of its being an outlier among such alternative universes. Once we account for the initial conditions stochastically, and thus discount their improbable features, we can carry out the rest of our explanatory work by whatever combinations of statistical and deterministic reasoning the science of the day validates. The partial configuration spaces within which deterministic explanation applies will all have their ultimate origin in the particular initial conditions of our universe.

If we are lucky, some of the laws, symmetries, and constants that we are able to establish will, by this reasoning, apply not just to our universe but to all the universes. They will be compatible with different initial conditions for each universe, with the choice of initial conditions determined stochastically. The combination of the super-universal laws with the accidental, universe-specific initial conditions will in turn generate the set of laws, symmetries, and constants applicable to that universe.

The reader may smile and imagine that he is reading a passage from Plato's *Timaeus*, in updated vocabulary. In fact, this statement simply explicates and extends the direction that much of contemporary cosmology and physics has taken in its approach to the explanation of the initial conditions of the universe. The point of such pseudo-scientific speculation is not to replace the standard ways of thinking stochastically and deterministically about the initial conditions, but rather to make up for their manifest inadequacy: to prop them up and make them seem more powerful than they are, given that we do not know how to replace these efforts at explanation.

The alternative basis on which to understand the initial conditions is the idea of succession, interpreted as an instance of the thesis of the singular existence of the universe rather than as an alternative to this thesis. The crucial point is that the formative trauma of the earliest moments of the universe remained at every moment within the realm of the finite: physically finite, albeit very large quantities; geometrically finite, although very small space; and universal time, flowing inexorably, the one reality in nature that deserves to be considered sovereign.

The significance of the finitude of each of the attributes of the events from which the present universe originated is that they permit a causal pass through from the universe before those events to the universe after them. They suggest a way of accounting for the strange and improbable initial conditions of the universe on the basis of a historical explanation.

The deepest complications presented by such an approach are twofold. The first complication is the difficulty of generalizing when the subject matter is a class of one: one universe, the one real world, passing through distinct stages or phases. The analogy to the physics of phase transitions fails because here we do not deal with a process that is local as well as recurrent, and indeed recurrent under laws and symmetries that are stable in the cooled-down universe even if they are not immutable. We reckon with events that, so far as we know, may never recur in the same way and with the same characteristics. The second complication is the difficulty of accounting for changes in nature so radical that they may be accompanied by changes in the laws, symmetries, and constants of nature. As a result of this second difficulty, we may need to abandon the assumption of a framework of natural laws that remains unchanged amid the changes of nature. Non-recurrence and change of the laws and other regularities define the form of universal history described by the conjecture of non-cyclic succession.

A historical explanation may help account for the relative homogeneity and isotropy of the post-traumatic universe. Conversely it may also help explain the way in which, as well as the extent to which, the universe displays asymmetrical and possibly non-Gaussian features.

The question then arises whether such an understanding of the initial conditions of the universe is a mere speculative conjecture or whether it can be developed, although from the starting point of natural philosophy, in a fashion that lays it open to empirical inquiry. It is a question vital to the force of our argument. I address it only in indirect and fragmentary fashion. (Lee Smolin addresses it directly and systematically.)

Two sets of observations and experiments would shed light on the respective merits of these two approaches to the initial conditions of the present universe. One is an approach relying on the conception of an infinite initial singularity, with its logic of infinite temperatures and densities, implied by leading interpretations of general relativity and of solutions to its field equations. Such an approach invites or allows elaboration by a cosmology of multiple universes, each of them causally closed and all of them causally unconnected to one another, with the qualifications implied by eternal inflation (if eternal inflation is accepted as part of the approach). The alternative approach reinterprets the formative events in the present state of the universe to ensure that they remain within the realm of the finite. It therefore fits with an idea of succession. (Earlier in this chapter (section entitled "Arguments for the singular existence of the universe"), I suggested reasons to prefer non-cyclic to cyclic succession as the working assumption of a research agenda.) In this second approach, the causal nexus between the formative events of the present universe and events preceding them is shaken and stressed, but not broken, by the violent origins of the universe in its present condition. The case for the thesis of the singular existence of the universe around us turns in part on the reasons to prefer this alternative view.

One set of experiments and observations pertinent to the choice between these two views goes to the impossibility of generating any physical process that leaves the domain of the finite to enter the realm of the infinite. The emergence of a physical process with infinite rather than merely just very large quantitative attributes is so unlike anything that has ever been observed, and suggests so radical a change in the workings of nature, that we must ask through what transitions such a leap could ever take place. It presupposes that the radicalization of a physical process – its quantitative accentuation – will eventually reach a threshold at which it jumps into another realm of non-finite quantities.

The quest for a perpetual-motion machine was abandoned as the result of a combination of a repeated practical failure with a theoretical argument about why such an outcome was impossible: the failure to make such a machine even under what seemed to be the most propitious condition worked together with the persuasive force of the argument about the impossibility of perpetual motion in a world of friction, resistance, and entropy. Similarly, we would need to persist in the experimental radicalization of certain physical processes to observe whether there is ever a sign in nature of anything that fails to lend itself to finitistic characterization and explanation. At some point, people will give up.

A distinct set of experiments and observations would address the physics of the stressed and shaken but not interrupted causal nexus between the events prior and subsequent to the extreme changes that lie at the beginnings of the present state of the universe. Observationally, we could achieve this goal by studying local circumstances in the post-traumatic universe that reproduce certain features of the traumatic events. Experimentally, we can try to mimic these conditions. A series of increasingly powerful particle colliders, conceived by some in the service of the project of the final unification of the laws of physics under the aegis of the non-finitistic view of the cosmological singularity that is suggested by general relativity, can be enlisted in the service of this goal. (At the present time they have already been used to generate a quark–gluon plasma, as a way of probing the strong interactions.)

Having observed or mimicked those fiery circumstances, we can seek confirmation of the two conjectures that are central to this argument about the initial conditions of the universe. The first conjecture is that the events and the structure subsequent to the trauma can be understood only in the light of the events and the structure prior to the trauma. The causal pass through is real and indispensable to the understanding of the later event.

The second conjecture is that the traumatic and extreme character of the formative events may not only change the laws, symmetries, and constants of nature, as the idea of non-cyclic succession implies; they may also generate a circumstance in which causal connections cease to exhibit a general, recurrent, and therefore law-like form. Because such observations or experiments will always deal with a localized part of the universe as a proxy for the universe as a whole (which we can neither observe nor tinker with), they will provide only fragmentary evidence. However, the accumulation of such evidence may throw light on a question of fundamental interest to cosmology. What combination of earlier states of the universe, prior to the formative events of the universe in its present state, as well as of changes in the laws, symmetries, and constants of the universe, can help explain the otherwise inexplicable features of the universe in which we find ourselves?

In this empirical effort, we shall find encouragement in the thought that one day our scientific equipment and ideas may enable us to discern more directly in our universe the vestiges of that earlier universe: nature in two moments of its violently discontinuous history, changing on a scale of time vastly disproportionate to the scale of our experience.

THE UNEXPLAINED CONSTANTS OF NATURE

A third controversy is the unexplained and precise values of a number of constants of nature, including the free parameters of the standard model of particle physics as well as of the standard cosmological model.

Consider first the problem in its most general form. We find certain constants or constant relations, which we can describe as the parameters of a wide array of physical theories, omnipresent in the world. However, we do not know why they have the values that they do. Their values appear to be brute facts, the unexplained furniture of the universe. Among them are the masses (and the ratio of the masses) of the elementary particles, the strength of the different forces or interactions, the cosmological constant (the energy density of space), the speed of light, Planck's constant, and Newton's gravitational constant. These values have so far defied all attempts to account for them on the basis of the laws of nature that we are now able to discern.

Three of the unexplained parameters – Newton's gravitational constant G, Planck's constant h, and the speed of light c – are intrinsically dimensional: to the extent that they fail to vary, we can take them to define the units by which we measure everything else – including time, mass, and energy. Their function as part of the equipment by which we measure the world may give them some exemption from the query about why each of them has one value rather than another.

However, the enigma of brute, irreducible facticity then attaches all the more strongly to the remaining unexplained parameters. These residual parameters are unitless or dimensionless ratios. The mystery of their having one value rather than another stares us in the face. If the dimensional parameters do vary, then the ratios of their values at different times are also dimensionless numbers, with the result that the mystery applies to them as well.

With regard to the standard model of particle physics, there are 28 dimensionless parameters in addition to the three dimensional parameters, G, h, and c, giving units for counting quantities in nature.

The problem presented by the pervasive presence of these surprisingly arbitrary values has two aspects.

The first aspect is the special-tuning riddle: were the parameters even slightly different there might not have emerged stable nuclei, stable molecules, longlived stars, or life. There seems to be no mechanism by which the teleological result could exercise causal influence: that is to say, no mechanism by which the indispensability of those particular values of the parameters to these outcomes can help explain why they are what they are. Some version of the anthropic principle – a reverse causal explanation appealing to the properties of a universe capable of accommodating us – then fills the vacuum left open by the failure of an adequate causal account.

The second aspect is the fine-tuning conundrum. Many of the parameters are a very small number. The constants are thus specified with a refinement that makes more disconcerting the absence not only of an adequate explanation but also of any serious and sustained attempt to explain them.

As mysterious as the parameters of the standard model of particle physics are the parameters of the standard cosmological model. This model works very well at least back to nucleosynthesis, which is the time when neutrons were created from protons and electrons in the plasma. However, it depends on additional unexplained parameters. Many of them also need to be finely tuned for galaxies and stars to form and for life to exist.

Consider again the larger problem, now in the context of the enigma presented by the precise but unexplained values of the parameters of both the standard model of particle physics and the standard cosmological model. These parameters surface as constants in the cooled-down universe. Did any or a few of them but slightly differ from what they are, everything would be different in the world. That the human race would never have appeared represents just another twist on this more fundamental enigma.

The problem of the unexplained parameters or constants presents the puzzles of contemporary physics and cosmology from a distinct vantage point: the perspective of underdetermination. The prevailing theories fail to show why the values of the parameters or the constants are what they are rather than something else. In this sense these theories underdetermine the outcome.

To such problems of underdetermination, there are in general three classes of solutions: the dialectic of chance and necessity, the view of our universe as one of a crowd of universes, and the appeal to historical explanation. Historical explanation includes the possibility that the laws of nature change, in the course of time, together with the phenomena that they govern. It is not just the content of the laws of nature that may change; it is also the law-like character of causal connection (according to the thesis of causality without laws).

The third class of solutions is merely undeveloped; part of the program of this book is to develop it. The first two are irreparably defective. Their flaws result from their failure to come to terms with the implications of the inclusive reality of time.

Thus, the completion of our argument about the uniqueness of the universe, and therefore of the advantages of succession, especially non-cyclic succession, over plurality in our cosmological thinking, depends on the argument for the reality of time, presented in Chapter 4 of this book. The theses of the singular existence of the universe and of the inclusive reality of time are intimately connected. It is nevertheless useful to sketch, even before working out the argument about time, reasons to reject both the dialectic of necessity and chance and the conception of multiple universes as acceptable solutions to the problem presented by the unexplained parameters or constants of nature.

We may be tempted to attribute the otherwise unexplained value of the parameters of nature to the effects of probability – the roll of cosmic dice. It is a solution that becomes increasingly less satisfactory as we expand the scope of the explanatory work that we expect it to carry out. This approach may be useful in helping explain certain physical and biological events. Expanded, however, into a cosmological thesis, it is so incomplete as to be unavailing. It is the half rather than the whole of an answer, and it makes little sense without the missing half.

To justify the metaphor of the dice, we must be able to say how such dice are put together, and how they are cast, and within what setting of changeless or changing reality the cosmic gambling goes on. No wager sets its own terms; a probabilistic explanation can work within a framework determined in another way, not when it is used to account for the most general framework of natural events. On this vast scale, to make use of probabilistic thinking is to replace one mystery by another.

A yet more fundamental objection to the use of probability is that on the assumption of the singular existence of the universe, there is only one case of a universe at a time. The unexplained parameters are enduring attributes of nature in this one case. Such a circumstance violates the first requirement for the applicability of probabilistic reasoning: probability demands a well-defined group.

A second basic approach to the theoretical underdetermination of the parameters or constants lies in the appeal to a plurality of universes. This appeal, we have seen, can take two main forms. In one, macro form, it is the idea of parallel or divergent universes (the multiverse), of which our universe represents one – the thesis of plurality. In another, micro form, it is the notion of multiple states, realized in different dimensions: a major theme of contemporary string theory.

The macro idea pushes the laws of nature many levels up, assigning them the role of governing what is common to the multiple or parallel universes rather than what is peculiar to the universe in which we find ourselves. The relation of the laws of nature to the unexplained parameters or constants would then resemble the relation of basic biochemical constraints and regularities to the relatively accidental, path-dependent details of natural history.

The micro idea takes the concept of plural worlds many levels down, into the multiplication of different ways in which the constituents of matter can interact. The way in which they do interact in the observed universe may then be explained as one of such possibilities: the possibility consistent with our own emergence. We shall then read the seemingly arbitrary constants in our own world as part of the indispensable background to our emergence – thus converting, to our satisfaction, arbitrariness into providence.

In either of these two modes, the invocation of multiple universes amounts to an evasion rather than to an explanation of the mysterious factual residue in the present world view of science, manifest in the unexplained constants or parameters as well as in the strange and finely tuned initial conditions of the universe. Such an invocation provides no account of why our universe is one of these many fanciful universes rather than another. The "anthropic principle," which presents the values of the parameters or constants backwards, as part of the condition for our rise, stands in for a missing explanation.

The intellectual sin of this latitudinarian perspective is the transmutation of a scientific enigma into an ontological fantasy: the notion of the multiverse. Under the weight of this transmutation, science sinks into allegory; the actual universe takes on some of the non-reality of the conjectural universes so that the conjectural universes can borrow some of the reality of the one that we are in.

The result is to rob the world of what, for science as well as for art, represents its most important attribute: that in all its present, past, and future particularity, it is what it is, or has been, or will be, given its all-decisive history. The real world is what it is, not something else. The more clearly we acknowledge this feature of nature, the deeper becomes, in our ideas about reality, the abyss between being and non-being. The conjectural worlds of the rejected allegory would provide the *tertius* between non-being and being and make the contrast less absolute.

The failure of these two ways of dealing with the factual residue, whether of the unexplained parameters and constants or of the initial conditions of the universe, drives us to a third position. According to this third view, there is facticity because there is history, because time is both real and inclusive. Here lies the connection between the singular existence of the universe and the inclusive reality of time. The phenomena change, and so do the laws. Causal connections exist in certain states of nature without the feature of recurrence and repetition that leads us to think of them as based upon laws of nature. The parameters that we observe in nature – some of them unexplained by the effective laws established by science – may, according to a conjecture suggested by such a view, be explained by the past of nature: the evolution of its regularities as well as of its structure.

Such a cosmology completes its turn into a historical science. To do so, however, it must abandon the war that physics, allied to mathematics, has long waged against full recognition of the reality of time. It must cease to rely on the idea of an immutable framework of natural laws.

4 The inclusive reality of time

THE PROBLEM PRESENTED: HOW MUCH OF NATURE EXISTS IN TIME?

Time is real, and everything that exists, or has ever existed, or will ever exist, takes place in time. From this thesis there results the idea that the laws of nature must in principle be susceptible to change. Like everything else in this one real universe, they have a history.

The inclusive reality of time is not a tautology or a truism. It is a revolutionary proposition. Rightly and therefore radically understood, it is incompatible with a major element in the dominant tradition of modern science, the tradition that goes from Galileo and Newton to the particle physics of today. In particular, it contradicts the "block-universe" picture of the universe as well as the application of the Newtonian paradigm – the explanatory practice that explores law-governed phenomena within a configuration space bounded by initial conditions – to the universe as a whole. It puts pressure on our conventional notions of causality. It compels us to reconsider our beliefs about the possible and the new in nature. It suggests that the laws of nature are mutable and that the relation between laws of nature and states of affairs varies. It gives us reason fundamentally to invert the relation between historical and structural explanation in natural science, so that we may come to see the former as more fundamental than the latter rather than as derivative from it.

The argument for the inclusive reality of time requires for its development, and generates as one of its results, a view of the nature of time. This view is in many respects incompatible with the prevalent understanding of what contemporary science has discovered about the universe and its history but it is not incompatible with what science has actually discovered. What physics has found out about the workings of nature must be laboriously separated from the metaphysical pre-commitments in the light of which the significance of these findings is commonly interpreted. The reasoning of this chapter, as of this whole book, suggests not only a reinterpretation of the discoveries of science about the universe and its history but also a redirection of the agenda of empirical inquiry and theoretical work in cosmology and physics.

The physics of the twentieth century undermined the view of space and time as an independent background to natural events. In so doing, however, it reaffirmed the idea of a backdrop of timeless (because immutable) natural laws. Poincaré and many others claimed that this idea was not only a matter of fact about the world but also a requirement of the practice of science. It is a thesis of this book that the conception of an unchanging framework of natural laws must also be overturned if science is to advance. We cannot, however, overturn it within the limits of a way of thinking that treats time as an insubstantial extension of space. To carry out this overturning, we must come to understand time as fundamental, non-emergent, and inclusive: nothing is outside it, not even the laws of nature. Time is not emergent, although space may be.

That such a view cannot be developed and supported without trauma to influential ideas and practices, within and outside science, can immediately be shown by a first impression of its consequences for some of our most widely held scientific and philosophical beliefs.

Consider how our conventional ideas about causality are incoherent for reasons largely unrelated to Hume's canonical criticism of them. Causal relations, unlike relations among logical and mathematical propositions, presuppose time. The cause must precede in time its effect. If time is illusory, effects are simultaneous with their causes. As a result, no deep distinction then exists between causal and logical connections. We commonly rely on the legitimacy and the distinctiveness of causal explanation. To that extent, we assume, in our everyday beliefs about causation, as about much else, the reality of time.

If, however, time goes all the way down, so that the laws of nature are within rather than outside it, then these laws must sooner or later be liable to change. As everything else in the world, they have a history. Causal explanations ordinarily rely on laws of nature, which serve them as warrants: the idea of recurrent and persistent connections among phenomena forms a major if not the predominant part of the ordinary understanding of what laws of nature are. However, if time is inclusive and the laws of nature are therefore susceptible to change, all our causal explanations rest on laws that are mutable even when they fail to change.

Moreover, an unprejudiced reading of what cosmology has already discovered about the history of the universe may suggest that nature has existed in forms radically different from those that it takes in the mature and observed universe, with its definite structure of elementary constituents, its clear distinction between states of affairs of laws, and its severe limitation on the range of what, given any state of affairs, can happen next. When nature fails to exhibit these attributes, causality may exist without laws. If it can exist without laws, it must be a primitive feature of nature. We do better to regard the laws that science is able to establish as codifications of recurrent causal connections in certain states of nature than to see them as the basis and warrant of causal connections.

The notion of causality without laws extends the reach of the idea of the mutability of the laws. It does so by making a historical claim about the universe: that the law-like causality we observe is a characteristic of certain states of the natural world rather than a permanent feature of nature. The development of this idea begins in the rejection of what I earlier described as the second cosmological fallacy.

Once we confront these problems, we must change how we think about causation. Our commonplace ideas about causation are confused. They assume that time is real – real enough to establish the distinction between logical and causal connections – but not so real that it threatens the stability of our causal explanations and the legitimacy of our explanatory procedures. This ramshackle compromise, however, fails to do justice to the truth of the matter. Time is real; the discoveries of science, embodied in the vision of a universal history, have given us, over the course of the last century, increasing reason to acknowledge its all-encompassing reality.

The thesis of the inclusive reality of time exposes the trouble in the picture of the world presented by contemporary science just as it reveals the incoherent character of our conventional beliefs about causation. In the central tradition of physics, time had no sure foothold. In his "scholium on absolute space and time," Newton famously wrote: "Absolute, true, and mathematical time, of itself and from its own nature flows equably without regard to anything external; and by another name is called duration ...." In Newton's own physics, however, no basis exists on which so to affirm the reality of time. Newton's laws of motion are time-symmetric; they supply no reason or occasion to distinguish between forward and backward temporal orderings of events.

The denial of the reality of time in this Newtonian tradition is not merely an implication of the reversibility of the laws of motion; it is also a consequence of the explanatory approach that we label the Newtonian paradigm. All phenomena are to be explained as if they took place on a trajectory of movement within a well-defined configuration space, described by initial conditions that are stipulated to hold for the purpose of that explanation. These same assumed starting points may figure as explained phenomena rather than as stipulated initial conditions for another instance of the same explanatory practice.

Given the initial conditions that define the configuration space and the laws that govern the events within this space, nothing is left to chance. It is possible, in principle, to infer both future and past events from present ones. Thus, present, past, and future can all be held simultaneously in the mind of the observer-theoretician. Such obstacles as may exist to this collapse of all moments into a single moment result only from the frailty of our minds and from the imperfection of our knowledge.

The observer stands outside the configuration space. He looks upon the events from the vantage point of time-symmetric laws expressed in the timeless propositions of mathematics. The regulative ideal to which his knowledge conforms (however inadequate in a particular circumstance his fulfillment of the ideal may be) is that of the knowledge that God has of the world he made (with such qualifications as may be required by the divine provision for human free will). For such a scientist, past, present, and future are simply now.

The Newtonian paradigm presupposes and reinforces a view of time from which physics and cosmology have never completely freed themselves in their subsequent history. According to this view, time resembles a film made of still photographs.

However, time is not an accumulation of slices. Insistence on speaking as if it were a film results from the widespread antipathy to an unconditional recognition of the reality of time. The deep problem revealed in the film image of time is the difficulty that all our non-causal categorical schemes – beginning with our logical and mathematical reasoning – have in dealing with the temporal continuum (a subject that we address in the development of the third of our central claims: about mathematics and its relation to nature and to natural science). The slice or film language is yet another way to subordinate time to the anti-temporal biases of these forms of reasoning. Indeed, this whole tradition – the commanding tradition of modern science – has trouble conceding to time the character of a continuum: not a continuum in the mathematical sense of the real number line but a continuum in the vulgar sense of unbroken flow, not subject to analysis into discrete elements. The adherents to this tradition return, despite themselves, to the conception of a series of still photographs.

In what respect has the subsequent history of natural science reversed the implications of these ideas and practices and laid a basis for recognition of the reality of time? The answer to this question requires us to confront a paradoxical fact, of immense interest to the concerns of this chapter and of this book. There is insufficient support, in the dominant theories of physics since Newton's day, for recognition of the radical and inclusive reality of time.

The bodies of ideas that have been regularly invoked as such a foundation – statistical mechanics (thermodynamics and hydrodynamics) and quantum mechanics – fail to provide a sufficient basis. It is worth pausing to consider why, for the reasons help elucidate the enigmas that I here consider. Thermodynamics and hydrodynamics are regional theories: they address parts or regions of nature. (Einstein made a similar distinction between what he called principle theories, which hold by virtue of general principles, and constructive theories, which depend on assumptions about the composition of matter. He cited thermodynamics as a prime example of a principle theory, but he would have seen hydrodynamics as a constructive theory.) Such local theories foresee, in the parts or regions of nature that they address, processes that are in principle reversible and that become irreversible only in the context of particular initial conditions and therefore of a particular history: not just the history resulting from the initial conditions but also the history resulting in the initial conditions. Despite their appeal to statistical rather than to deterministic causation, they apply only through the practice of defining initial conditions and specifying a configuration space of law-governed phenomena. For the reasons earlier adduced, this practice cannot be legitimately generalized to the explanation of the whole world (rejection of the first cosmological fallacy).

The attempt to ground a view of cosmic and preferred, irreversible, continuous, and non-emergent time on the quantum-mechanical description of the smallest present constituents of nature is misguided for a different reason. The structural outcome of a historical process, which is the subject matter of quantum mechanics, can provide fragmentary clues to the understanding of such a process but it cannot support a general theory of the transformation that produced the outcome. Time evolution in quantum mechanics is thus reversible. It is often said that the reality or necessity of time results, in quantum mechanics, from the phenomenon described as the collapse of the wave function. However, such time as results from this picture of the workings of nature at a fundamental level is time in small installments, not time as universal history.

Thus, neither of the two theoretical foundations for the reality of time that the science of the last hundred and fifty years is credited with having produced in fact give adequate grounding for the recognition of that reality. They do not contradict such a recognition. They are consistent with it. They suggest how, once the reality of time is established on a different, more general foundation, it can be reconciled with what we have discovered to be true about the world. However, they do not, and cannot, accomplish the time-confirming work that they are widely but mistakenly believed to perform.

If the parts of contemporary science that are often alleged to provide a sufficient basis for the recognition of the reality of time in fact fail to do so, the theories with the broadest cosmological scope have worked against such a recognition. Nowhere did this impulse to discount the reality of time take more powerful and influential form than in Einstein's physics, under what have been its most influential interpretations. (I discuss in the next section, "The argument in science and natural philosophy," the relation of the thesis of the inclusive reality of time to special and, above all, general relativity: the point of greatest tension between the argument of this book and the reigning ideas in physics.)

Time is absorbed into the geometry of space, captured in the idea of a spacetime continuum, and made accessory to the disposition of matter and motion in the universe. Its geometrical representation provides the key to the understanding of time. The spatial metaphor describing time as the fourth dimension is the popular rendering of this explanatory move. It reveals, in proto-scientific language, what is at stake in this movement of ideas: the spatialization of time.

In all these respects, the central tradition of physics since Newton has diminished or devalued the reality of time when it has not entirely denied it. Nevertheless, this same scientific tradition has also supported and developed the idea that the universe has a history. We know enough to assign the present universe an age, of about thirteen billion and eight hundred million years. We can infer from the study of the present universe and of the vestiges, or delayed representations, of the universe, what the universe at different moments in the past must have been like. We can project from our knowledge of the present and past of the universe features of its future. Contemporary cosmology is almost entirely consumed by debates about this history as well as about its relation to fundamental physical theories.

How can the idea of a universal history be squared with ideas and assumptions that diminish, if they do not deny, the reality of time? The thesis that the universe has a history is not extraneous to the development of science; it is one of its most formidable achievements. The tension between theories (or their associated explanatory practices) that limit the reality of time and an idea of universal history that makes little sense unless time is real – more real than we have generally been willing to allow – draws a fault line within the established body of scientific ideas. The conception that the universe has a history, a decisive, irreversible history, is the single most important expression within those ideas of the thesis that time is real. It is therefore also the most important source of the trouble that recognition of the reality of time makes for the time-devaluing traditions of science and natural philosophy.

The prosecution of an argument for the unqualified reality of time thus forces us to address confusion outside science and disharmony within it. Outside science, it exposes the incoherence of some of our most influential conventional beliefs about how nature works, such as our beliefs about causality. Within science, it requires us to deal with the relation between the time-dependent idea of a universal history and foundational theories that have yet to provide a general and sufficient basis for acknowledgment of the reality of time.

For the reality of time to be accepted without qualification, we must radically revise those conventional beliefs and distinguish between empirically validated insight and supra-empirical speculation in those foundational theories. We must accord to the idea of a universal history the prerogative that it deserves. We must consider to what extent resistance to this prerogative results from prejudice rather than from knowledge: from a certain interpretation of what science has discovered about the universe rather than from those discoveries themselves.

THE ARGUMENT IN SCIENCE AND NATURAL PHILOSOPHY

The conception of the inclusive reality of time and of the mutability of the laws of nature presented in this chapter and Chapter 5 forms part of a larger set of ideas and arguments. I now mark out this larger terrain, later to address a particular part of it. I mark it out by emphasizing less the detailed content of the view that I here present and defend than the reasons to develop such a view.

Once we place the conception in this larger context, it becomes clear that it proposes a reorientation of the agenda of cosmology, not simply a reinterpretation of prevailing scientific ideas. It rests, though for the most part indirectly, on an empirical basis. It is rich in implications that lay it open to empirical confirmation or falsification. As with any set of comprehensive ideas in science or natural philosophy, it faces the tribunal of the facts of the matter less proposition by proposition than in the aggregate. It confronts empirical test nonetheless along a wide periphery of consequences and presuppositions of its tenets.

In the course of developing and establishing these views, we must overcome a number of metaphysical prejudices that inhibit their understanding and acceptance. We must also distinguish, in some of the most influential scientific ideas of the present, what is scientific discovery about the workings of nature, supported by observation and experiment, from what is a metaphysical gloss on these findings. This divorce between discovery and speculation is nowhere more important than with regard to the place of general relativity in cosmology.

* * *

The universe has a history. Reckoning with its history is the common element in almost all important cosmological discoveries made over the last hundred years. Moreover, according to the view that now enjoys abundant and increasing empirical support, this history began in a supercondensed and very hot state. From that state there developed, by steps, the cooled-down universe that we observe, with its discriminate structure and its stable regularities. We know how old the universe is, or, more precisely, the present universe or its present phase since expansion from the hot and condensed plasma that it once was: the statement that the universe is about 13.8 billion years old – now widely agreed among cosmologists – lacks meaning outside a historical account. We can, by inference from observation and theory, form a view of the early history of the universe and infer the steps by which the infant hot and condensed universe became the relatively cool and much larger universe that we observe. We can see images of its early history in the distant sky.

Any cosmological or physical idea that fails to do justice to the historical character of the universe must, on that account alone, be found wanting. By this standard – empirical rather than merely speculative – the now dominant cosmological ideas are defective.

The discovery that the universe has a history may be thought to have little consequence for the basic ideas now ruling in cosmology and physics. At least such a conclusion may result if the history of the universe is represented as governed by changeless and timeless laws of nature and if it leaves untouched the elementary structure of nature, as described, most notably, by particle physics.

However, any such restraint on the implications of the historical character of the universe faces two objections, each of them developed later in this section and in this book. One objection has to do with the specific content of the history that cosmology and large-scale astronomy, for close to a century, have begun to disclose. The other objection goes to the idea of history.

It would be a fiction to suppose that the initial conditions of the universe and the making of the basic constituents of nature before nucleosynthesis can be inferred from the now leading physical theories. At best, certain aspects of this history can be reconciled with these ideas, for example with the string-theory development of particle physics. The trouble is that these ideas (to stay with the salient example of string theory) can also be reconciled with a vast number of universes other than the one that we actually observe. It is this radical underdetermination of observation by theory that helps motivate the fabrication of imaginary universes in the multiverse conception, discussed in Chapter 3.

The dominant theories break down in their application to the earliest moments in the history of the universe. A characteristic sign of this breakdown, addressed further ahead in this section, is the inference of an initial infinite singularity from the field equations of general relativity.

Even on the present hegemonic theoretical ideas, it is hard to see how the laws of nature, as we now understand them, could have applied at the beginning of the present universe. One reason why it is hard is that these laws have as their central subject the interactions among the basic constituents of nature, in particular as they are represented by the standard model of particle physics. These constituents, however, including particles and fields, are themselves protagonists in this history, rather than part of an eternal backdrop to it. They emerged in the course of the historical changes, in real time.

It is true that computer models, expressing our present understanding of the laws of nature, can "predict" many aspects of the evolution of the universe, including its chemical evolution, converging with what we know observationally. It is undisputed that the laws (the recurrent, formulaic causal connections), symmetries, and dimensional or dimensionless constants of nature have been stable since early in the history of the universe. It is tempting to infer from the stability of the laws and other regularities of nature, as well as from our preconceptions about science, that the laws must be immutable.

Consider, however, four reasons for which we should hesitate to infer the immutability of the laws of nature from their stability.

A first and most important reason is that the identification of the stability of the laws with their immutability leaves not only unanswered but also unanswerable the question: Where do the laws and the initial conditions of the universe come from? Historical explanation makes the question at least in principle susceptible to an answer. It does so, however, on several provisos: that there not be an absolute beginning, or a beginning out of nothing, such as is implied by the idea of an infinite initial singularity; that the history be at least in principle and indirectly open to empirical investigation, if only by the traces it may leave on subsequent states of affairs; and that temporal reality or the succession of universes or of states of the universe is indefinitely old although we have reason (as I earlier argued) not to conclude that it must therefore be eternal. We are unable to look into the beginning of time. That we explain the *after* by the *before* without ever reaching an initial moment is a limit to our understanding. However, it is not a circularity or a contradiction in our reasoning about what we can collectively, and over time, hope to discover about nature.

An additional challenge to this alternative, historical approach is that it is incompatible with what we call the Newtonian paradigm, with its characteristic distinction between initial conditions and a bounded configuration space of law-governed phenomena. Given that here we deal with the cosmological rather than the local, we should count this incompatibility as a strength rather than as a weakness.

A second reason to resist supposing that the laws of nature may be immutable because they are stable is that what we already know about the history of the universe suggests, as we later argue, that both the laws and the elementary constituents or structure of the universe must have been very different at the beginning. This inference suggests that the laws and elementary structure of nature may change sometimes quickly and at other times slowly or not at all – a concept familiar as the punctuated equilibrium of neo-Darwinian evolutionary theory. However, if we are to make good on the cosmological application of the idea of punctuated equilibrium, we must develop a way of thinking about how the laws and the structure co-evolve. In cosmology there is even more reason to undertake this task than in natural history:

we cannot in cosmology, as we may in natural history, reassure ourselves with the prospect that some other, more fundamental or comprehensive discipline will do the work.

A third reason to avoid inferring the immutability of the laws from their stability is that, among the features of the universe and of its history, there are some that are directly related to the irreversibility of the history of the universe. Prominent among these features is that the universe remains out of thermal equilibrium. We shall argue that to attribute this irreversibility either to improbable initial conditions or to a special law or principle of irreversibility is just another way of saying that explanation in cosmology must be historical before it is structural.

A fourth reason not to translate the stability of the laws into their immutability is that the laws, such as we know and express them, radically underdetermine the observed universe. Computer simulations of the evolution of the universe may beguile us into thinking that, after the mysterious origins, everything else is accounted for by the established understanding of the laws of nature. In fact, there is no such gapless synthesis of the now standard cosmological model and of the standard model of particle physics. Particle physics accommodates our universe only by also accommodating countless other universes, in the existence of which we have no other grounds for belief.

The reasons to think that the historical character of the universe may have far-reaching consequences for cosmology are not limited to the implications of this history for the premise of the immutability of the laws of nature. They have to do as well with the idea of history: with what the claim that the universe is historical means. If history were determined by structure, rather than the other way around, and if the determination of history by structure were expressed by eternal laws of nature, the result would be to eviscerate the sense in which there is a history at all. Laplacean determinism, and its yet more radical successor – the block-universe understanding of the cosmological implications of general relativity – annul the significance of time. They place the end in the beginning. That is not history; it is the negation of history. Its opposition to history ends in a denial of the reality of time. It places the scientist in the position of God in the Semitic monotheisms, for whom there is only an eternal now because he sees the end in the beginning.

It is a problem familiar, at least for the last hundred and fifty years, to students of human history and of social theory. Belief in the existence of permanent laws governing the succession of institutional and ideological regimes in the history of humanity was justified, among other ways, by the claim that without such belief historical explanation would be left with no basis whatsoever and descend into a narrative agnosticism. Yet the loss of that faith has been followed by new forms of causal explanation, dispensing with the notion of timeless laws of historical change. So it must be, and can be, in cosmology.

* * *

The question then arises whether any aspect of nature is exempt from this history: that is to say, from susceptibility to change. Under the views that have long been dominant, much of fundamental importance is exempt from change, and, in this sense, is unhistorical. What is most unequivocally left out of the history of the universe are the laws, symmetries, and supposed constants of nature. These regularities, according to long and widely held conviction, from which only a few scientists have dissented, are unchanging. Indeed, many philosophers and scientists have mistakenly regarded the idea that all the laws, symmetries, and constants of nature may change as nonsensical. Also left out of openness to change, according to the prevailing views, is the elementary structure of nature.

These exemptions from the reach of history and of change are so entrenched in established scientific thinking that they have survived successive revolutions in cosmological and physical theory. The assumptions underlying them have set their mark even on theories that put the idea of a succession of universes in place of the idea of a plurality of universes. Such cyclic universe theories (for example, now in the early twenty-first century, those of Penrose and of Steinhardt and Turok) commonly suppose that the successive universes exhibit the same structure and conform to the same laws.

These prevailing ideas about the distinction between what changes and what fails to change in the history of the universe have a twofold basis. One basis is the stability of the laws, symmetries, and constants in the observed cooled-down universe. From stability it is common to infer immutability. Whether this inference is justified, however, depends on the facts of the matter about the history of the universe. The current cosmological models imply that the universe may in some states exist in a condition of extreme density and temperature. These states may be incompatible even with the elementary structure described by the standard model of particle physics. They may fail to conform to the regularities – laws, symmetries, and constants – found in the cooled-down universe.

In other states, however, nature exhibits such a structure and appears to obey such regularities. To regard these law-obeying and structurally differentiated natural states as the sole regime of nature (the second cosmological fallacy) is then to misrepresent the universe and its history. The stability of the laws, symmetries, and constants of nature finds its most ready explanation in these historical facts about nature rather than in the speculative idea of unchanging laws, symmetries, and constants.

Another source of the unjustified inference of the immutability of the laws (and symmetries and constants) from their stability is a pair of notions, at once methodological and metaphysical. According to one of these notions, the idea of changing laws is confused, self-defeating, or senseless: we can discern change only by reference to something that does not change. It is, on this view, only because part of nature, including its basic structure and laws, fails to change that we can hope to understand the part that changes. This objection, however, would count only against the view that all laws change at once, or in the same way. Nothing about nature or its history suggests that laws could change in such a fashion, if indeed they do change. This false objection is, however, related to a real problem: the one that we address in this book under the label of the conundrum of the meta-laws.

The second methodological and metaphysical notion deployed in the attempt to infer immutability from stability is the claim, made by Poincaré and many others, that science cannot dispense with immutable laws. This objection, however, amounts to an attempt to entrench a particular explanatory practice as a requirement of science. It is contradicted by the methods of the life and earth sciences as well as by those of social and historical study.

The conundrum of the meta-laws suggests one of several reasons for resistance to the idea that the laws and other regularities of nature may change. The idea that the laws may change, notwithstanding their stability in the cooled-down universe, seems to present us with an unacceptable choice between saying that such change is law-governed, with the result that the high-order or fundamental laws then benefit from an exemption from history that we deny to the lower-order or effective laws, and saying that the change is uncaused, with the implication that it is arbitrary or uncertain in reality, not merely unknown to us.

* * *

The view that the universe has a history, amplified by the conjecture that everything in the universe, including its rudimentary structure and its laws, changes sooner or later, and has indeed already changed in the past, can be stated radically and comprehensively only if we accept that there is a preferred cosmic or global time. A simple description of what such time means is that everything that has happened or that will ever happen in the history of this universe, or in the history of the universes that may have preceded the present universe, can in principle be placed on a single time line. When I subsequently use the term cosmic or global time without further qualification I mean it in this sense as preferred cosmic time.

Without a strong, non-arbitrary conception of preferred cosmic time, the reality of time and the historical character of the universe would be limited and compromised for reasons that each of us later discusses in greater detail. These reasons can be summarized as follows.

First, if there is no cosmic time, there can be no overall history of the universe, only a series of local or fragmentary histories. Such histories are history in a diminished sense: they treat history as subordinate to structure. Structure, or structural explanation, may be universal, whereas time and historical explanation may be only local. Moreover, if even events that are causally connected cannot be lined up in unique temporal succession, the link between causality and time is broken or substantially modified. It becomes hard to see how time can be fundamental or non-emergent. It must be derivative from something else.

Second, if there is no global time, the non-existence of such time in the mature universe must relate to events in the very early universe in one of two ways, either of which is troubling. On one view, there may have been a unified time at the very beginning, when the universe remained a superhot and supercondensed plasma within which light was trapped rather than a gas within which light could travel freely. Then, however, as the universe began to expand, cosmic time fell apart, giving way to many-fingered time. It then seems that the disintegration of time is an event within a larger history rather than a deep feature of time. On another view, cosmic time never existed even at the beginning. If so, it becomes unclear in what sense we can say that the present universe has a history at all, or make empirically grounded claims about its age. It becomes senseless to state, as cosmologists now generally do, that it is about 13.8 billion years old.

Third, if the reason for the non-existence of a preferred time is the inseparability of time from space, or more specifically its shaping by the geometry of space and the disposition of matter in the universe, as the leading interpretations of general relativity imply, then the non-existence of global time results in a substantial qualification to the reality of time altogether.

The combined and cumulative significance of these considerations is to suggest that we cannot rid ourselves of cosmic time without at least diminishing the sense in which time is real at all as well as the sense in which the universe has a history. Any discoveries about the incidents of this history may require more or less radical reinterpretation if they are to be reconciled with the theoretical denial of a unified time, expressed in a unified history. The end of cosmic time, as a legitimate, factually based, scientific concept, represents the beginning of the end of time itself.

The existence of a preferred cosmic time seems to conflict with general relativity. The ample empirical and experimental support for general relativity may therefore appear to strike a fatal blow against the idea of such a preferred time. I discuss this apparent contradiction in two steps: first here as part of the development of the idea, and then later in this section with emphasis on the implications of my argument for the future agenda of cosmology.

In general relativity, the relativity of simultaneity is enlarged into a general freedom to choose the time coordinates on the spacetime manifold. From the standpoint of the theory, such a choice is arbitrary. There are many cosmic time lines, not one. In that strong sense, of one preferred cosmic time, there can be no preferred time: no one time line describing exclusively the history of the universe.

The non-preferred time suggested by the relativistic way of thinking is a feature of its way of representing the universe, not just a contingent aspect of some classes of solutions to its field equations. It is incidental to our arbitrary slicing of the spacetime continuum, described as a four-dimensional semi-Riemannian manifold. Under the aegis of this idea, we can always slice the cosmological solutions of general relativity into an evolving succession of spacetime states. Almost every cosmological solution under the leading interpretations of general relativity can be represented as a one-parameter succession of states. It must conform to the assumption of a gauge invariance rendering the choice among such representations devoid of physical significance. Moreover, we can slice the spacetime continuum in a mathematically infinite number of ways.

A defender of this eviscerated view of global time against any proposal of a preferred cosmic time may object that whereas the former benefits from the ample empirical support for special and general relativity, the latter requires a conjecture for which there is no corresponding validation. This objection amounts to an illegitimate shifting of the burden of proof in the debates needed to illuminate the future path of cosmology. Neither relativistic time, understood in the now traditional way and therefore subject to arbitrarily chosen spacetime coordinates, nor a preferred cosmic time is directly observable. Nevertheless, the latter may figure, as the former already does, in theories generating a host of implications susceptible to empirical confirmation or challenge as well as to experimental inquiry. The question is which of the conflicting bodies of ideas, taken as a whole, generate the most fertile research agenda for science and best stand up, cumulatively and over time, to empirical and experimental test.

To vindicate the existence of a preferred cosmic time, it does not suffice to show, as I earlier argued, that the thesis of the inclusive reality of time and of the thoroughly historical character of the universe require it. Such an argument would be merely speculative and circular. In the end what are decisive are the facts of the matter about the universe, its constitution, and its history. Is the universe, or is it not, so evolved and arranged that it allows for a preferred cosmic time?

The answer is that the concept of a preferred cosmic time can be translated into the idea of a preferred state of rest of the universe, but only at a cosmic rather than a local scale. The claim that a preferred state of rest exists at local scales would contradict special relativity, for which there is abundant empirical evidence.

The existence of a preferred state of rest gains operational significance, and a chance for empirical validation, from features of the universe that support preferred or fundamental observers: observers for whom the universe appears to be expanding and changing equally in all directions and from all directions. Their position is, in this sense, neutral.

That there is more than one way of selecting such observers, and that the observers selected by these alternative criteria are the same despite the plurality of criteria for selecting them, are powerful signs of the reality of the phenomena that these criteria select. One criterion is the range of frames of reference from which the galaxies recede at the same speed. The second criterion is the range of frames of reference to which the cosmic microwave radiation background of the universe presents itself at the same temperature.

A factual condition for the existence of such preferred observers is the relative homogeneity and isotropy of the universe. We cannot infer this condition from theoretical or a priori considerations. Our inability to do so supports the point that the most important feature of the universe is that it is what it is rather than something else. Its features may nevertheless have historical explanations.

The applicability of these two overlapping set of criteria for choosing preferred observers of a preferred state of rest in the universe is no mere speculative conjecture. On the Earth, these criteria apply, to a close approximation, so long as we control for the consequences of the movement of our planet in space.

Reconsider in the light of these remarks the objections to the existence of a preferred global time that claim the authority of general relativity. The concomitant effect of these two sets of objections is to shift the burden of proof in favor of the many-fingered time of the prevalent understanding of general relativity and against the thesis of a preferred cosmic time.

A first set of objections rejects such a time because it is incompatible with the notion of the spacetime continuum and its spatialization of time. I later argue that this notion embodies a supra-empirical ontology. The ontology does not deserve to benefit from the authority granted by the classical and post-classical tests of general relativity. On the contrary, we have many reasons to strip from the hard core of general relativity this metaphysical accretion. Of these reasons, the most important are those that concern the inclusive reality of time and the historical character of the universe and the requirements of a way of thinking and of a research agenda that can do justice to them.

A second group of objections denies that the features of the universe that we see as making possible a preferred state of rest and preferred observers have any such significance. It dismisses them as accidental features of the history of the universe, without larger consequence. These conditions would, according to this way of thinking, be without theoretical significance. They are not explained by laws and symmetries of nature, nor do they suggest, so far as the present state of scientific knowledge can discern, such symmetries and laws. At best, according to the ideas expressed in the idea of a multiverse, the set of these initial conditions expresses one of the indefinitely large possible variations of nature, given that our mathematical representations fail to select the universe that we actually observe.

This objection, however, is self-discrediting. It reveals the dangerous alliance between the practice of the Newtonian paradigm (the first cosmological fallacy) and the spirit of metaphysical rationalism. The view of mathematics as the oracle of nature and the prophet of science joins the same coalition. Grant me that the most important fact about the universe is that it is what it is rather than something else; that the distinction between stipulated initial conditions and timeless laws, which defines the Newtonian paradigm, lacks legitimate cosmological application; and that we cannot infer the truths of nature from metaphysical preconception or mathematical abstraction. At least we cannot infer them except insofar as mathematics helps make explicit the implications of empirically validated discoveries about the workings of nature. On these assumptions, we have no good reason to discount, as bereft of significance for cosmology, characteristics that the one real universe in time actually has. Natural science must seek to explain these characteristics. So much the worse for our present theories if they fail to do so; we then need to come up with better ones.

* * *

Up to this point, I have discussed the meaning of the idea of preferred cosmic time, explored its importance for the development of the thesis of the inclusive reality of time and for a historical approach to the universe, and considered objections to its existence. I conclude this introduction of the notion of preferred global time by addressing another way in which preferred cosmic time may be introduced into the canon of our cosmological conceptions. Once again, a crucial point is the apparent irreconcilability of such time with general relativity.

The reinterpretation of general relativity in the light of other ideas may reconcile the preponderant part of general relativity – and especially its legacy of empirically validated discoveries – with a notion of preferred cosmic time. One such set of ideas is shape dynamics, the character and implications of which Lee Smolin discusses in his part of this book. For the moment, it suffices to say that whereas in general relativity (according to its predominant interpretations) size is universal and time is relative, in shape dynamics time is universal and space is relative. The propositions of general relativity can be converted, without loss of empirical success, into this different language.

The path by which shape dynamics reaches an idea of preferred cosmic time is a gauge-fixing condition achieved through constant-mean curvature slicing. In addition to reconciling general relativity with preferred cosmic time, shape dynamics has the advantage, from the perspective of the agenda defended here, of eliminating singularities. As a result, it can be useful to a cosmology that is determined to lift the impediment that an infinite initial state (a singularity in the familiar technical sense) imposes on its historical approach to the universe.

Nothing, however, in the following argument turns on the merits and prospects of shape dynamics as a scientific theory in its own right. The allusion to it here serves the purpose of suggesting that the relation of general relativity to the idea of preferred cosmic time is more complicated and ambiguous than it may at first appear to be and cannot be adequately described as a simple and insuperable contradiction. Later in this section, I explore another response to the objections raised by the ruling accounts of general relativity to preferred cosmic time: the separation of the empirically supported core of general relativity from what I argue to be a metaphysical gloss on its discoveries. The major part of this gloss is the idea of a spacetime continuum and the spatialization of time that this idea has served.

The translation of general relativity into the language of a theory such as shape dynamics weakening or undermining the obstacles to the acceptance of a preferred cosmic time and the divorce of general relativity from the supra-empirical ontology of the spacetime continuum represent alternative routes to the reconciliation of general relativity with preferred time. The second route is more radically revisionist than the first. These two paths may not diverge but one requires more, by way of reconstruction in the agenda of contemporary physics and cosmology, and may go further, than the other.

* * *

The historical character of the universe was discovered and confirmed after both special and general relativity had been proposed and had begun to be vindicated by empirical findings. Quantum mechanics was already in the early stages of its development. In different ways, both these theories, as they came to be understood and worked out, lent support to the diminishment if not to the denial of time. The most acute contradiction was with general relativity, for general relativity, under its predominant interpretations, was incompatible with acceptance of the idea of a preferred cosmic time as distinguished from many-fingered time: the array of alternative spacetime coordinates, choice among which remains arbitrary from the perspective of the theory. Without making use of the concept of preferred time, we cannot give full and radical meaning to the historical character of the universe or to the reality of time.

Before saying more about the significance of this real or apparent contradiction with general relativity, it is useful to remember an important feature shared by special and general relativity, quantum mechanics, and all the most powerful movements of twentieth-century physics. When they rejected Newton's vision of space and time as an unmoved backdrop to physical events, they reaffirmed the conception of an immutable framework of natural laws.

The discovery that the universe has a history – that it is not unchanging and eternal – thus took place in an intellectual setting hostile to the full recognition of time. In fact, in the move from classical mechanics to general relativity and quantum mechanics, physics had become more rather than less resistant to acceptance of the reality of time. The discovery that the universe has a history had to be cut down to size: it had to be reconciled with ideas limiting the reality of time. One way in which these ideas qualified that reality was by rejecting the concept of preferred cosmic time.

Although Newton's laws are in principle time-reversible, their time-reversibility might be annulled by any of the frictions abounding in nature. The subsequent commonplace grounding of the direction or "arrow" of time in entropy might be seen as another example of the effort to rescue the direction and the reality of time by establishing them on the basis of asymmetries in the workings of nature.

Absolute time and absolute space remained in classical mechanics the stage on which natural events take place rather than an integral feature of the events themselves, as a relational view, such as Leibniz's, would require. Only in this diminished sense does Newton's physics make room for a notion of preferred cosmic time. It empties this notion, however, of content as a result of the time reversibility of the laws of classical mechanics, qualified solely by the unexplained facts that may render the operations of nature irreversible in the circumstances of the actual universe. Morever, the explanatory approach – what we call the Newtonian paradigm – that it shares with subsequent physics cannot deliver any conception of a history of the regularities and of the structure of nature. Newton's physics assumes a universe without a history.

With general relativity, under its standard interpretations, the situation changes. It is vital to ask whether the argument about the inclusive reality of time, including the existence of preferred cosmic time, that we present here contradicts general relativity. To the extent that it contradicts general relativity, we must further consider whether what it contradicts is the core of empirically confirmed theory or only the association of this empirical idea with a metaphysical vision from which the empirical core may be divorced.

From the outset, the cosmological consequences of general relativity were subject to dispute; its inventor was himself uncertain. The crucial point is so simple that it can easily be forgotten. General relativity was a theory about the structure of the universe, not about its history. It was formulated within a tradition of science for which physics, and cosmology viewed as a part or an extension of physics, are structural, not historical, sciences. To the extent that this tradition makes any room for history at all (and it is doubtful that it does), it subordinates history to structure.

The subsequent discovery that the universe has a history therefore presented a problem for the elucidation of which such structural theories had not been designed. It is unsurprising that the implications of these theories for an understanding of the history of the universe were so controversial. The most influential early class of solutions to the field equations of general relativity – the Friedmann–Robertson–Walker–Lemaître (FRWL) solutions – suggested that the universe must have begun in a singularity in which the gravitational field, and hence the values of density and temperature, must have been infinite. Penrose and Hawking proved the past incompleteness of any solution to the field equations of general relativity that is compatible with a small list of conditions describing our universe, of which the two most important are that the energy density of matter be positive and that there be a three-dimensional spacelike surface on which the universe is expanding everywhere. This proof was widely understood to support the inference of an infinite initial singularity from the field equations.

Part of the significance of the notion of an initial infinite singularity was to suggest a theoretical basis for the idea of an absolute beginning of the universe, a beginning as it were out of nothing, and thus as well for the emergent or derivative character of time. When these interpretations of general relativity were combined with the relativity of local simultaneities, established by special relativity, the result was to imply that the concept of cosmic time had no legitimate place in physics and cosmology. However, this suggestion remained inconclusive until a third element was added to the other two, the relativity of simultaneity and the initial infinite singularity. This third piece of the time-limiting picture was the representation of time as inseparable from space in the language of spacetime, represented as a four-dimensional semi-Riemannian manifold.

Right from the start, it was understood by many that the appeal to an initial infinite singularity, rather than representing the inference of an actual state of affairs, amounted to an acknowledgment that general relativity failed to apply, or broke down, when brought to bear on the extreme events of the very early universe. (Among those who shared the view that such a singularity could describe no physical state of affairs were Einstein and Lemaître. Despite his association with the FRWL solutions, Lemaître clearly indicated, as far back as 1927, his preference for what he himself described as a cyclical account of the history of the universe.)

To this idea of the infinite there corresponded, and could correspond, no picture of the workings of nature. The infinite, tamed by nineteenth-century mathematics, has no presence in nature. Yet the conception of mathematics as a shortcut to insight into how nature works, criticized in Chapter 6, cast a semblance of legitimacy on the appeal to the infinite, representing what was in effect a mathematical conceit as if it were a physical conjecture. A major consequence of this appeal to the infinite was to establish an absolute bar to causal inquiry into what happened before or beyond the moment when the values of phenomena pass from the finite to the infinite.

However, it is not for this reason alone that the outcome is unsatisfactory. Under this view, the laws of nature hold, regardless of time, until a certain instant, close to the formation of the early, super-condensed universe. Then, at some unknown moment, they cease to hold. What then does hold in their stead is not only unknown, but is also, insofar as it crosses the threshold of the infinite, unknowable.

An empirical scientist could be forgiven for regarding such a combination of conjectures as a ramshackle compromise in the explanation of the universe and of its history: at best, an approximation to some yet unformulated and more comprehensive idea. What deserves emphasis is that the more general view would have to be more general with regard to periods or moments in the history of the universe, notwithstanding the resistance to consider time as separate from space.

The FRWL view of the cosmological implications of general relativity was never the only one. On one side, there were also Gödel's solutions to the field equations, which resulted in a universe in which there are closed and cyclic timelike worldlines. On the other side came suggestions that general relativity might be interpreted in a way allowing for a preferred state of rest in the universe and thus for a preferred cosmic time. (Lee Smolin discusses later in this book one of these interpretations, in the variant of shape dynamics.)

The significance of these differences in the interpretation of general relativity for our argument is that a combination of special relativity with the FRWL and related solutions to the field equations fails to exclude preferred cosmic time or, more generally, to deny or diminish the reality of time altogether. These time-denying conclusions follow only when we add the third element: the conception of spacetime, as well as the spatialization of time that is implicit in this conception, with its revealing spatial metaphor of time as a "fourth dimension." This conception in turn yields the "block-universe" view in cosmology: the view of space and time as a single reality, or a single system of causal events, so that an event taking place in time can be represented as a point in spacetime.

The block-universe view can be immediately recognized as a successor to Laplacean determinism, except that it implies even more strongly than does Laplacean determinism the essential unreality of time, not just the non-existence of cosmic time. The block universe does more than place the early history of the universe beyond the reach of explanation, as happens with the inference of an initial infinite singularity. It contradicts altogether the idea that the universe has a history. It does so by translating historical change into non-historical structure: moments in time into points in spacetime.

This third piece in the prevalent understanding of general relativity and of its cosmological implications differs from the other two – incorporation of the confirmed insights of special relativity and appeal to an initial infinite singularity – in its character. Special relativity is a scientific theory about the relativity of inertial frames, richly confirmed by empirical and experimental evidence. We can best understand the initial infinite singularity, for its part, as a mathematical inference from the field equations of general relativity, suggesting the limits to their domain of application. The spatialization of time – the third part of the dominant approach to general relativity – in turn combines two distinct elements. They enjoy radically different degrees of empirical grounding.

The first element is the subsuming of Newtonian gravitation within a theory that explores and explains the structure of space. Space ceases to be an independent and invariant backdrop to physical events. Gravitation becomes a name for the reciprocal interactions between matter and space.

The second element is a metaphysical vision according to which time is best regarded as an extension, or an integral feature, of space, varying with space and having no independent reality of its own. Its existence and its flow are, on this account, illusions, as the block-universe development of general relativity makes clear. The exclusion of cosmic time works as only the first step in the exclusion, or radical diminishment, of time altogether.[1]

[1] Einstein himself rejected efforts to represent general relativity as committed to the spatialization of time. In his review essay, "À propos de La Déduction Rélativiste de M. Émile Meyerson" (*Revue Philosophique de la France et de l'Étranger*, **105** (1928), 161–166), he wrote: "Furthermore, Meyerson correctly stresses that many presentations of the theory of relativity incorrectly speak of a 'spationalization du temps'. Space and time are indeed fused into a unified continuum but this continuum is not isotropic. Indeed the character of spatial contiguity remains distinguished from that of temporal contiguity by the sign in the formula giving the square of the interval between two contiguous world points. The tendency he denounces, though often latent in the mind of the physicist, is nonetheless real and profound, as is unequivocally shown by the extravagances of the popularizers, and even of many scientists, in their expositions of relativity."

A mathematical conception expresses the metaphysical vision. It represents spacetime as a four-dimensional pseudo-Riemannian manifold. The manifold can be sliced in an infinite number of ways by alternative global spacetime coordinates. None of these coordinates enjoys priority. None describes a preferred cosmic time: the time of a universal history.

That we should not confuse this second element – the metaphysical vision and its mathematical expression – with the first, but should understand it as the suprascientific proposal that it is, can be shown by considering the wholly different relation of each of these two elements to what are taken to be the empirical supports for general relativity. These supports bear directly only on the structure of space. They bear on the structure of time not at all, unless it is already assumed that space and time are inseparable and that time is simply an extension or modification of space.

Such an assumption, however, has no basis in any of those empirical tests of general relativity. Its association with general relativity is the product of a metaphysical bias: a bias against the inclusive

He reiterated his protest twenty years later in his letter to Lincoln Barrett of June 19, 1948. There he wrote: "I do not agree with the idea that the general theory of relativity is geometrizing physics or the gravitational field."

It is clear from these and many other statements that Einstein's main interest in general relativity was a unification affirming the equivalence of gravitational and inertial force and treating both space and time as inseparable from the disposition of matter and motion in the universe rather than as a distinct and absolute background to physical phenomena. Along the way, he insisted, as we do, on the empirical vocation of mathematical ideas: "... [I]n the end geometry is supposed to tell us about the behavior of the bodies of experience .... [T]his association makes geometry a science of experience in the truest sense, just like mechanics. The propositions of geometry can then be confirmed or falsified, just like the propositions of mechanics." (Draft of an article, left unfinished and unpublished, for *Nature*, 1919/1920.)

The defense in this book of a preferred cosmic time, in the setting of our broader claims about the inclusive reality of time and the mutability of the laws of nature, cannot be reconciled with some of Einstein's theoretical proposals. It contradicts even more his methodological preferences. He would not have approved the radical divorce, for which I argue here, between the empirically validated core or residue of general relativity, as well as the local relativity of simultaneity in special relativity, and the supra-empirical ontology of a four-dimensional semi-Riemannian manifold.

Einstein nevertheless avoided and resisted the transformation of general relativity into a full-blown metaphysic reducing time to space or treating the former as emergent from the latter. He was saved from this "extravagance" by, among other precautions, his refusal to treat mathematical conceptions as a surrogate for physical insight. Our arguments contradict his theories less than they oppose the views of many of his interpreters and successors.

reality of time. This bias has accompanied, with certain exceptions, much of the history of modern physics. It has been inspired by a view of the relation of mathematics to nature and to science that we later criticize.

None of the classical empirical tests of general relativity – the perihelion precession of Mercury, the deflection of light by the Sun, and the gravitational redshift of light – have any direct or proximate relation to time or any reliance on the notion of Riemannian spacetime and to Minkowski's geometrization of time. They are all fully accounted for by the geometrical or spatial reinterpretation of gravity and by the relation between the placement of matter in the universe and the geometry of space. They need not be interpreted as contradicting either the independent reality of time or the existence of a preferred cosmic time.

Of the post-classical tests of general relativity, such as gravitational lensing, frame dragging, and observation of binary pulsars, only one, Shapiro's time delay test, confirmed by very long baseline interferometry, has any close relation to Riemannian spacetime. Even this test, however, with its prediction of so-called time dilation in the movement of photons close to the surface of the Sun, can be accommodated by the part of general relativity bearing on the interaction between the physical nature of light and the gravitational potential of our star. It requires no reliance on the impulse to render time spatial. This impulse betrays a metaphysical idea from which the hard empirical core of general relativity can and should be rescued.

It is true that the description of these observational tests is commonly couched in the language of geodesics of a spacetime continuum, connected with the suprascientific program for the spatialization of time. In every instance, however, even the aspects of these tests involving time dilation can be fully accounted for by the relativistic way of thinking about matter, motion, gravity, and space and the effects of velocity and gravity (in conformity to Einstein's principle of the equivalence of gravitational and inertial masses) on the movement of everything, including clocks or cells in human bodies, our biological clocks. There is nothing in these tests, or in what they test, that requires us to make the extra leap of regarding time as a merely local extension of space or as an extra and inseparable "dimension" of a spacetime continuum.

The identification and criticism of this element of supra-empirical ontology in general relativity (or in its leading interpretations) is consistent with acknowledging that the theory invoking this ontology may be very successful in its core domain of application, with the result that the supra-empirical leap may seem to be supported by empirical observation. However, it is no more true that the successful application of general relativity to this domain vindicates the metaphysical conception of Riemannian spacetime than it would be justified to say that the similar success of classical mechanics in its central realm of application vindicated Newton's picture of a world of interacting forces and bodies against an independent background of space and time. In each instance the experience of reaching the limits to the domain of application precipitates a characteristic question for natural philosophy: when, and to what effect, the marriage between the empirical substance or residue and the metaphysical vision to which it is wedded should end in divorce.

This line of reasoning becomes easier both to understand and to accept once it is demonstrated that the framework of general relativity can be restated in an equivalent vocabulary such that the choice between the two vocabularies is neutral with regard to the empirical observations. (For example, Julian Barbour's shape dynamics, related to Juan Maldacena's gauge/gravity duality, trades the many-fingered time gauge invariance of general relativity for a different gauge principle: that of local changes of scale.) It is not crucial to this proposal to divorce the empirical content of general relativity from the project of spatializing time that any particular such equivalent formulation be true or successful as a scientific enterprise in its own right. What is decisive, for the immediate purpose, is that the discussion of such formulations reveals the gap between what we know for a fact and how we choose to interpret, frame, or represent our empirical findings.

I am now able to begin answering the question about the extent to which the argument of this book contradicts general relativity, with regard to full acknowledgment of the historical character of the universe, to the inclusive reality of time in general, and to the existence of a cosmic time. The answer is that the argument contradicts general relativity as general relativity is often represented. However, it does not contradict a version of general relativity that has been subject to three sets of clarifications. I have sketched in the preceding paragraphs the reasons for these clarifications.

First, the local relativity of simultaneity must be qualified by recognition of the existence of a preferred state of rest in the universe and thus, as well, of preferred observers. Such observers might in principle see the universe, averaged over a suitably large scale, from a vantage point undistorted by local inhomogeneity and anisotropy in the disposition of matter. They could in principle determine whether an event happening in Andromeda was happening before or after, on the cosmic time line, an event taking place in NGC 3115, and they would have reason to credit their observations as in correspondence with the cosmic time line. That there can be such observers depends entirely on certain facts about the constitution of the universe such as its relative homogeneity and isotropy; it cannot be inferred from a priori considerations. That we on the planet Earth fail perfectly to figure among such preferred observers in no way contradicts these limits to the relativity of simultaneity or their implications for the existence of cosmic time. We can in fact correct for the movement of our planet and approximate the circumstance of preferred observation.

Second, the inference from the field equations of general relativity to the notion of an initial infinite singularity must be understood as a mathematical revelation of certain temporal limits to the domain of application of general relativity. It must not be interpreted, instead, as a conjecture about an actual state of the universe, either in its earliest formative moments or in its subsequent evolution.

Third, and most importantly, general relativity must be reformulated without the addition of the Riemannian spacetime conception, the disposition to spatialize time, or the block-universe view, none of which are vindicated by the empirical and experimental evidence adduced in favor of general relativity. The metaphysical gloss must be lifted from the empirical theory, with the result of suggesting a redirection of the agenda of cosmology.

These multiple qualifications represent no small adjustment to the common understanding of general relativity. Yet they are all justified by what cosmology has discovered about the universe and its history. They are justified as well by the need to resist the seduction and corruption of physics by mathematical ideas (e.g. about the representation of the infinite) that have no place in the workings of nature. These justifications are no mere grab bag of unrelated findings and ideas. Rather they comprise different aspects of the same basic insight into the historical character of the universe, to which even the most ambitious structural theory of nature must accommodate if it is not to misrepresent its subject matter.

* * *

I now address more briefly the relation of our argument about time and the historical character of the universe to quantum mechanics. With respect to the quantum theory, as with regard to general relativity, the issue is whether what the theory has discovered about the vicissitudes of nature contradicts the claims that we make here. My discussion will be more compressed both because quantum mechanics bears less directly on this part of our argument than does general relativity and because, to the extent that it does bear on the argument, Lee Smolin will address its significance later in this book.

Like general relativity, quantum mechanics is a structural rather than a historical theory. It presents no account of transformation in natural-historical time. It is true that the time-dependent Schrödinger equation, when taken in tandem with thermodynamic principles, creates a basis for grounding transformation and time in the minute workings of nature and for connecting change at the atomic scale with change in the macroscopic world.

That this set of implications is nevertheless insufficient to serve as a basis for the understanding of temporal processes on the scale of the history of the universe was long ago shown by the Wheeler–DeWitt equation. This equation suggests that, once applied to the whole universe, quantum mechanics has no place for time: its cosmological application results in a quantum-mechanical equivalent to the block-universe view inspired by the most prestigious interpretations of general relativity. Barbour and others have developed the implications of this view for denial of the reality of time.

Quantum mechanics and its crowning achievement, the standard model of particle physics, have been richly confirmed by empirical and experimental results. Can they be legitimately invoked as an objection to the view of time and of universal history that we advance? They cannot because quantum mechanics is essentially incomplete, and therefore approximate, in two distinct but connected ways.

In the first place, it is incomplete because its account of its subject matter, the operations of nature in the cooled-down universe at the most rudimentary level, fails to yield any determinate picture of these operations. Not only must it deploy stochastic rather than deterministic reasoning but it must also appeal, in this reasoning, to the idea of the infinite: infinite possible configurations rather than the initial infinite singularity that has been inferred from the field equations of general relativity.

However, here as always in science, the infinite is not a physical reality. It has no place in nature. The infinite is "the measure of our ignorance." The appeal to stochastic reasoning, under cover of the mathematical (but not physical or natural) idea of the infinite, was the legitimate target of Einstein's objections to the quantum theory that he had helped create.

The effort to redress this ignorance has set some on the search for the "hidden variables" that could help replace probability with determination and conjectural infinity with a view of what happens when and where at the atomic scale. The hidden-variables theories capable of accounting for the nature and effects of these "hidden variables" could only be relational views: they would show how the particles, fields, and forces studied by quantum mechanics relate, through reciprocated action, to features of the universe that have remained outside its scope of inquiry.

Here, however, we come to the second way in which the quantum theory is incomplete and thus restricted in its domain of proper application. It is incomplete because it is unhistorical. One might argue that quantum mechanics deals with all of nature, albeit at a certain level, the level of the most basic natural processes, rather than with some patch of nature. To that extent, it would be untainted by what I earlier called the first cosmological fallacy: applying a method suited to explore part of nature to the study of the whole universe. However, even if one accepts this response, there is no adequate defense against the accusation of committing the second cosmological fallacy: quantum mechanics is also incomplete because it treats as the intrinsic and eternal furniture of the universe the law-obeying and discriminate structure that the cooled-down universe displays.

What cosmology has already discovered about the formation of the universe suggests that the elementary structure described by the standard model of particle physics did not always exist. For example, a crucial moment in the formation of the universe, the moment called decoupling, occurred when the atoms became stable. If this structure once did not exist, or took different form, it may later in the history of the universe also differ: in regions of the cooled-down universe, such as in the interior of black holes, or in the remote future. Our experimental technology – including our particle colliders – may develop to the point of allowing us to simulate aspects of these extreme states of nature.

Quantum mechanics, as it is now understood, would then become the theory of the elementary workings of nature in one of the phases of nature, if we make the analogy at cosmological scale to the local physics of phase transitions. In this inquiry, as in every department of science, to understand phenomena is to grasp what they have become, or can become, under certain circumstances or provocations. Structural understanding is ultimately subordinate to historical insight.

The hidden-variables theories needed to continue the work of quantum mechanics, and replace its reliance on stochastic reasoning and on the mathematical infinite, must therefore be temporal as well as relational. The science of which they would form part would be informed, through and through, by a temporal naturalism.

Until we come into the possession of such a science, we can have no confidence in the cosmological applications of quantum mechanics. These applications will, on account of the twofold incompleteness of the underlying theory, continue to point in opposite directions: either supporting or denying the reality of time, and either accommodating or resisting the discovery that the universe and all of its regularities and structures have a history.

* * *

It is common to suppose that both the reality and the direction or "arrow" of time need no basis of support other than the long-established principles of thermodynamics. It is true that the special and seemingly unlikely circumstances making an entropic process irreversible may play a role in the evolution of the universe: great homogeneity and consequent low entropy in its very early history. It is also true that the temporal asymmetry of these macroscopic processes may be anchored or prefigured in particle physics by the so-called CP (charge, parity) violations of the weak interactions.

Nevertheless, thermodynamic principles are insufficient to provide a basis for foundational thinking about time for three reasons.

The first reason is that an account of the basis for an arrow of time cannot suffice as a response to ideas that deny the reality of time altogether or that treat special and general relativity as insuperable obstacles to the affirmation of preferred cosmic time.

The second reason is that statistical mechanics is, by the character of its theories, equations, and procedures, a study of local realities in the universe. Like classical mechanics, it works by applying invariant laws to changing phenomena within a configuration space defined by stipulated initial conditions. The cosmological application of such a theory takes it to a level at which no such distinction between initial conditions and timeless laws makes sense. The configuration space becomes the entire universe; there can be nothing outside it (other than the imaginary universes of the multiverse conception). The initial conditions are simply the state of the whole universe at some arbitrary point in its history: the same as the configuration space, only at any earlier time. Such a cosmological extrapolation of the local theory represents an instance of the first cosmological fallacy.

The third reason is that entropic processes are not in fact irreversible except in certain circumstances. Conversely, the laws of motion in classical mechanics become irreversible as soon as we add certain real-world features of events to the circumstances that they are used to explain. The contrast between reversibility in classical mechanics and irreversibility in statistical mechanics is circumstantial and relative.

Any understanding of how a diminution of entropy at a cosmological scale could come to be irreversible presupposes the reality of time and the historical character of the universe more than it explains them. One way of explaining irreversible cosmological entropy invokes special circumstances in the very early universe. If these circumstances are not hidden behind the impenetrable screen of an initial infinite singularity, they demand historical explanation: a previous state of affairs must explain how they came to be what they were. Such an explanation invokes time and, at least at the formative moment, a line of preferred cosmic time, even if (by the prevailing interpretations of general relativity), such unified time later gave way to many-fingered time.

Another explanation appeals to a directional law that, at least at a cosmological scale, we would have to add to the standard repertory of thermodynamics the better to ensure an asymmetry between past and present. Such a "law," however, is only another name for time rather than an account of its emergence out of putatively more fundamental physical realities.

We require a deeper and more general basis for our understanding of time than any that is offered by statistical mechanics and by the attempt to give it a cosmological application.

* * *

Thus far, I have considered reasons to believe in the inclusive reality of time and in the existence of a cosmic time that have to do with particular features of the observed universe as well as with the discovery that the universe has a history. I have discussed whether these ideas about time conflict with established empirically supported science: primarily general relativity, the body of ideas with which our claims seem to be in greatest tension, and secondarily special relativity and quantum mechanics, with regard to which the conflict is both less obvious and less acute.

These are not, however, the only reasons to adopt the views about time that we propose. There are also reasons of a more general character. Proposals in natural philosophy and in comprehensive scientific theories should be evaluated by the totality of their advantages and defects, including the intellectual opportunities opened up by the research agendas that they inspire, as well as by the explanatory power and the empirical foundation of their component pieces. To deserve influence, such general considerations should retain a connection to empirical challenge or confirmation. The power of comprehensive views needs to be assessed as a whole and by comparison to rival research programs.

Among general reasons of this order, it is in turn important to distinguish between those that have to do with some understanding of the nature of the phenomena studied by the research agenda and those that rely on assumptions about science itself.

We should entertain this second class of general reasons – reasons about the requirements of scientific practice – with parsimony and suspicion. There exists no single uncontested and universally applicable way of doing science. Method must follow vision rather than the other way around. The defense of a program of inquiry on the ground that it, and it alone, embodies the correct practice of science can easily serve to mask metaphysical prejudice and to inhibit the dialectic between vision and method.

An example of immediate pertinence to our argument is the commonplace idea that science cannot dispense with the idea of a framework of immutable laws of nature. Whether there is such a framework is a matter of fact, regardless of how difficult and perplexing the matter of fact may be to investigate. A more complicated example is the attempt to entrench a particular view of causal connection as intrinsic to human understanding, as Kant did against the background of Newton's physics.

Nevertheless, there is a legitimate residue in such general reasons having to do with scientific practice. It is the preservation of openness to empirical and experimental test, however varied and oblique the testing may be. An objection to the invocation of an infinite initial singularity as something more than a mathematical *reductio* is that it places a range of phenomena beyond the reach of further inquiry. An objection to the idea of a plurality of causally disconnected universes is that it multiplies entities that cannot, even in principle, be investigated, confirmed, or excluded by empirical test. It disposes of an explanatory embarrassment at the cost of creating a much larger one.

By contrast, the argument in natural philosophy and in cosmology that we develop here is open on all fronts to empirical consequence and challenge. As in any such ample theoretical view, some parts of the system of ideas are much closer to empirical implication than others. We nevertheless count it as a fatal objection to a view that it be, especially in principle, immune to investigation of the fact of the matter.

This remark leads into the first and more legitimate set of general reasons to prefer an agenda of theory and research to its rivals: reasons that have to do with the phenomena that it addresses. An advantage of the ideas about the inclusive reality and the existence of global time that we propose is that they fit much more easily than do their rivals with other views about the singular existence of the universe and the selective realism of mathematics. There are, we argue, independent reasons to hold these other views, about the solitary character of the universe and the imperfect relation of mathematics to nature.

The singular existence of the universe, the inclusive reality of time, with its implications for the mutability of the laws of nature (despite their stability in the cooled-down universe), and the selective realism of mathematics are intimately related and overlapping conceptions. It is not easy to adopt any one of them without accepting the other two.

The thesis of the inclusive reality of time is connected to the notion of the singular existence of the universe because it makes it possible to put the idea of a succession of universes or of states of the universe, without break of causal continuity, in the place of the idea of a plurality of causally unconnected universes. If everything that has ever happened, or that will ever happen, can be placed on a single time chart – the import of the concept of preferred cosmic time – not only will events in the present universe fit on this chart; so, in principle, will events in every universe that preceded it, if such universes have existed.

If, according to another attribute that we claim for time, time is not emergent, its flow will be uninterrupted by a succession of universes, or of states of the universe, if indeed there is such a succession. There will be no interruption of causal continuity even if, for example, successive universes undergo periods of extreme density and temperature. Time will flow unbroken even if such periods are marked by a conflation of the laws of nature with the states of affairs that they govern. It will continue its uninterrupted progress through other periods in which the universe cools down and expands, acquires a discriminate structure, and sees a distinction established between states of affairs and laws governing them.

Expressed in this abstract and comprehensive form, as the natural-philosophical basis for a cosmological program, these ideas may at first seem to be entirely speculative and beyond the reach of empirical test. Yet they can be falsified by a single observation or experiment suggesting the disappearance, emergence, or interruption of time, in any of the states in which nature presents itself, local or cosmic.

These conceptions and arguments do indeed fail to take issue with theories that deny the reality of time altogether, or that reduce time to a series of frozen configurations of the universe. However, they fail to take issue only because the time-denying accounts make themselves invulnerable to empirical test. Such theories set themselves against empirical challenge by redescribing in time-free language all the observations and experiments that we perceive as taking place in time. By the same token, they render senseless the discovery that the universe has a history, for if history means anything it means causal succession in real time.

The quasi-empirical view of mathematics – the view of mathematics as a simplified representation of the most general aspects of nature, which then takes off on its own even when its inventions have no counterpart in the natural world – denies to mathematics any privilege to reveal the intimate nature of reality. This deflationary approach to mathematics is similarly connected with affirmation of the reality of time. This connection remains fundamental to the understanding of both mathematics and nature.

The relations among mathematical propositions are timeless even when, as in the calculus, they are used, or were even devised, to represent temporal events. If everything in nature, even the laws of nature, is time-bound, a gap opens up between what is natural and what is mathematical. At a minimum, there can be no mathematical object that is homologous to the natural world. More generally, the abstraction of mathematics is closely related to its timelessness. Its immediate subject matter is not the real natural world but an imaginary proxy for that world, bereft of both time and phenomenal particularity. Disregard for phenomenal particularity is closely related to denial of time.

These views about mathematics also serve an agenda of empirical and experimental reason for a straightforward reason. They oppose the legitimacy of any attempt to use mathematical reasoning as a substitute, rather than as an instrument, for empirical inquiry. They render empirical the question of whether any given mathematical conception is or is not realized in nature. They help explain how mathematics can be so useful despite its divergence from nature and because of that divergence.

The argument composed by our three central ideas supplies the substance of the temporal naturalism that we advance. Such a naturalism supports the continued transformation of cosmology into a historical science that becomes central rather than marginal to physics. It informs conceptions that are hospitable to a fuller reception of the discovery that the universe has a history and that everything in the universe changes sooner or later.

* * *

These special and general considerations justify overturning all supposed barriers to the full reception of the most important cosmological discovery: the discovery that the universe has a history. Everything in the universe was once different than it is now. Everything in the universe will change sooner or later.

There is no good reason to suppose that anything is outside time, which means that anything is not susceptible to change.

To this principle, however, the predominant cosmological and physical ideas make two closely related exceptions. They make a first exception in favor of an immutable framework of laws, symmetries, and constants – the values of certain fundamental relations or ratios in the workings of nature. They make a second exception in favor of a picture of the most elementary particles and fields of which nature is constituted.

The two exceptions are connected. The laws, symmetries, and constants have as their subject matter the interactions among a particular stock of constituents of nature and would have no significance apart from them. The stock is an arbitrary bric-a-brac until considered in relation to the laws, symmetries, and constants calling them to order. These two exceptions are not warranted by our cosmological discoveries or required for the practice of science.

Under our present understanding of the history of the universe, however, neither the regularities nor the stock could have been the same at the beginning of the universe as they subsequently became in the cooled-down universe, with its discriminate structure. If they were different then, they may become different again.

Here are four reasons – all of them mistaken – why we might insist on this twofold exception to the reach of time. Although the list is not exhaustive, it includes all the most influential classes of argument for exempting the ultimate structure and regularities of the universe from its history.

The first reason is the view that there is no time, or that there is time only in some greatly restricted and qualified sense. Then we cannot accommodate the discovery that the universe has a history, and all the observations and experiments with which this history is associated, except by radically reinterpreting their significance. I later address two classes of philosophical objections to the reality of time, whether inclusive or not.

The second reason is that science cannot continue to do its explanatory work without presupposing such an immutable framework and stock. This second reason is sheer metaphysical prejudice. It is contradicted by the existence of many versions of scientific inquiry, particularly in the life and earth sciences and natural history, relying on no such presupposition. It is true that we do not know how to do physics without invoking permanent laws and constituents. We need to learn how. The history of modern science offers many sources of guidance. Cosmology and physics without unchanging laws and natural kinds present the genuine enigma that we call the conundrum of the meta-laws.

The third reason is that the earliest events in the history of the universe would be, by virtue of their inclusion or proximity to an infinite initial singularity, beyond the reach of scientific inquiry. However, this reasoning mistakes a mathematical inference (from the field equations of general relativity) for an image of the workings of nature. Moreover, it draws a picture that must be unsatisfactory to the scientist: a structure that is immutable until, at some uncertain point, it falls off into the abyss of the infinite. It would better accord with both the discovery that the universe has a history and the practice of specialized and local branches of physics (e.g. the physics of phase transitions), to suppose that there is over time some set of discontinuous transformations, including transformations of both the regularities and the stock, that we can come progressively to investigate and to understand.

The fourth reason is that, in the cooled-down universe we observe, both the regularities and the stock are in fact very stable. It is tempting to infer their eternity and their necessity from their stability. (I earlier enumerated reasons to resist this temptation.) However, a comprehensive cosmological view, one accommodating the historical character of the universe, must be able to reconcile stability at some times with transformation at others. That such a reconciliation can form part of a successful research program in science, we know for a fact because it is one of the main tenets of the now dominant neo-Darwinian synthesis in evolutionary theory, as amplified by the conception of punctuated equilibrium. It also has counterparts in the investigation of lifeless nature in geology as well as in the study of human history and society.

A lesson of these analogous research agendas is that the understanding of stable laws and types may be decisively reshaped by insight into their transformability even if they are in fact stable over extended periods and in particular domains. At its heart, scientific insight is always insight into transformation: a structural understanding is never deeper than its grasp of the conditions under which the phenomena that it addresses change into something else, and have changed to become what they are.

* * *

These arguments, developing the cosmological program of a comprehensive temporal naturalism, both invoke and justify an idea of time. They are arguments about change. However, in being arguments about change, they are also arguments about time, for time and change are internally related concepts. Both are in turn internally related to causation.

Time is the fundamental aspect of reality – of all nature – by virtue of which everything changes. Because everything is connected, directly or indirectly, with everything else and what it is is the sum of such relations, to say that everything changes is to say that it changes with regard to these other things.

Such an understanding of reality conforms to three minimalist postulates. The first is the postulate of reality: there is something rather than nothing. The second is the postulate of plurality: there is more than one phenomenon or being. The third is the postulate of connection: the plural things that exist are connected.

These three postulates might be described as forming a proto-ontology, were this label not likely to arouse a misunderstanding. This proto-ontology can inform a view, like the temporal naturalism developed in this book, that rejects the idea of a permanent repertory of natural kinds and of law-like regularities governing their interactions. To define such a system of beings and laws or principles was the aim of classical ontology. Thus, the system jointly formed by the postulates of reality, plurality, and connection might be better described as an anti-ontology, or as the beginning of one.

Time, understood in this way, is also susceptible to differential change. Because things change differently or unevenly, the change of some can be used to measure or clock the change of others.

To say that time is inclusively real is to hold that nothing in nature is exempt from the susceptibility to change, including change itself. Susceptibility to change reaches all laws, symmetries, and supposed constants as well the kinds of things, the natural kinds, that there are, including the most elementary constituents of nature.

We develop this concept of time into a particular conception. Fully to elucidate the significance of the discovery that the universe has a history, such a conception must treat time as non-emergent, global or cosmic (in the strong sense of preferred time), irreversible, and continuous. Much of the rest of this chapter is devoted to developing this conception of time and to discussing its cosmological and historical bases, uses, and implications.

Of these four attributes of time, the first – non-emergence – is the most fundamental. The second – that it exists in the form of preferred global time as well as in the form of local times – is the most controversial, given the most influential interpretations of general relativity. The third – that time is irreversible – is the proximate premise of universal history. It must be affirmed cosmologically, as a view of the universe and its history, or not at all. We cannot derive it securely from any theory of local phenomena such as statistical mechanics. The fourth – that time is continuous – is the most mysterious, in light of the anti-temporal bias of the mathematical imagination.

Such a conception of time may at first appear to contradict the assumptions of a relational approach. We may be tempted to interpret it as the description of the attributes of a thing or substance, in the manner of a metaphysical system like Aristotle's or Spinoza's. The outcome of such an interpretation would amount to a return to an absolute idea of time like Newton's.

We may be seduced into taking this turn by the spatial quality of our intuitions. So we may think of time as a medium in which all phenomena move, like the ether of late nineteenth-century physics, or like an empty moving film on slices of which physical events set their mark.

Time, however, is no such thing. It is an integral part of the way in which everything is what it is. Everything is what it is only because it can become something else. Only because it can become something else can we hope to understand what it now is or was.

Like causality, from which it is inseparable, temporality is a primitive feature of nature and of how it works. Contrary to Hume and to Kant and to the traditions that they inaugurated, causation is only secondarily a requirement of the understanding. It is about the world before it is about us.

* * *

These ideas suggest a cosmological view in which the laws of nature, as well as the symmetries and supposed constants, may change. Nevertheless, the mutability that in principle they enjoy must be reconciled with their stability in the cooled-down universe.

The need to reconcile their mutability with their stability is forced on us by the facts of the history of the universe as they begin to emerge from the findings of large-scale astronomy as well as from the ideas of theoretical cosmology. At different moments in the history of the universe, nature presents itself in radically different forms. Only the development of cosmology as a historical science and the formulation of a cosmological equivalent to the physics of phase transitions can provide the theoretical basis for such a reconciliation between the stability and the mutability of the laws of nature.

It is a development that must be undertaken with full acknowledgment of the limits of the analogy to the physics of phase transitions. A first such difference is that this physics deals with local realities and can thus legitimately deploy the Newtonian paradigm. A second such difference is that the phases addressed by the physics of phase transitions are enduring states of nature, in accordance with stable regularities.

From the outset, such a project faces the dilemma that we call the conundrum of the meta-laws. If change of the laws is law-governed, we have solved the problem only by reinstating, in favor of such higher-order laws, an exemption from the reach of time, that is to say, from susceptibility to change. If change of the laws is not law-governed, it may seem to be arbitrary or unexplained, without benefit of any theory accounting for the nature and limits of such uncertainty. Progress in the solution to this conundrum is crucial to the future of cosmology.

Later in this book, each of us explores responses to this conundrum. A beginning lies in treating causality as more fundamental than laws of nature rather than in treating laws as the indispensable warrants to causal judgments, as we generally suppose. It lies as well in recognizing causal connection as a primitive feature of nature, rather than simply as a presupposition of our apprehension and understanding of nature, according to Kant's influential approach. Not just what causes what in time but how it causes what it causes, as well therefore as the character of causal connections, cannot be mere matters of theoretical prejudice. They are proper subjects for empirical inquiry. In cosmology, they take on their widest and most important form.

If, however, no progress were yet possible in the solution to these problems, our failure to solve them would give us no excuse to resist acknowledging that the mutability of the laws of nature is both implied by the idea of inclusive reality of time and suggested by what we already know about the history of the universe.

* * *

Before I address philosophical objections to the reality of time, I pause to look back on the preceding arguments in this section and to summarize the main reasons to pursue the cosmological agenda that they suggest and support. If we set aside all technical refinement and complication, the better to grasp the essential content of these ideas, we can see that they provide three main reasons to complete the transformation of cosmology into a historical science, founded on recognition of the inclusive reality of time.

The first basic reason to pursue this agenda is to widen the intellectual space in which we can seek answers to the question: Where do the laws and symmetries as well as the initial conditions of the universe come from? History does not succeed to rationalist metaphysics: it does not pretend to tell us why the laws and symmetries must be what they are. It fails to dispel, at the wave of a metaphysical wand, what, together with its temporality, is the most important feature of the universe, its factitiousness: that it is what it is rather than something else. However, a historical approach to the universe, like the view for which we argue in this book, broadens the room in which the central practice of science, causal inquiry, can work.

In so doing, such an approach also attenuates the contrast between the unwarranted appearance of the necessity of the laws and the seeming strangeness, unlikelihood, and arbitrariness of the initial conditions. Both the initial conditions and the laws emerge and evolve, under this view, in the course of universal history. To the history of the kinds of things that there are, including the kinds that there were in the early universe, there corresponds a history of the laws, symmetries, and supposed constants of nature – its regularities.

To act on this first reason requires that we lift the bar on the extension of causal inquiry imposed by the inference of an infinite initial singularity from the field equations of general relativity. The justification for lifting that bar is, as I previously argued, both particular and general. The particular ground is that the inference of an infinite initial singularity is best interpreted as a sign of the breakdown of the theory when carried beyond its proper domain of application rather than as the description of a physical state of affairs. The force of this argument rests, however, in large part on a more general ground: the absence of the infinite from nature. Behind the concept of a cosmological singularity stands the mathematical idea of the infinite, benefiting from the dangerous seductions as well as from the matchless power of mathematics.

The second major reason to follow this agenda is that it allows us to avoid the unsatisfactory position to which cosmology has been driven in the absence of such an agenda. Our successive discoveries concerning the evolution of the universe lead us, increasingly, to conclude that everything in the arrangements of nature – its elementary constituents and its organization – was once different from what it now is and will be different again at another place and another time in the universe. How can we make sense of the coexistence of unchanging laws, symmetries, and apparent constants of nature with change in the elementary composition and organization of nature?

We might make an analogy to the local physics of phase transitions and say that the laws explain the change in the constitution of nature just as they govern phase transitions. Moreover, we might seek support for this claim in the undisputed success of simulations of the evolution of the universe, back to relatively early moments in its history, on the basis of the established laws. The trouble is that the laws and other regularities of nature fail to explain change beyond or before nucleosynthesis. Beyond or before that time, the composition and organization of nature, and even the extent to which causal connections displayed law-like regularity, may have been radically different from what they subsequently became in the cooled-down universe.

What we call the initial conditions of the universe is a name for part of this early constitution. It is the part defined by the combination of two criteria: that we can distinguish it from the regularities of nature and that the regularities fail to explain it. Failure to explain the initial conditions of the universe is just another name for the failure of the structural laws established by contemporary physics and cosmology completely to explain the evolution of the universe.

Two very different responses to this predicament are then possible. One response is to suppose, in the spirit of the now dominant ideas and practices in cosmology, that as science advances, we shall be able to reproduce, on a cosmological and universal-historical scale, the success of the physics of phase transitions. This physics explains in law-like fashion changes in the local constitution of nature.

For the cosmological analogy to the physics of phase transitions to be successful, we would have to explain this evolution on the basis of an adjustment to the established catalog of laws. Given the radical changes that appear to have taken place in the course of the history of the universe, we would, at a minimum, have to enlarge our view of the regularities of nature. To do so, we would need to appeal not just to more, or adjusted, laws and symmetries but to a different kind of law-like explanation. If nature in those formative moments of the universe consisted in singular events and failed to display law-like regularities, we could not hope to solve the problem by simply extending or adjusting the present stock of our explanatory ideas. It would be necessary to change the way in which we explain as well as the content of our explanation.

The alternative response is to proceed on the working assumption that if the structure changes, even radically, so may the laws and symmetries. The familiar model for this approach is the one that we find in natural history and in the life and earth sciences, with their characteristic appeal to a co-evolution of the laws and the phenomena, and their recognition of the mutability of natural kinds as well as of the pervasiveness of path dependence. Under this view, we shall no longer have to ask ourselves what the laws of nature were doing when the states of affairs that they supposedly govern did not yet exist. We shall have room for a thoroughly historical view of the universe, one in which we need not rely on the idea of a permanent constitution of nature or treat the laws and symmetries of nature as commanding but unmoved bystanders to universal history, exempt from time and change.

The third important reason to redirect cosmology to this agenda is the need to offer a consequential alternative to a combination of ideas that have acquired increasing influence over the last few decades: string theory, in its cosmological implications and applications; the multiverse, eternal inflation, and anthropic reasoning. Although these theories have distinct origins, motivations, and proposals, they have been combined to produce an ominous turn in the fundamental science of the present day: a turn away from the empirical and experimental discipline of science. Together with this lapse from empirical challenge has gone a squandering of the treasures of science, which are the riddles presented to it by nature in defiance of prevailing ideas.

The influence of these ideas can be effectively resisted only if there is an alternative to them. One of the aims of this argument about the inclusive reality of time and more generally of the whole argument of this book is to offer, in the discourse of natural philosophy, the sketch and the bases of another agenda. The cosmology needed to achieve this alternative as science, not simply as natural philosophy, does not yet exist, or exists only in fragmentary form. Much of the empirical material and conceptual equipment needed to begin developing it is, however, already at hand.

The turn in cosmology has consequences for the future of physics and indeed of all fundamental science. Our understanding of any part of nature depends on our view of the whole to which these parts belong. Such assumptions are no less decisive when they remain implicit as they often do.

When we come to approach the universe historically and affirm that both the constitution of nature and its regularities are susceptible to change, we further strengthen the sense in which an understanding of part of nature relies on an understanding of the whole universe and its history. A region of the universe or a period of its history can then no longer be counted on to reveal the most basic truths of nature, even when we study it in its connections with other regions.

In the dominant traditions of physics, universal truth is revealed locally. Local discoveries and insights are scaled up into theories of the universal structure of nature and of the laws and symmetries governing this structure. Physics has resorted to such scaling up throughout its modern history. Scaling up underlies the use of what we have called the Newtonian paradigm, the first cosmological fallacy.

A view subordinating structural to historical explanation and refusing to exempt either the constitution or the regularities of nature from the reach of time and change, robs scaling up of legitimacy. It does so by suggesting that the most significant features of a state of affairs are those resulting from its placement in the history of the universe and from its consequent relation to the changing repertory of natural kinds and of laws and symmetries of nature. Cosmology, converted into a historical science, ceases to be peripheral. It displaces particle physics as the most encompassing and fundamental study of nature.

* * *

In the history of thought many philosophers have argued for the unreality of time, not just, as the defenders of the standard interpretations of general relativity do, for the non-existence of preferred cosmic time. I put aside conceptions of the shallow and ephemeral character of the phenomenal world in the history of philosophy, by contrast to underlying unified and timeless being, to address only arguments against the reality of time that have been especially influential in recent thought. These arguments deserve attention because their authors have developed them against the background of an engagement with cosmology and have often explored their cosmological implications. The summary consideration of these views here cannot do them justice. It may nevertheless help elucidate, by contrast, the content and the bases of the temporal naturalism that we embrace.

The objections to the reality of time that I have in mind fall into two main groups. One family of objections exploits the consequences of the contrast between time as an objective feature of nature and time as the human agent experiences it and talks about it. These objections belong to the history of modern idealism. The other set of arguments emphasizes our inability to give a clear and precise account, in particular a mathematical statement, of what we mean by saying that time is real and by supposing that it flows interruptedly. The intended effect of both sets of arguments is to make it seem that the idea of the reality of time is incoherent, contradictory, or incurably vague. We do not really know what we mean when we say that time exists. Time, therefore, cannot exist.

A notable instance of the first set of objections is McTaggart's argument for the unreality of time. It is an argument that trades on the confusions of an agent who recognizes that time must involve change but who cannot make sense of the idea of time from the standpoint of his agent-oriented experience and his language about past, present, and future. If all that we have is before and after, we do not possess real time, unless we can also fix the references of our talk of future, present, and past. However, this talk is incoherent: every event must be, and yet cannot be, simultaneously past, present, and future. If we say that it is future, present, and past at different times – McTaggart's "obvious objection" to his own argument – we simply assume the reality of a time that we have failed to clarify, allowing ourselves to be caught in a vicious circle or an infinite regress. Once we have discarded both our "tensed" and our "tenseless" discourse about time, we are left with a timeless ordering of events.

Consider, however, the idea, supported by the cosmological discoveries from which we take our point of departure, that the present universe has a history, with a defined beginning and a likely end. In such a universe, change happens through causation. The aspect of reality by virtue of which it is susceptible to uneven or differential change, within a world of plural and connected phenomena, is what we call time.

Whether or not the universe has a preferred state of rest, and can therefore accommodate a preferred cosmic time, we shall be able to speak of early and later moments in its history. If time is only many-fingered, as the predominant interpretations of general relativity propose, any global time that may have applied in the early universe will have given way to multiple times, inseparable from the placement of matter and the geometry of space. However, any observer, depending on his place in the universe, can relate his experience to at least one of those times.

If the constitution of the universe is such that throughout its history it can support preferred observers who can discern the passage of cosmic time (notwithstanding the inability of other observers to escape the relativity of simultaneity), any event, anywhere in the universe, fits on a single cosmological time line. Our position on Earth enables us to approximate, with certain corrections, the circumstance of such preferred observers.

The way in which our temporal experience lines up with preferred cosmic time – or with its dissolution into the alternative cosmic times of the prevailing interpretations of general relativity – will not be determined by the idiosyncrasies of our experience or the syntax of our language. It will be ultimately determined by the facts of the matter about the history of the universe and by our place in this history. We can expect to have difficulty in translating our agent-oriented vocabulary, developed to suit the scale of human activity, into the discourse of a historical cosmology, or in translating the latter into the former. We have no reason to be either discouraged or disoriented by such difficulties: they exact a small part of the price that we must pay for exercising the prerogative to think far beyond our station.

These considerations serve as the beginning of an answer not only to McTaggart's objection to the reality of time but also to all objections leveled from the perspective of idealism. They exemplify the standpoint of a naturalism that sees both causality and time as internally related and primitive features of nature. They interpret our linguistic equivocations and ambiguities as a sign of the limitations of natural language in representing the nature and history of the universe rather than as evidence for the nonexistence of realities that we have many other reasons to credit.

A second class of contemporary objections to the reality of time argues that we are unable to state with clarity and precision what we mean when we claim that time exists. In particular, we cannot make sense, especially mathematical sense, of the flow of time. More generally, we cannot distinguish time clearly and coherently from other aspects of reality. If we cannot either explain it or distinguish it, we may well conclude that it does not exist.

There is an element of validity in this family of objections. However, the objections to the reality of time to which I now refer turn the true significance of these problems upside down. The seat of the greatest difficulties to the elucidation of time and of its flow is mathematics: the indispensable but dangerous tool of physical science.

In its use to account for the nature and attributes of time, natural language is accommodating but vague. Mathematics is precise but recalcitrant. In Chapter 6 I argue that both the power and the limitations of mathematics in the representation of nature have to do, in large part, with the trouble that mathematics has with time. The relations among mathematical or logical propositions are timeless.

All mathematics can best be understood as an exploration of a simulacrum of nature: a version from which time and phenomenal particularity (inseparable in nature from each other) have been banished. Having begun in the study of the most general aspects of nature – shape and plurality or number – mathematics soon found its chief inspiration in itself and its secondary inspiration in the tasks presented to it by natural science. Its radical simplifications are the source of its power but also of its limitations. Given its trouble with time as well as with phenomenal distinction, it is more useful in understanding some parts of nature than others. Its trouble with time implies, as well, trouble with history and with any historical understanding of natural processes. It lends itself more readily to the development of a structural science than to the elaboration of a historical one.

An important aspect of the quarrel of mathematics with time is the difficulty that it has (also discussed in Chapter 6) in representing the flow of time. The mathematical idea of the continuum, traditionally modeled on the real number line, can represent flow only as a series of slices. The dominant, modern program of discrete mathematics is inadequate to the task of representing the vulgar (non-mathematical) idea of the continuum as uninterrupted, non-discrete flow. The sequence of real numbers remains a series of steps: the steps fail to melt into a flow by virtue of being uncountably infinite. Analogous difficulties beset the mathematical representation of other attributes of time and natural history.

It does not follow, however, from such difficulties that we should reject what mathematics can state only imperfectly, or state only by some approximation to the corresponding realities. Mathematics is the most powerful instrument of science, but not the infallible one. It is not equally suited to all the tasks that science has reason to undertake or to all the phenomena that it has reason to study. A physical picture, inspired by physical intuition, and developed with the help of physical equipment, may turn into a theory that is disciplined by observation and experiment without being exhaustively described by its available mathematical expressions.

Instead of regarding these problems of the mathematical representation of time as a sign of the unreality of time or of its attributes, we would do better to take them as a reminder of the difference between mathematics and physics.

* * *

There remains another source of suspicion of an argument in favor of the reality of time, including the existence of cosmic time. It may seem that in the effort to uphold that reality we surrender to our perceptual experience when much of the progress of science consists in revolt against the illusions of experience.

We experience the Earth as flat, and discover that it is not, and so it happens, the argument goes, step by step, with every aspect of our perception of the world.

Consider how incomparably more central the reality of time is to our pre-scientific understanding of nature and of ourselves than is the flatness of the Earth. That the Earth is round rather than flat, and that despite its roundness we do not fall off it into space, is a curiosity and a marvel. Having discovered it, we can retrospectively reinterpret in its light many aspects of our perceptual experience that our ancestors had misinterpreted. However, the roots of this fact in the general shape of our experience are narrow and superficial.

By contrast, no aspect of our experience is more pervasive to all our experience, and more far-reaching in its consequences for how we understand the workings of nature and our place in the world, than is the reality of time. No aspect of our existence or of our perception is untouched by it. The concepts with which we understand the manifest world, and the words with which we state our understandings, are all penetrated by implicit reference to time.

Take a homely example of a concept that crosses the divide between natural language and the language of science: the concept of a ruler for measurement. It is part of the concept of a ruler that it not change, not at least in its local environment, with regard to other things that do change. In this respect, a ruler is like a clock. It refers implicitly (whereas a clock refers explicitly) to time. From this simple instance of time reference we ascend to instances that are central: first and foremost, our understanding of ourselves as mortal organisms moving, in real and irreversible time, toward death.

If time is unreal, all bets are off in our perception and in our understanding of what is what with respect to the manifest world around us and well as to ourselves. Our perception of reality is then radically disconnected from the real. It becomes questionable how we can continue to do science at all. We may perform certain operations – for example, provoking particular particle collisions – and treat our ideas – about particle physics in this case – as the pragmatic predicate of such capabilities. However, our causal language, as well as our experimental practices, will always be time-related.

The expulsion of time does not amount to the correction of an isolated element in our perception, which, once corrected, we can continue to relate to the remainder of our experience. It dismantles the whole of our experience. It does so to such an extent that it undermines the procedure, on which science has always depended, of departing from perception without severing its link with perception. There remains in even the most revolutionary science a dialectic between our direct experience of nature and its theoretical correction. Denial of the reality of time threatens to suppress this dialectic.

It would then be as if we were in the hands of a malevolent demon like the one Descartes evoked, or as if the powers of perception and understanding with which natural evolution has endowed us were so narrowly suited to the goal of short-term survival that nothing of cognitive value can be gained by their exercise. Why then would we have reason to trust any speculative conception, such as the idea of the unreality of time, that is unsupported by direct, operational confirmation? And what could such confirmation look like, given that operational confirmation is bound up with change and causality, and causality and change with time?

The philosophies, in ancient India and Greece as well as in modern Europe, that denied the reality of time did so as part of a metaphysical program. They supplied an alternative account of reality as it is and as we experience it, and connected this account with an approach to the conduct of life. Denial of the reality of time is not an idle speculation to be pursued as if we could pull out this master thread of our experience and leave the rest in place.

A striking feature of the worldwide history of philosophical ideas about time is that neither of the most influential traditions of thought about these matters can be reconciled with what science has already discovered about how the universe evolves and about how nature changes. One tradition, associated with what we might call the project of classical ontology (exemplified by Aristotle's metaphysics and by most strands of subsequent Western philosophy) upholds both the reality of time and the existence of a permanent repertory of natural kinds or types of being. Another tradition (represented by many of the philosophical systems of ancient India and by Western philosophers as different as Spinoza and Schopenhauer) denies both the reality of time and the existence of any such permanent catalog of beings. It treats our experience of temporal change and of phenomenal distinction as an illusion obscuring the reality of unified and timeless being.

The truth suggested to us by science is, however, incompatible with both these contrasting views. There is no permanent list of natural kinds precisely because time is inclusively real. The mutability of natural types as well as of all regularities of nature is the corollary, not the opposite, of the reality of time. Such a view, the closest to the truth on the showing of modern science, speaks with the weakest voice in the history of philosophy.

These remarks go to the reality of time, not to the existence of a preferred cosmic time. It may be objected that the same considerations that weigh against renouncing the idea of the reality of time fail to count in favor of reluctance to abandon the notion of global time. Indeed, there is nothing necessary about the existence of such time. It depends on features that are better described as factitious than as contingent: that the universe happens to be constituted, and to have evolved, in a fashion that allows for a preferred state of rest and for preferred observers placed in a neutral position with respect to that state. Time may exist in some qualified but significant sense even if preferred cosmic time does not. We would then disaggregate the idea of time, treating it as a bundle of notions that can be disassembled: some of them to be retained but reinterpreted, others to be rejected outright.

Nevertheless, for the reasons that I adduced at the beginning of this section, the rejection of preferred cosmic time might well be regarded as the beginning of the end for the reality of time. It is enough to recall that if no such time exists, it is no longer clear how, or in what sense, the universe can have a history. Only the local or partial histories of many-fingered time would then exist. Such histories would be inseparable, under the leading interpretations of general relativity, from the disposition of matter in the universe and the physical geometry of space. They would be incapable of fitting together to form a master narrative. As a result, cosmology could not complete its turn into a historical science.

* * *

I do not execute in detail the intellectual program sketched in this section. Its execution would require cosmological work beyond the limits of this essay in natural philosophy. Lee Smolin will address many of these themes and develop many of these ideas, exploring their significance for the theoretical and empirical agenda of cosmology.

Within this broad terrain, the remainder of this chapter and Chapter 5 undertake four tasks. Taken together, they represent the beginning of a pathway in the direction that I have just described. They offer a natural-philosophical prolegomenon to a cosmology that could carry out this intellectual program.

The first task is to describe what it would require for time to be inclusively real. No conception of the inclusive reality of time can be useful, or stand a chance of being vindicated, unless we can specify its meaning and the properties that it ascribes to time. We can do so only with the degree of precision that is allowed by the vagueness of natural language and by the temporal recalcitrance of mathematics.

To say what the inclusive reality of time means is not to show that time is in fact inclusively real. However, unless we can further develop the idea of its inclusive reality, we cannot hope to distinguish a cosmology or a philosophy of nature informed by these views from rival conceptions. We fail to fulfill one of the conditions for an agenda of work that theory, observation, and experiment can hope to pursue.

The second task is to consider the role that a conception of the inclusive reality of time can play in a cosmology that sees itself as a historical science because it defines the history of the universe as its central subject.

The third task is to discuss the implications of this view of time and of the discovery that the universe has the particular history that it does have for the mutability of the laws, regularities, and supposed constants of nature. Such a discussion must show on what terms we can reconcile the mutability of these regularities of nature with their stability in the cooled-down universe.

The fourth task is to address the major problem to which this view of the mutability and the stability of the laws of nature gives rise: the problem that we name the conundrum of the meta-laws. We have reason not to deny either that any transformation of the laws is itself law-governed, in derogation of the inclusive reality of time, or that it is uncaused. It may be caused without being law-governed.

A beginning of a solution to the conundrum of the meta-laws lies in regarding causal connection as a primitive feature of nature, like time itself, with which it is bound up. It lies as well in appreciating that the recurrent and law-like form of causality may not be manifest in all expressions of nature. It may be most characteristic of the cooled-down universe, the traits of which we should not mistake for permanent features of nature, according to the argument against the second cosmological fallacy.

This more selective intellectual program serves the larger program outlined in this section as a preliminary. It also reveals the character and some of the implications of the temporal naturalism that we propose.

TIME AS THE TRANSFORMATION OF TRANSFORMATION

That time is related to change is a proposition that has been a persistent theme in thinking about its nature, both in philosophy and in science. However, the change that is implied by time has often been represented as movement: that is to say, in spatial terms. The result is to favor a form of thought affirming the primacy of space over time. It is a slant evident in views as far apart as Aristotle's metaphysics and Einstein's general relativity, as general relativity is commonly understood. However, it is a bias that we cannot accept if we are to grasp what an affirmation of the radical and inclusive reality of time implies.

Time is indeed about change. Its relation to change, however, need not and should not be represented in terms that make time seem to be an appendage to space.

Time is the contrast between what changes and what does not change. More precisely, it is the contrast between what changes in a particular way and what either does not change or changes in some other way. It is the relativity or the heterogeneity of change.

Within this view, time is intimately and internally connected with change. Change is causal. Time is change. In the spirit of these propositions, we should take inspiration, not discouragement, from Mach's remark: "It is utterly beyond our power to measure the change of things by time. Quite the contrary, time is an abstraction at which we arrive by the changes of things." (If we fail to recognize the intellectual opportunity bound up with the resulting conceptual confusion, we may be futher discouraged, as Thomas Hobbes was when he wrote in *De Mundo* that "time has always been whatever anyone has wanted it to be.")

The significance of the association of change, and thus of time, with causation remains circumscribed so long as we continue to think that causal connections are only instances of unchanging laws of nature. In fact, as I argued in Chapter 1, causation is a primitive feature of nature, and there may be states of nature in which causal succession fails to assume law-like form.

If change were uniform in pace, scope, direction, and outcome, it would not be change, other than in a greatly diminished sense. We could not notice it or register it. There could be no clocks because clocks are devices that measure one set of changes by reference to another set of changes. That is so whether the clock is a device that we build or a part of nature.

Might we not still remember an earlier state of affairs even if all states of affairs change uniformly and in concert? Even in principle, we could not. There must be a form of consciousness, or its physical expression in the brain, that changes on its own – by its own procedures, as well as at its own pace – for time to be perceptible.

The thesis that time has to do with the uneven character of change immediately suggests the view that time can exist to a greater or lesser extent. The deeper and more comprehensive the unevenness of change, the greater becomes the reality of time. It is most real if change itself changes. To claim that change changes is to say just what the words mean: the ways in which phenomena change also change.

Consider a simple example, to which I later return in my discussion of the conundrum generated by the idea of the mutability of the laws of nature. The forms of genetic recombination, or the supervening regularities of sexual selection, did not exist before the emergence of the phenomena that they shape. They were new types of changes, associated with new types of phenomena. Neither the kind of things involved, nor the types of change, had existed before. When they arose, change changed. The change of change was accompanied by a change in the kinds of things that exist in the world. On this view, the full reach of the reality of time is revealed in the transformation of transformation, which thus becomes another way of defining what time is.

What science has already discovered about nature, in the course of its revolutionary history, confirms abundantly that in this one real world change changes. Nevertheless, the implications of this fact for the understanding of time are almost never fully drawn. A crucial test for the acceptance of such an understanding turns out to be the temporal status of the laws of nature.

Many have argued, mistakenly, that the assumption of the timelessness of laws of nature is indispensable to the work of science. Yet this assumption appears on its face to contradict the idea that change changes. To hold that change changes amounts to allowing that the laws of nature may change. The mutability of change can be reconciled with the immutability of the laws only if we resort to the view that the change of the laws is itself law-governed. However, this move resolves the contradiction only by qualifying the idea that change changes: the higher-order laws would then be proof against time.

A simple thought experiment helps clarify both the grounds and the implications of the idea of the reality of time. The content of the thought experiment is nothing other than the circumstance that is conventionally labeled Laplacean determinism. The world is governed by a single, unchanging set of laws. Any variety in the pace, scope, and direction of change is itself law-governed. The laws of nature fully determine everything that has ever happened, or will ever happen, in the world until the end of time. They do so minutely, by what we know as causal determination. We may not have discovered the laws of nature to the extent necessary to see how they completely shape the behavior of all particulars in the world. But they do: infirmities of knowledge are not to be mistaken for disorder in the world. If we knew enough, we would be able to infer from the present moment all past and all future events: not only those that have happened, and are therefore irreversible, but all those that will happen. Indeed, in principle, from the state of affairs at any moment it is possible to infer the states of affairs at all other moments. Chance and catastrophe – including the production of vast reversals out of relatively small disturbances – are ruled out. So is genuine novelty: what may appear to be new is the working out of what existed before.

In such a world, time would exist only in vastly diminished sense. For the mind with sufficient – that is to say, complete – knowledge the whole of the history of the world is recapitulated, or foretold, in the present moment. Such a mind would be the mind of God, or of natural science insofar as it takes such divine insight as the regulative ideal whose fulfillment it progressively approaches. Under the Laplacean regime, the reality of time is compromised and circumscribed; the deeper our level of insight, the less weight we have reason to grant to the reality of time.

In these circumstances, the difference between the causal sequence of events and the connections among logical or mathematical propositions would shrink: the relation of consequences to their causes in our understanding of nature would more closely resemble the relation of conclusions to their premises in our mathematical and logical reasoning. There would be more reason to think of mathematics as an ante-vision of the workings of nature.

Our world, however, is not the world described by this thought experiment. Here are three reasons why it is not.

A first reason is that we have increasing reason to think that the range or depth of the adjacent possible – the scope of what can happen next, given a certain state of affairs, is not constant. It varies in the history of the universe. On the basis of the cosmological models that are now in the ascendant, it must be imagined to have been much greater in the formative moments of the present universe than it has, on the whole (but not always), become in its subsequent evolution. (Recall the discussion of the second cosmological fallacy in Chapter 1.) The combination of probabilistic explanation with "hidden-variables" theories may seek to subsume the periods of the universe and the forms of nature that are marked by greater degrees of transformative opportunity under the aegis of Laplacean determinism. However, the more that nature approaches the condition I earlier called causality without laws, as the distinction between laws of nature and the states of affairs that they govern fades and as natural phenomena fail to present themselves as a differentiated structure, the less plausible the attempt to uphold Laplacean assumptions will be.

A second reason is that the way of representing regularities in nature that the thought experiment portrays has, in the history of science, proved effective only in the context of the Newtonian paradigm: the explanation of parts of nature on the basis of the stipulated initial conditions that define both a configuration space and laws governing phenomena within those boundaries. The explanation of the initial conditions is always relegated, in this explanatory practice, to another explanatory account; the stipulated initial conditions of the former become the explained phenomena of the latter. Because of this feature, as well as because of the other considerations I have invoked, the Newtonian paradigm cannot be generalized to the world as a whole.

A third and more fundamental reason is that the obstacles to the application of this Laplacean view are not merely the results of transitory deficiencies in knowledge. They result from the grounds we have to believe, on the basis of what science has already established, that change does indeed change and that such transformation of transformation forms a major part of the reality of nature.

That change changes should not be interpreted as an invitation to conclude that we are therefore liberated from causality: that the universe is open, or creative, or possessed of any of the attributes that would enable us to believe that it is somehow on our side, offering us solace for the indifference of nature to our concerns. A view of what happens to causation – and to causal determinism – when change changes must be a major part of a conception of the reality of time.

ATTRIBUTES OF TIME: NON-EMERGENT, GLOBAL, IRREVERSIBLE, AND CONTINUOUS

Consider, on the basis of this initial idea of time as the transformation of transformation, four sets of additional attributes that time must have if it is to be radically and inclusively real: to be irreducible to anything else and to admit no exemption to its rule. The description of these attributes develops the conception of time implicit in the previous section. What science has already discovered suggests that in this one real world, time does have these attributes. Or – to make the same claim in different words – the view that it possesses these attributes is more readily reconciled with our established knowledge of nature, all things considered, than the view that it does not. The resulting account of time nevertheless conflicts with theories that have exercised great influence in contemporary science. What we must decide is whether we have better reason to affirm the reality of time or to cling to such theories, even in those respects in which they contradict the full acknowledgement of that reality.

* * *

The first attribute of time is that it is not emergent. Time is what remains, quipped Richard Feynman, when everything else leaves. We can conceive of a world in which time would emerge: for example, a world in which change fails to change, or in which the emergence and disappearance of time result from changing configurations of space. That is precisely the case in ruling interpretations of general relativity. That time is non-emergent implies that it cannot adequately be represented in the dimensional language often associated with non-mathematical formulations of general relativity. It is not an appendage to space.

The non-emergent character of time forms part of what it means for time to be inclusively real. If time were emergent, it would by definition derive from other, more fundamental realities such as timeless and even object-less laws of nature. The special and general reasons to believe in the inclusive reality of time considered in the previous section are consequently also reasons to regard time as non-emergent.

* * *

The second attribute of time is its cosmic or global character. There is a time of the history of the universe that is not simply a collection of local times. Its existence enables to say, for example, that the universe is about 13.8 billion years old. At least in our universe (as distinguished from earlier or later universes or from earlier or later states of ours), we can translate the speculative idea of cosmic time into the more precise and restrictive notion of a preferred cosmic time. Such a time can be registered by preferred observers whose neutral position in the universe fulfills requirements that I earlier specified.

In the history of twentieth-century physics, every view that began by disputing the existence of such a global time ended by discounting or denying the reality of time altogether. Thus, one might say that with respect to the thesis of the inclusive reality of time, the denial of the global character of time has regularly served as a first step in the formulation of a view robbing time of some or all of its reality. Resistance to this first step, however, may seem to contradict much that the physics of the twentieth century revealed about the workings of nature. It may appear a reactionary concession to our perceptual experience of time, no more justified than resistance to correcting our sense that the Earth is flat. To understand in what sense the affirmation of the global character of time does and does not conflict with the legacy of twentieth-century physics, it is necessary to say more about this attribute of time than about any other.

The thesis of the global unity of time is easily misinterpreted. The chief difficulty in understanding it is the tendency to interpret it from the standpoint of a specter that continues to stalk our commonplace beliefs about nature: Newton's idea of absolute time. The life of this idea is perennially renewed by its proximity to our unreliable perceptual experience.

Cosmic time must, instead, be interpreted from the perspective of a relational view. The hallmark of such a view is that time be understood in terms of both relative change and causal connection. In such an understanding, the ultimate grounds of the causal unity of both the universe and its history are the variability, discontinuity, or irregularity of change within this nevertheless unified and historical universe. Time, on this view, is global because the singular universe has a single history. Global time is perceptible and in principle measurable because change is not uniform over the universe and its history and because how change happens is itself subject to change.

In the ontological program informing Newton's physics, the real is equated with the absolute: it is real by virtue of standing outside the chaos of change and motion. This chaos amounts to an illusion, hiding the reality of nature. We, on the contrary, argue in this book that phenomena are real precisely to the extent that they belong to the network of evolving relationships constitutive of the world. It is not an external reference that defines change. It is the dynamical network of relationships. Real time is the reality of such changes.

The temptation to misinterpret the thesis of global time from an absolute rather than a relational perspective is aggravated by a strange and striking feature of the history of these ideas: the absence of any adequate statement of the relational view. This absence puts a heavy burden on those of us who would invoke a relational conception.

There are two comprehensive statements of the relational vision in the history of natural philosophy: that of Leibniz and that of Mach. Neither of them serves the purpose.

The trouble with Leibniz's version of the relational approach is that it forms part of a philosophy that hollows out the meaning of causation and thus, as well, of time. It does so not just by the metaphysical machinery of his monadology but also, more generally, by the reduction of causation to logic (or mathematics): an impulse prophetic of the direction that physics would take.

The flaw in Mach's variant of the relational position is that it confusingly combines a relational perspective with a very traditional, indeed Newtonian, view of causation. The relational idea comes into its own only when we revise our ideas about causality and about the relation of causal connections to laws of nature. It is only when we understand causation as more general and more basic than laws of nature, therefore as antecedent rather than as subsequent to laws of nature and as a feature of the world rather than as a mere construction of the mind, that the radical character of the relational view becomes apparent. Mach's phenomenalism prevented him from moving toward any such conception, or indeed from giving a clear account of what Einstein came to dub "Mach's Principle."

Three questions immediately arise with regard to the idea of global time and its meaning in the argument of this book. How does cosmic time under a relational view differ from cosmic time under an absolute (Newtonian) view? In a relational conception how can the thesis of the global character of time be reconciled with the undisputed finding of variation in the passage of local time, or of the relativity of local times? If global time exists, how can it be measured? I give greater consideration to the second of these three questions. It lends itself to the vital task of disentangling science from metaphysics: what twentieth-century physics actually discovered, as distinguished from the meta-scientific lens through which the hard core of its empirical findings have been read. (By cosmic or global time I mean always a single, preferred, and universal time rather than the infinite number of alternative spacetime coordinates allowed by the most widely accepted interpretations of general relativity.)

The chief distinction between cosmic time under a relational and an absolute view is that, for a relational account, the global character of time resides in the unbroken and inclusive web of connections in the singular universe and in its singular history rather than in some independent place: the place of the invariant background or of the eyes of God.

It may be objected that this statement simply alters the meaning of words, by calling change and causation time. However, something of great consequence is at stake. Nothing is for keeps in nature: no typology of being (as described by particle physics and by the periodic table), no set of laws of nature, no ways by which some things change into others. Only changing change endures.

When we consider change from the standpoint of its variability – the way that some things remain unchanged while others are changing, or change in some ways while others change in other ways – we call it time. When the seat of this ceaseless transformation is the whole universe, viewed in relation to its own history or to the local changes that go on within it, as more than the sum of its parts, we call it global time.

The prime empirical referent of the concept of cosmic time is the discovery that the universe has a history, against the background of the assumption that both the universe and its history are causally united. In invoking the concept of global time, we bet that this history can be told as more than an open-ended collection of local histories that remain incapable, even in principle, of being combined. Cosmology needs a conception of global time if it is to become a historical science.

The denial of cosmic time, and the view that this denial follows from the empirical discoveries of twentieth-century science, contradicts the view that science can properly investigate only patches of spacetime. Such a cosmological minimalism invites two responses.

No one can deal with a part without making assumptions, albeit loose assumptions, about the whole. The assumption that the history of the universe is just a pack of local histories does not exempt us from this requirement. The universe, we know since Lemaître and Hubble, had a beginning about 13.8 billion years ago. Our understanding of the local histories will be informed by assumptions about universal structure, about universal history, or about both.

If cosmology is not to be a historical science, it must be a structural science, like chemistry. The problem now is that it is neither. The fundamental objection to turning cosmology into a structural science is that everything structural changes sooner or later.

The relational revision of the meaning of global time, suggested by the preceding considerations, goes a long way to answering a second question provoked by affirmation of the thesis of global time. How can variation in the local passage of time, according to the disposition of matter in the universe, abundantly confirmed by the empirical findings that we associate with special and general relativity, be reconciled with the existence of cosmic time? Under this view, no obstacle exists to such a reconciliation so long as the idea of global time can be given observational significance and laid open, once again in principle if not yet in practice, to measurement (the third question).

There is abundant evidence that time passes more slowly in some parts of the universe than in others: for example, close to the stars, as opposed to intergalactic space. The universe appears to be so ordered that causal connections accelerate locally under certain conditions and slow down under others. Their hastened pace under such circumstances is what we mean by the more rapid passage of time.

When might we think that there is an insuperable difficulty in reconciling cosmic time with such local variation? We might think so if we were to misunderstand global relational time as global absolute time.

We might also think so if we were to mistake, as we generally do, the empirical discoveries associated with special and general relativity for the ontology to which Einstein's successors, more than Einstein himself, were attracted. Call it the Einsteinian–Riemannian ontology, marked by the project of rendering time a shadowy department of geometry: the spatialization of time. Its characteristic mathematical statement is the representation of spacetime as a four-dimensional pseudo-Riemannian manifold. It was Gödel who, toward the end of his life, most clearly grasped the time-denying implications of this ontology, and carried them to the extreme in his reflections on general relativity.

No necessary, one-to-one relation exists (as I argued in the previous section) between the Einsteinian–Riemannian ontology and the hard empirical content of general relativity. We can keep the empirical residue while dispensing with the ontology. We can do so, however, only by "rescuing the rational kernel from the mystical shell" and by beginning to work it out. We are not required to pay homage to the ontology, if we have reason to resist it, until we have a fully formed alternative view. The arguments of this book suggest that we have many reasons for such resistance.

A persistent feature of the history of science is the association between empirical discoveries and ontological programs. The most ambitious theoretical systems bridge – and conceal – the gap between the former and the latter. It is crucial, however, to remember the difference between them. It is never more important to do so than when a science is tempted to redefine its successes as failures, through special pleading and allegorical fabrication. Present-day physics and cosmology have succumbed all too often to this temptation.

The history of science is in part the history of such ontological programs: in physics, most notably those of Aristotle, Newton, and Einstein. They bewitch. To struggle against this bewitchment is often the condition of new insight. For Einstein, the metaphysical commitments were seductive enough to have conflicted his relation with the quantum mechanics that he had done so much to create. Against the seductions of Einstein's ontology, we should heed Einstein's advice: pay attention to what scientists do (which I interpret to signify: what they discover), not to what they say about what they do.

How, under the view for which I have argued, can cosmic time be measured, and what would it mean to measure it? Here childlike simplicity may help, by laying bare both the intellectual difficulty and the intellectual opportunity.

There are three kinds of clocks: local, foreign, and global. Local clocks measure the passage of time. Their movements behave in the fashion of the movements of the things around them. All that they can measure is one change by comparison to another change in that place.

Foreign clocks measure the passage of time in one site of the universe from the distance of another part of the universe. Their advantage is that they are outside the region of the universe for which we use them to measure time. Because they move according to different constraints, the constraints resulting from the distribution of matter in their neighborhood, they can better exploit the variability of change in the universe. That is, of course, also their disadvantage: information signals to and from them take time, muddying the clarity of what it is exactly that they record. Any use of such foreign clocks must contend with the intractable problems of simultaneity to which special relativity introduced us.

A global clock presents problems of a different order. It seems that such a clock could be neither outside nor inside the universe. What and where could it be? For such a clock to be observable, the universe must be so arranged that it allows for preferred observers. Their position, set by the mean distribution of matter in the universe, must enable them to watch the clock without any distortion resulting from the anomaly of their position.

Such observers cannot themselves be the clock. They can only be watchers of the clock. The existence of such a clock is made possible by factitious traits of the universe, such as the trait of having a preferred state of rest.

The clock that is neither inside nor outside the universe must be the universe itself. What this means is that certain general, quantifiable, and slowly changing features of the universe, such as the recession of the galaxies in all directions, and the color temperature of the photons in the cosmic microwave radiation background, must be used to time some other change, either in the universe as a whole or in part of the universe.

* * *

The third attribute of time is its irreversibility. The irreversibility of time is directly related to its being non-emergent and to its being global. The time reversal that has occasionally been imagined (for example, by Gödel, as a way of interpreting the implications of general relativity) is always local time reversal. The reversal of global time lacks a foothold in any empirical finding or theoretical proposition of contemporary science. Although irreversibility cannot be inferred by pure analysis (logical or mathematical) from non-emergence or non-emergence from irreversibility, they are closely related. It is by virtue of its global irreversibility that time resists reduction to any other phenomenon. It is on the basis of that irreversibility that it touches – and threatens – everything.

Moreover, the irreversibility of time should not be represented as simply a consequence of certain specific physical processes, such as the entropic processes studied in thermodynamics and hydrodynamics. Nor can it be said to be conditional on conformity to specific physical limits such as the limits imposed by the speed of light on the power of light – or of a light cone – to convey information. It is rather the other way around: the irreversibility of time appears in countless manifestations, among which are the dynamics of entropy and the light-borne conveyance of information about the recent or the distant past. In fact, entropic processes, unlike time, are reversible under certain specific initial conditions. It is a matter of fact, rather than of logical or mathematical analysis, to establish that those are not the conditions that hold in this one real universe of ours.

The irreversibility of time is the foundation of the asymmetry between past and future. The past is closed. The impossibility of changing it is, for the scientific investigation of the world, a requirement of causal continuity. For the shape of human experience, our powerlessness to change the past is the condition of tragedy. The future is not closed in this sense, although the precise sense and extent to which it is open – open to surprise, to novelty, and, in the domain of those realities we are able to influence, to transformative projects – depends on what the facts about how change changes turn out to be.

* * *

The fourth attribute of time is its continuity. The continuum is, for the representation of time, much more than a heuristic device. It is an unadulterated expression of the nature of time.

For a physicist, the idea of a continuum is associated with the real number series. The advantage of this association is to give the idea a precise mathematical formulation. We gain this advantage, however, at too great a cost: we deny ourselves a concept by which to designate that which in nature may be unqualifiedly continuous – that is to say, not susceptible to discrimination into separate parts, however small. This is the vulgar rather than the mathematical idea of the continuum: the continuum as a flow that can be broken into separate parts only by arbitrary discrimination, of the kind, for example, that we use to measure time. (See the discussion of the mathematical representation of the continuum in Chapter 6.)

There is at least one aspect of nature that satisfies the vulgar idea of the continuum: time. The application to time of the mathematical concept of the continuum seems, however, to support the view that time is in fact, in the vulgar, everyday sense, discontinuous, like the real number series. Whether there is a mathematical representation of the vulgar idea of the continuum better than the real number series is a challenge laid down to mathematics, not a reason to mistake what is susceptible to mathematical representation for the reality of nature. However, mathematics has intrinsic limitations in its ability to represent uninterrupted flow, given its roots in the counting of discriminate entities and its time-denying perspective on reality.

Suppose that we could devise no such adequate mathematical representation. We would still not be justified in adjusting our scientific insights to our mathematical capabilities on an issue central to the understanding of nature. Any contrary view of the implications of our mathematical powers rests on a misunderstanding of mathematics and of its relation to nature and to science, a misunderstanding that Chapter 6 of this book is written to oppose.

That such a non-mathematical concept of the continuum fails to apply to energy and matter in the relatively mature and differentiated universe is the import of much of what science, culminating in quantum mechanics and in the standard model of particle physics, has discovered about the structure of the world. In its formative moments of greatest (but not infinite) density, when it had not yet developed a discriminate structure, or any of the discrete components that we now observe and theorize, the distinction between the laws of nature and the states of affairs that they govern may not have been clearcut. Many more degrees of freedom may have been excited than is now the case; if so, the concept of the continuum was a far closer approximation to the facts of the matter about space than it has since become. It may turn into such a closer approximation again in a future contracted, superdense, and extremely hot universe.

To time, however, the extra-mathematical idea of the continuum applies directly, always and without qualification. The application of the concept of the continuum to time is not itself time-dependent. It is an inherent trait of the non-emergent, global, and irreversible reality of time.

This claim – the claim implied by the statement of the fourth attribute of time – is far from being a matter of definition or of logical and mathematical analysis. On the contrary, it is a contentious conjecture about natural reality. It excludes a view that is pervasive in the discourse of science: the film image of time, the idea that time should be understood and represented as a succession, or a combination, of discontinuous and static moments, or slices of spacetime, just as a film is composed of a rapid succession of still photographs.

The continuous nature of time cannot be discounted as a mere matter of definition. It could be invalidated by discoveries about the world: in particular, by observations and experiments that show time to have a quantum nature, as it has sometimes been suggested to possess. Such a result would run counter to ordinary experience of time, which is wholly consonant with its representation as a continuum. If, however, time is indeed a continuum, just as we experience it, rather than an accumulation of slices – each of them a static state of the world – then we have yet another reason to affirm the distinctiveness of time and its irreducibility to other aspects of nature.

Non-emergent, global, irreversible, and continuous time resembles, in many respects, time as described in Newton's famous scholium. It differs from Newton's time, however, because, from the standpoint of the temporal and relational naturalism that we espouse, it is not a thing, or stage, or backdrop separate from the phenomena. It is a fundamental and primitive feature of nature: the susceptibility of all nature to differential change. Moreover, in Newton's physics there seems to be no place for time as so defined. One reason is the time symmetry of the laws of classical mechanics. Another reason is the yet more fundamental evisceration of the reality of time that is implied by the explanatory practice that we call the Newtonian paradigm. The subsequent development of physics, whether in the direction of general and special relativity or in the direction of quantum mechanics, has on the whole narrowed rather than widened the space for the reception of a view of time as non-emergent, global, irreversible, and continuous.

The most familiar exception to the observation that post-Newtonian physics has limited rather than broadened the role of time in nature is statistical mechanics and the theory of entropy. For the reasons previously stated, however, this exception cannot – many statements to the contrary notwithstanding – provide a sufficient basis to affirm the non-emergent, global, irreversible, and continuous character of time. At least it cannot do so without the combination of two moves: one, illegitimate; the other, a confession of its own incompleteness. The illegitimate move is its transposition from a regional to a universal explanation: from the explanation of particular phenomena, in a bounded configuration, to the explanation of the universe as a whole (the first cosmological fallacy). The confession of incompleteness is its need to rely on certain initial conditions, the presence of which such a view is unable to explain, in order to obtain the result of irreversibility. (Since the time of Boltzmann and of the Ehrenfests it is understood that entropy is not universally irreversible in statistical mechanics. Such irreversibility is characteristic only of special, low-entropy initial conditions.)

Now, however, we have a problem. It is not the problem of the unification of general relativity (or of gravity as reinterpreted by general relativity) with quantum mechanics, which, under the label of a final unification, has become the highest ambition of contemporary physics. It is the problem posed by the contrast between the implications of the idea that the universe has a history and a tradition of science that is unwilling fully to acknowledge the implications of the universe having a history. We have already learned enough about its history to know that at some point in that history the relation of laws of nature to the states of affairs that they govern and the elementary structure of nature may not have been what they subsequently became. Universal history suggests not only that time is real but also that it has the attributes I have just enumerated, for each of them is implied in the notion of having a history. The inclusion of the laws of nature within this history lends support to an idea of time that relates these attributes to the transformation of transformation.

The universe cannot have a history unless time is real. For time to be fully real, it must be non-emergent, global, irreversible, and continuous. For time to be inclusive as well as real, there must be nothing outside it, nothing safe from its ravages and surprises, not even the laws of nature. However, such a view of time and of universal history cannot easily be reconciled with some of the central tenets and major assumptions of the prevailing ideas in physics. Either these ideas must be revised or reinterpreted, or the idea of a universal history and the affirmation of the reality of time must be qualified in accordance with them.

The commanding project of contemporary physics – the grand unification of the fundamental physical forces and, in particular, the reconciliation of general relativity with quantum mechanics – takes a new direction once reconsidered in the light of the more fundamental implications of the idea of universal history.

THE PROTO-ONTOLOGICAL ASSUMPTIONS OF THIS VIEW OF TIME

Before I restate more systematically the thesis of the reality of time, consider what more general assumptions about the world are assumed by such a view of time. The question posed is whether there is implied in this conception a fundamental ontology: a view of the basic structure of the world that would, at its periphery of implication, remain subject to empirical confirmation or disconfirmation.

The answer to the question is both a qualified no and a radical no. There is no fundamental ontology implied by this view. What there is, instead, is a proto-ontology: a set of broad but connected assumptions about the way the world is and the conditions on which we can grasp it. The statement of these barebones assumptions amounts to the qualified no.

However, we should not understand such a proto-ontology as an ontology in the making: as if its failure to develop into a full-fledged philosophy of being were a transitory or secondary weakness. The thesis of the full and inclusive reality of time dooms the classical project of ontology running through the history of Western metaphysics: the project of establishing any view that represents the world to have an abiding structure, complete with a list of the kinds of things that there are and of the ways in which they eternally interact. If time is to be fully and inclusively real and the idea of a universal history is to have a meaning undiluted by our time-resisting reservations, history must trump structure. As a consequence, we must reject as misguided in principle the program of a fundamental ontology, ever the core of classical Western metaphysics. That is the sense in which the no is radical rather than qualified.

The proto-ontology is not a picture of the world. It is not a crude forerunner to science, a kind of minimalist scientific theory. It makes no claims of fact. It offers an account of the minimal assumptions about nature that the view of time and nature presented here must make. These assumptions can be justified only by the advantages of the scientific and philosophical program that they make possible.

The first idea in this proto-ontology is the idea of reality. There must be something rather than nothing. The cosmological expression of this idea is belief that something cannot come out of nothing. Nothing will come of nothing. A moment of greatest density and temperature – a formative moment in the earliest history of the present universe – should be assumed to have a prior history. That is the import of placing succession – the succession of universes – in the place of plurality – the plurality of universes – as a working assumption.

In this sense, it must be an operational premise of our understanding of nature that the world – not the present universe but the succession of universes, or the successive states of the universe, or the causal continuation of reality, stressed but not broken – is indefinitely old and has a prospect of indefinite continuance. Its indefinite age and longevity, enlarging the realm of causal inquiry, is not to be equated with eternity. To affirm the eternity of the world, as Avicenna and others did, would be to admit the infinite into nature, for eternity is temporal infinity.

There are at least two ideas that could take the place of this postulate of the indefinite antiquity of nature. One idea is that the world is not indefinitely old because it was invented or created by a force or a being who the believer calls God. The other idea is that the world is not eternal because at some moment, at the edge or the beginning of time, something came of nothing. Then time would indeed have emerged but its emergence would be a one-time event, coeval with the emergence of being: that is, of something out of nothing. These two alternatives to the eternity of the world – the divine making of the world and its unassisted emergence out of nothing – are conceptually equivalent because they are equally miraculous and mysterious, and equally incapable of translation into our study of the world that we in fact encounter.

The second idea of the proto-ontology that is required to affirm the reality of time is the idea of plurality. There must be many things rather than one thing. By many things we mean some form and measure – any form or any measure – of difference, of variation, in reality. On the view presented earlier, time is non-emergent, and its nature lies in the uneven character of change and in the transformation of transformation. It follows that even in moments of greatest density and temperature, when the universe did not yet have, or (in the course of the succession of universes) had ceased to exhibit, a definite, law-governed structure, there could not be a uniform one thing. There would still have to be variation within the one: differential change leading to further differentiation. No change could occur without occurring in particular parts of the one and without changing the relation among its parts.

Notice, however, the uncompromising minimalism of this view. It does not presuppose any particular structure of the world: any enduring set of types of being. It does not even require that there be a differentiable manifold, with distinct kinds of being and with a clear contrast between laws of nature and states of affairs. All it demands is that there be a susceptibility to variation or differentiation within the one: the otherwise single and uniform reality.

The third idea is the idea of connection. Everything in the world is somehow connected. That there are things, or states of affairs, with regard to which there arises the issue of connection results directly from the combination of the first two ideas: reality and plurality.

Of the three ideas of the proto-ontology – reality, plurality, and connection – this third one is the most difficult to grasp and to circumscribe. The nub of the difficulty, however, lies in the somehow rather than in the connected. In principle, connection requires the transmission of information: by light or other means. Moreover, the transmittal of information takes time, and therefore presupposes time, as the example of its conveyance by light shows. That the elements of this proto-ontology must be presupposed by science is shown by reliance of the concept of information on the postulates of reality and of plurality as well as of connection. There is no information without organized contrast.

However, once space – the disposition of plurality – emerges, there is an additional mode of connection – the geometry of space – that may be instantaneous and require no time. We have reason to believe, from the facts of universal history such as science has begun to discover them, that this non-temporal form of connection is supervening: it supervenes on connections that exist only in time. Space may be emergent and its emergence compatible with the eternity of the world, if the world is eternal. Whether space is emergent or not, its geometry evolves in time, as a feature of the history of the universe.

The reason why the somehow in the statement that everything is somehow connected represents the nub of the difficulty presented by the idea of connection is straightforward. No closed and permanent set of forms of connection exists in the world. Like everything else, the forms of connection change, sooner or later. Like everything else, they have no exemption from time.

The three ideas that comprise this proto-ontology are far from being empty or tautological. There is much that they exclude. Insofar as they inform an agenda of natural science, they become, though indirectly, subject to empirical test, notwithstanding their extreme abstraction, generality, and minimalism.

It should now be clear that these ideas of reality, plurality, and connection cannot serve as the germ of an ontology – a doctrine of the kinds of things that there are in the world – in the sense that turned ontology into the core of the metaphysical tradition of the West. Instead of exemplifying the aspiration to supply a foundational philosophy, they amount to a proto-ontology only in the sense of also being an anti-ontology. According to this view, the world has no abiding structure; it has only a structure some of the features of which change very slowly and other features of which may change rapidly and decisively at certain times: that is to say, discontinuously within the continuum of time.

There is no permanent list of natural kinds, the types of being that exist in the world. Nor is the nature of the way in which things – phenomena or states of affairs – are distinct from one another itself permanent. The history of the universe does not merely witness the emergence of new types of things; it also witnesses the birth of new ways in which the things that there are differ from one another. The individuals of a species do not differ from one another in the way in which species themselves differ. Species do not differ from one another in the way in which, say, sedimentary rocks differ from igneous rocks. And igneous rocks do not differ from sedimentary rocks in the way in which protons differ from electrons. Protons and electrons do not differ from one another in the way in which, prior to their emergence, the different parts of the superhot and superdense early universe differed.

The same impermanence besetting the types of things that there are and the character of their distinctive being affects as well the ways in which they interact: the forms of connection, even the one that appears least time-bound, the geometry of space.

What then is permanent? Only one aspect of reality: time. The condition of the radical and inclusive reality of time is the impermanence of everything else. Time is internally related to change and causal connection: equally fundamental and primitive features of nature.

The project of a fundamental ontology is therefore misguided in principle. Its subject matter, the permanent structure of the world, does not exist. As this project has been forever the centerpiece of Western metaphysics, notwithstanding all skeptical assaults on its pretensions, the tenets of that metaphysical tradition deserve to be repudiated. The reason to repudiate them is not the limitation of our insight into the world – our inability to gain access to the nature of things in themselves – as Kant and many others held. It is, on the contrary, the outcome of the insight into the nature of the world that, especially through the work of natural science, we have nevertheless gained.

The implications of this line of thought for our approach to both physics and mathematics are far-reaching. Throughout its history, at least from the time of Newton, the canonical subject matter of physics has been the permanent structure of the world. The history of physics since the late nineteenth century has entrenched this privileged subject matter even further. It has done so by giving pride of place to the study of elementary constituents of nature.

It has done so as well by undertaking this study in the light of an agenda that puts the static question – how do these constituents combine and interact? – in place of the dynamic question – how does one state of affairs result from another? When the explanatory attitude enacted in the exploration of nature at the quantum scale then extends to the study of the cosmos, the result is to conceive the history of the universe as subordinate, in explanatory significance, to a non-historical analysis of the fine structure of nature. Cosmology then becomes a marginal specialty rather than the most comprehensive part of physics: the part with the best credentials to represent its master discipline.

The anti-ontology that I have outlined suggests reasons why the facts of the matter about the world make this approach misguided. Those who defend the present dominant agenda in physics may then respond by acknowledging the ultimate impermanence of any seemingly abiding structure of nature but then go on to dismiss this concession as largely irrelevant. They may argue that once the universe cooled down and took shape as the discriminate manifold that it now is, governed by laws clearly distinguished from the states of affairs to which they apply, its structure can be regarded, for the purposes of the work of science, as permanent and even timeless.

However, this response is inadequate for two reasons. The first reason is that the renovation of the repertory of natural kinds or types of being, of ways in which they enjoy distinctive being (and therefore of the sense in which they are distinctive), as well as of their modes of connection, is susceptible to novelty and change at any moment in the history of the universe. The emergence on Earth of life and its laws exemplifies this truth. The second reason is that for time-bound, historical phenomena, such as all natural phenomena, structural explanations must be ancillary to historical ones.

We cannot understand what something is without grasping what, under certain conditions, it can become. Moreover, we cannot understand what it can become unless we understand how it came to be what it is. What is ultimately at stake in this contest of approaches is the vital link, in the development of science, between insight into the actual and imagination of the possible: not the fanciful possible of an outer horizon of possible worlds or possible states of affairs (which we are in fact unable to discern) but the real possible, which is the adjacent possible – what theres we can reach from here.

If the rejection of the project of classical ontology has implications for the agenda of physics, it has no less significant consequences for our understanding of mathematics and of its relation to nature and to natural science. For the moment, it is enough to remark on the close relation between the idea that the world has a permanent fundamental structure and the hope that mathematics can offer a shortcut to the discovery and the understanding of that structure. A requirement for the legitimacy of this hope is that there not be a fundamental difference between the relations among parts of nature and the relations among mathematical propositions. Otherwise the latter could in no way be homologous or fully representative of the former.

However, there is such a difference. The relations among mathematical propositions are wholly outside time, even when designed and deployed to represent movement in time. The relations among parts of nature are squarely within time.

THE IDEA OF THE INCLUSIVE REALITY OF TIME RESTATED

In the light of this development of a conception of time, as well as of the proto-ontology on which this conception rests, the thesis of the radical and inclusive reality of time can be restated in the form of a single proposition. *Everything that is real in the world is real in a particular moment, belonging to a flow of moments.*

A corollary of the idea that everything real is real in a particular moment is that the laws of nature may change. Their mutability is the theme of Chapter 5 of this book.

The thesis of the inclusive reality of time gains clearer meaning by contrast to a number of standard positions in contemporary philosophical thinking about time. Among such positions are presentism and eternalism.

According to presentism, only what exists in the present moment is real. The now has special significance for science. It is all that really exists. The past was real but no longer is. Of the past, we have only vestiges, although, given delay in the transmittal of information through light, the afterglow of the past may live brightly, all around us, in the now. The future has no reality and is uncertain.

According to eternalism, all present, past, and future events (to the extent that time is real and that real distinctions exist among the present, past, and future) are equally real. They are all necessitated by the structure and regularities of the universe. The now matters for human experience. It enjoys, however, no special significance for science.

The contrast between presentism and eternalism is thus connected with the question of whether the distinctions among present, past, and future are features of reality rather than just traits of our experience as situated agents, marking, from our perspective, our relation to other people, events, and phenomena.

The share of truth in presentism is that in each present moment we can ask in principle what is real in that moment throughout the universe. The whole of reality is weighed on the scales of the present moment. We may, however, be unable to complete the weighing, given the difficulty of retrieving information and of establishing simultaneity – joint participation in the now – among places in the universe. In this sense, the now counts for science as well as for human experience.

The weighing of reality on the scales of the present moment, when combined with the thesis of the existence of a preferred cosmic time, ensures that present, past, and future are characteristics of natural reality in time, not in our experience alone. The present moment is the now, separating past and future, in the cosmos as well as in our experience. This temporal character of nature would persist in our absence as it existed before our emergence.

However, what is real does not fit within the confines of the present moment. To that extent, presentism is untenable. Reality fails to remain within the now, regardless of the extent to which we believe the future to be determined by the past or by the regularities and structure of nature.

Everything in the universe is always becoming, or ceasing to be, and changing into something else, more slowly or more quickly. The structure and regularities of nature may be in a condition of stability or of rapid and radical change. Change itself is changing. Time is the differential susceptibility of everything, including change, to change.

This process of becoming and of ceasing to be is not only real, it is more real than anything else in nature. It is real if anything is. It forms the subject matter of science and therefore as well of the science of largest scope, which is cosmology. It cannot, however, be accommodated within the now because the now is instantaneous.

In this sense, the now has no special significance for science despite its centrality to our experience. The now may be all that we can ever possess, for it is only in the now that we can live, but it fails to define the perimeter of scientific inquiry. We do science to outreach, in our understanding of nature, unaided perception and uncorrected experience.

If nature were organized on the basis of a permanent structure of elementary constituents, governed by unchanging laws, symmetries, and constants, the failure of the now to capture the real and to provide science with its subject matter might be less striking than it is. All of nature would now be what, fundamentally, it always remains. The now would thus be representative of the whole. However, it is just this idea that we deny when we affirm the inclusive reality of time and the historical character of a universe in which everything changes sooner or later. Anyone who accepts our ideas and arguments must reject presentism.

The part of truth in eternalism is that reality goes beyond the present moment. However, eternalism implies that past, present, and future are determined. If they are all determined, as Laplacean determination supposed and as today the block-universe view in cosmology proposes, they are, in a sense, simultaneous. A thoroughgoing determinism implies a time-destroying simultaneity of past, present, and future. They are simultaneous in the godlike eyes of the scientist, independently of other objections, such as those arising from the predominant interpretations of general relativity, to the reality of time. Tell him the initial conditions, and he can in principle see to the end of time, which means that he sees the future as if it were the present.

It may be objected that a determined future must still be acted out just as an actor handed his script must still speak his lines and just as a patient in the terminal state of an incurable and fatal disease must still die. To this extent, time may exist despite determinism. However, a comprehensive and intransigent determinism eviscerates the significance of this qualification, leaving no room for anything such as the chance of the actor to forget or to improvise or of the patient to be unexpectedly cured or to die for another reason. The treatment in the physics of the last hundred years of time as an emergent phenomenon, accessory to the disposition of matter and motion in the universe and consequently always local, never cosmic, or cosmic only in the sense of arbitrary spacetime coordinates, completes the destruction of the idea of reality as a flow of moments. It helps render insubstantial the distinctions among past, present, and future.

The future of the universe is a legitimate topic for science. It does not follow from the future being largely unknown that it is entirely undetermined and unknowable. If, however, there are no immutable laws, symmetries, or supposed constants of nature, and no permanent repertory of types of being or elementary constituents of nature, if the regularities of nature change together with its structures, in a fashion not commanded by any closed set of higher-order laws, and if historical explanation is therefore paramount over structural analysis, rather than the other way around, the most extreme forms of universal determinism cannot be right.

A cosmology untainted by the twin cosmological fallacies, receptive to the ideas of natural history – the pervasiveness of path dependency, the mutability of types, and the co-evolution of laws and phenomena – and insistent on the inclusive reality of time cannot be reconciled with eternalism. It must argue that the universe cannot be as closed as eternalists suppose.

There may be room for novelty in the history of the universe – the unscripted new rather than the false new that is just the enactment of a script pre-written in nature. Room for the new will be widest in those periods in universal history, or those extreme states of nature, in which stable structures are dissolved and causation ceases to exhibit law-like form. In the course of time, the deepening of our insight and the strengthening of our powers may enable us to influence the history of the universe as well as the history of our planet. If we can already conceive the idea of exploding asteroids before they hit the Earth, why should we assume that there are insuperable obstacles to other, more wide-ranging interventions in the future?

This criticism of eternalism fails to imply that the history of the universe is open in the sense or to the degree that human history is open. Nor does it mean that nature is on our side and hospitable to our plans and values. We do not yet know how open the history of the universe is. As a result, we cannot yet confidently discern what portion of eternalism deserves to be salvaged. What we can say is that eternalism must be greatly deflated, when not entirely rejected, if the view of time and the approach to universal history for which we argue are to be upheld.

FROM BEING TO BECOMING

The thesis of the inclusive reality of time comes fully into its own when stated and developed in relation to a more general way of thinking about nature. Call it the philosophy of becoming. It stands in contrast to what one might call the philosophy of being. It affirms the primacy of becoming over being and of process over structure. It informs an approach to reality in which the proposition that time is not emergent, although space may be, makes sense.

The centerpiece of the philosophy of becoming is the view that nothing in nature or in reality lasts forever, nothing except the changing flow itself, which we call time. Everything else changes sooner or later. Consequently, we do better to understand the varieties of being as the mutable and emergent products of becoming – that is to say, of the changing flow, than to understand becoming as merely an occasional modification of being – that is to say, of a structure possessed of unchanging elements or subordinate to unchanging laws. Anaximander said it long ago, at the origins of both Western philosophy and Western science: "All things originate from one another . . . in conformity with the order of time."

I do not here seek to infer the inclusive reality of time from the philosophy of becoming, or to defend this philosophy on metaphysical grounds external to science. Rather I argue that the trajectory of science, when viewed as a whole, gives us reason to prefer the philosophy of becoming to the philosophy of being, in physics, from which it remains largely excluded, as well as in the earth and life sciences and the social and historical study of mankind, where it has long been dominant. I argue as well for the advantages of making explicit, under the heading of the philosophy of becoming, the wider presuppositions and implications of the thesis of the inclusive reality of time. They are, in the first instance, advantages for this thesis itself: such an explication helps create a discourse in which our assumptions about nature cease to contradict these ideas about time. They free us from the need to develop the notion of the inclusive reality of time in the climate of metaphysical presumptions that are wrongly mistaken for empirical science. These presumptions prevent science from fully appreciating the significance of its own discoveries.

The relational view of time and space, invoked earlier in this chapter, represents an indispensable part of the argument of this book. So long as we cling to the absolute view of time and space, regarding them as entities separate from the phenomena that they would somehow house, rather than as orderings of events, we cannot do justice to the inclusive reality of time. We misinterpret the meaning of real and inclusive time. We misunderstand the global character of time in a manner that places it in unnecessary conflict with the core of undisputed empirical insight in special and general relativity.

The relational view, however, does not suffice to provide the background that we need to make sense of real and inclusive time. An isolated relationalism – one that fails to form part of a philosophy of becoming – remains compatible with the ideas that the universe has permanent elementary constituents and that the workings of nature are governed by timeless laws. It is only when we place relationalism in the context of a philosophy of becoming that we begin to think in a manner that can fully accommodate the singular existence of the universe, the inclusive reality of time, and the selective realism of mathematics. Relationalism – the relationalism that we need – is thus best understood as a fragment and as a step in the development of such a larger conception. It is therefore important to form as clear a view as possible of the content and consequences of this conception. One of the ways to do so is to understand it by contrast to the ideas that it opposes.

The philosophy of becoming has been present in the history of Western philosophy from the pre-Socratics to today. Among its exponents are Heraclitus, Hegel, Bergson, and Whitehead. The list suggests three features of this line of thought.

A first characteristic of the tradition of the philosophy of becoming is that, in Western philosophy, it has always remained a dissident view. The most influential philosophies have been one of the variations on the philosophy of being that I later enumerate. By contrast, the philosophy of becoming has been the dominant tendency elsewhere: for example, among the schools of ancient Indian philosophy.

A second characteristic is that it fails to represent a system, or even a well-defined tradition. It proposes an approach to reality that has been expressed in markedly different forms and combined with widely divergent ideas. It has been at best a family of philosophical beliefs and attitudes, a largely rejected and misunderstood direction.

A third characteristic is that its bearing on the practices of the particular sciences remains indistinct and undeveloped. Like many philosophical doctrines, it appears to float above the sciences, neither drawing life from them nor giving life to them. This disembodiment reinforces its apparent vagueness.

Nevertheless, the idea of the priority of becoming over being and of process over structure is more widely accepted than it may appear to be to those who approach it from the vantage point of physics. It has long been banished from the inner sanctum of modern physics. However, it is the largely unrecognized orthodoxy in natural history, in the life and earth sciences, and across the whole field of social and historical study. The communion, however, between the philosophical expressions of that approach to reality and its embodiment in the procedures of these disciplines has remained fitful and undeveloped. As a result, the philosophy of becoming continues to float more than to guide, and to propose by dogma rather than to learn, through science, from experience.

Consider now the set of ideas that the philosophy of becoming opposes. We can bring them under the general heading of the philosophy of being. It has been forever the hegemonic position in Western philosophy as well as in basic science, especially in physics as it has been understood from the time of Galileo and Newton to today.

The philosophy of being affirms the priority of structure to process and of being to becoming. It asserts that nature or reality has an ultimate structure, which it is the aim of philosophy, science, and even art to reveal. It treats change as the more or less localized modification of structure.

Like the doctrine of becoming, the philosophy of being has spoken through versions differing so widely that the unity of its core idea may be obscured. It has had three main instances in the history of thought and of science. When we see what has happened to it in each of these instances, we find, to our surprise, that its situation is far more precarious than its persistent and even overwhelming influence might lead us to suppose. The time is ripe for a change of direction, in our general approach to reality as well as in the particular sciences.

The first instance of the philosophy of being is the conception of a system of natural kinds in the world: of kinds, or types of being. Among the types available to us in our perceptual experience are those studied by the earth and the life sciences and in social and historical study. It is the observed division of manifest nature, as we encounter it in our daily experience, into enduring types of lifeless things and of living organisms as well as into types of economic, political, and social organization. Such is the inspiration of what I earlier called the project of classical ontology, canonically exemplified by Aristotle's metaphysics.

Consider what has become of this first and most characteristic instance of the philosophy of being. It survives in school philosophy in certain parts of the world: countless metaphysical texts profess to carry on the program of classical ontology, as if little or nothing had happened in the history of science. However, these exercises resemble the apparitions of ghosts: the basis for the execution of this program in our organized understanding of manifest nature and of social life was long ago destroyed.

In the earth and life sciences, the principles of natural history (discussed in Chapter 1 of this book) undermine the requirements for the enactment of the philosophy of being: the mutability of types, seen against the background of the path dependence and of the co-evolution of laws and states of affairs. Neither a permanent structure nor permanent laws obtain. Everything is historical in the subject matter of the naturalist: the kinds of the things that there are and the ways in which they change. Darwinian evolution is not the sole basis of this predominance of process over structure, and of becoming over being. Otherwise, these facts would be confined to the realm of life whereas they extend to lifeless nature as well: to the historical typology of rocks, for example, viewed in light of the evolution of the Earth.

When we turn to the institutional and ideological structures of society, the failure of the approach recommended by the philosophy of being is even harder to deny. The formative institutional and ideological regimes of a society amount to a frozen politics; they represent the outcome of a temporary interruption and a relative containment of conflict over the terms of social life. They change in character as well as in content: that is to say, in the extent to which they either insulate themselves against challenge or lay themselves open to revision, either disguising or exhibiting their nature as frozen politics. Although they change under constraint and rise or fall according to the advantages and attractions that they offer, no historical laws govern their succession.

In society, even more than in the manifest nature studied by the naturalist, there is no permanent set of indivisible types of organization, and no unchanging system of forms of change, on which a social theory faithful to the spirit of a philosophy of being can fix. The deep-structure social theories of the nineteenth and early twentieth centuries, with their characteristic belief in a law-like succession of indivisible institutional systems, have been discredited.

The hopes of the philosophy of being, dashed in the domains of natural history and of social life, then pass to the study of the fundamental and hidden constituents and laws of nature: the subject matter of physics and of its allied sciences. Before recalling why those hopes must there too be disappointed, consider one more instance – the second in this list – of the philosophy of being in the history of thought.

This second instance is speculative monism: the idea that there is only one being. All the distinct beings that appear to exist in the world, including our own selves, are, on this view, illusory, derivative or superficial. If they are not unreal, they are at least less real than the one hidden being that underlies appearances. The most determined expressions of this point of view in the history of Western philosophy are Parmenides (the fictional Parmenides of Plato's dialogue of the same name as well as the real Parmenides of the barebones fragments),

Spinoza, and Schopenhauer. In many other thinkers, speculative monism appears in more or less qualified form. For the most radical versions of speculative monism, the plurality of beings is an illusion. For the qualified versions, it is a shallow and ephemeral reality.

The relation of speculative monism to the philosophy of becoming is ambiguous and paradoxical. In one sense, it is the opposite of this philosophy. In another sense, it can be just another version of it: they can share in common the idea that everything changes, or turns, as Anaximander wrote, into everything else under the dominion of time.

Two factors determine whether a particular speculative monism represents the antithesis of the philosophy of becoming or an expression of it. The first factor is whether and how the doctrine acknowledges the reality of the kinds of things that exist at any given time and the distinctive way in which these natural kinds change. The reverse side of a view of the distinctions among phenomena must be an account of how they change and of how change changes.

The second, yet more basic factor is whether the doctrine affirms the inclusive reality of time. A speculative monism that denies or discounts the reality of time, as most versions of speculative monism in Western and non-Western philosophy have, stands in stark contradiction to the tenets of a philosophy of becoming. A speculative monism that affirms the inclusive reality of time moves in the direction of such a philosophy.

Even, however, when monism moves in that direction, it suffers from a defect that ought to be judged fatal: the lack of any account of how the one becomes the many, or comes to appear under the illusion of the many, and of how the different forms of the many – or of the one – change into one another. It lacks purchase on the manifest world and engagement with the methods of inquiry and the organized disciplines by which we seek to grasp the workings of nature and of society. It too floats: hence the reason to designate it as speculative.

There is a family of views, closely related to speculative monism, that may at first seem to offer a solution to this problem. It goes under the label panentheism. It sees God – or being, mind, the one – as constitutive of a world that the one being nevertheless exceeds. It struggles to form a loose view of how this one becomes expressed or embodied in the distinct types of being that we encounter in nature. It sees the many as emanations of the one.

The theory of emanation, however, has never been formulated in a way that would draw it close to the practices and sciences by which we can probe the world around us, and discover something of how it works. Under the panentheist dispensation, monism has thus invariably continued to be speculative, whether in the hands of a Plotinus, a Nicholas of Cusa, or a Heidegger.

For all these reasons, speculative monism represents a dead end in the history of thought: a dead end to which the philosophy of being may find itself confined if its two other main instances fail.

Earlier in this chapter I presented a proto-ontology: a series of postulates – of reality, plurality, and connection – listing the minimal assumptions about the real that a science faithful to the theses of the singular existence of the universe, of the inclusive reality of time, and of the selective realism of mathematics must make. This proto-ontology could be better labeled an anti-ontology. It eschews the project of classical ontology, with its commitment to describe a permanent typology of being.

An additional function of the proto-ontology or anti-ontology outlined earlier now becomes clear. It is to exclude the option presented by speculative monism, at least in the radical variant of the monist conception: the variant denying the reality of the many. The proto-ontology excludes this variant by virtue of its postulate of plurality: that there are many things rather than one. The postulate of plurality denies the defining thesis of a radical speculative monism. The significance of the postulate of plurality for science is made clear by the postulate of connection: the many connect. These connections are causal, presupposing time. They may or may not take recurrent and law-like form; thus, they are antecedent to any laws of nature rather than derivative from them. The study of causal connections in nature is the fundamental subject matter of science. The study of laws of nature is its occasional subject matter.

The third instance of the philosophy of being, alongside classical ontology, with its typology of natural kinds, and speculative monism, with its insistence on the oneness of being, is the one established in the dominant tradition of modern physics. The failure of the philosophy of being in this domain would leave this philosophy with no refuge other than speculative monism, given that it has already failed in its other domain: the effort to make sense of the manifest world, of natural history and social life, as we encounter it on the scale of human action and perception.

The twin mainstays of this third instance of the philosophy of being are the ideas of a permanent structure of ultimate constituents of the world and of an immutable framework of laws of nature governing the interactions among these constituents. There is one fundamental set of beings, the building blocks of all the others, and one unchanging system of change. To reveal this set and this system is the task of science.

Nature, according to this view, is a differentiated structure, subject to change on the basis of a fixed stock of elements. This stock is described, at two complementary or superimposed levels, by particle physics and by chemistry. The workings of this differentiated structure conform to recurrent and changeless regularities: the laws, symmetries, and constants of nature. These regularities underlie all causal connections and justify all causal explanations.

In such a view of nature, time cannot be fully real. If it is real, it is not inclusive: the basic elements and fundamental laws are changeless. They have no history, other than the inscrutable jumping off point of an infinite initial singularity at the origins of the universe. Where there is no change, there cannot be variation of change. Hence time cannot be inclusively real. Part of reality, if not all of it, lies beyond the reach of time.

If nature, at its most fundamental level, has such a character, mathematics is its privileged expression. The differentiated structure is a differentiable manifold. The existence of changeless forms of changing, connecting the phenomena, can be registered by laws of nature, stated as mathematical equations. That the relations among mathematical propositions are timeless whereas the phenomena that they represent are time-bound ceases to be perplexing if the timelessness of mathematics is at least partly seconded by limits on the timeliness of nature.

It is this view of what nature is ultimately like, once we see beyond the middle range at which natural history works, that the argument of this book seeks to challenge. In this contest, the issue of the reality of time plays a major part.

Prompted by its own most important discovery, the discovery that the universe has a history, cosmology must become a historical science. It cannot do justice to its own findings within the boundaries of the approach to nature that I have just summarized. Through the reorientation of its agenda and its assumptions, it must cast its lot with the philosophy of becoming, at least in the same relatively inexplicit sense in which the physics of Newton and of his successors tied itself to the philosophy of being. It must do so out of explanatory convenience and constraint, not out of philosophical prejudice.

The argument must take place at many levels, all the way from the internal disputes of physics and of the other specialized sciences to the large and loose contrast between these clashing metaphysical programs. In between the scientific and the metaphysical stands natural philosophy. It enjoys no prerogative to obtain for its arguments and proposals an exemption from the empirical discipline to which science is subject. It must sometimes do its work by seeking to dissolve the marriage of factual findings and ontological pre-commitments that marks any established system of scientific ideas. It seeks to renew the dialectic between empirical discovery and theoretical imagination, by enlarging the range of available conceptions and by allowing us to see with new eyes what we mistakenly think that we have already understood.

5 The mutability of the laws of nature

CHANGING LAWS

Do the laws of nature change? Many philosophers and scientists have claimed that the immutability of the laws of nature is a premise of the work of science. In pressing this claim, they reify a particular idea of science: an idea that takes the central tradition of physics, from Newton to Einstein, as the model of science. For it is only in this tradition that the notion of changeless laws of nature has had a secure place. Nevertheless, only very few physicists, Dirac and Feynman first among them, have explicitly questioned the immutability of the laws of nature and suggested that they must have been different in the early universe.

There are other branches of science in which the notion of unchanging laws does not immediately occur to a practicing scientist unless he is anxious to show how his scientific practice can be made to conform – or to appear to conform – to the supposed master science, modern physics. We commonly think of the explanatory force of the regularities of natural evolution that are enshrined in the contemporary Darwinian synthesis as having developed together with life. This joint transformation of the phenomena and of the regularities that they exhibit is not a one-time phenomenon; it keeps happening. For example, our account of the workings of the Mendelian mechanisms in the course of evolution is modified by the arrival of sexual reproduction.

A weak reductionism may assert that these effective laws can all be reduced to the supposedly fundamental and immutable laws of physics. However, such a claim amounts to little more than an empty genuflection to the model of explanation established in the dominant tradition of physics. It has no consequence for explanation in the earth and life sciences.

The idea of the unchanging character of the laws of nature has an especially close connection to the Newtonian paradigm. Changeless laws are invoked to explain change within a configuration space the contours of which are defined by unexplained initial conditions. A working assumption of this explanatory practice is that the laws applicable to the phenomena within the configuration space are held constant; the theoretician-observer remains in a timeless and godlike position outside the configuration space, wielding, as his instrument, the immutable laws of nature. It is precisely the universe as a whole, however, rather than any particular part of it, that is in question when we assert or assume the laws of nature to be unchanging.

This style of explanation – we argue – cannot be legitimately applied to the whole of the universe (so to apply it amounts to the first cosmological fallacy); its province is the explanation of local realities, bounded by stipulated initial conditions. The enthronement of physics as the exemplary science is followed by the extrapolation to cosmology of an explanatory strategy that has no legitimate cosmological use.

Those who claim that the immutability of the laws of nature is a requirement of any science also make another mistake, alongside their baseless entrenchment of a particular way of doing science. They treat a question of natural fact as if it were an indispensable presupposition of thought or, at least, of science. To succumb to this temptation is unjustifiably to circumscribe the self-subversive and revolutionary potential of science. Our interest is exactly the opposite: to reformulate assumptions that we are accustomed to treat as unavoidable requirements of empirical thinking in such a way that they too become open to empirical challenge. Much of what we are inclined to treat as necessary assumptions are in fact only the petrified byproducts of earlier efforts at scientific inquiry or the expressions of unjustified and unacknowledged philosophical prejudice.

I have suggested reasons to reject the claim that the immutability of the laws of nature is an indispensable assumption of science. I now offer reasons to suppose, consistently with what we know of the workings of nature as well as with a promising practice of scientific explanation, that the laws of nature *may* indeed change. To say, however, that they may change is not to say that they do change. Whether they change, when they change, and how they change are all facts to discover in the course of the investigation of nature.

There are three main reasons to think that the laws of nature may change. The first reason comes from the general view of the singular existence of the universe and of the inclusive reality of time that we defend and develop. It amounts to a philosophical conception, but only in the sense of a natural philosophy that seeks to work as the front line of natural science. One of the goals of natural philosophy is to rid science of the incubus of inherited metaphysical preconceptions inhibiting its advance.

The second reason comes from the picture of the history of the universe that is in fact emerging, in fragmentary and sometimes contradictory form, from the discoveries of cosmology. The striking disparities (often by several orders of magnitude) between what some established ideas predict to be true about the constitution of the universe and what we actually find should not discourage us. On the contrary, they should inspire us to formulate more clearly and boldly the most general meaning of the cosmological findings of the last several decades, in the hope that the formulation will help inform an agenda of research and theorizing that can overcome those disparities.

The third reason is the explanatory advantage of such an approach. The conjecture of the mutability of the laws of nature promises to expand the range of causal inquiry. It does so by suggesting that we might be able to explain historically why the laws are what they are rather than consigning them to the factitious residue of the unexplainable. Moreover, it makes this suggestion without surrendering to the metaphysical rationalism of the principle of sufficient reason or failing to recognize that science will never wholly explain why the world is what it is rather than something else.

The three reasons – the first, from above, from an idea about the onlyness and the timeliness of this one real world; the second and third, from below, from an account that refuses to sacrifice our strange empirical discoveries on the altar of our conceptual biases – converge. They converge toward a view in which the conjecture of the mutability of the laws of nature occupies a central place. This conjecture in turn suggests an enlargement of our vision of what scientific explanation can and should be like now.

* * *

Non-emergent, global, irreversible, and continuous time admits of no exception to its rule. Nothing remains outside it. It is an irreducible feature of reality. That is a conception – a philosophical conception – of what it would mean to take the idea of the reality of time to the limit, affirming it without qualification. Such a view of the radical and inclusive reality of time seems to agree with our experience of the manifest world. It also gives the fullest, least restricted meaning to the idea that the universe has a history.

The issue remains whether this conception of time is in fact realized in nature. A tenet of this book is that what science in general and cosmology in particular have already discovered gives us more reason to believe that this conception of time is realized in nature than to believe that it is not.

The crux of the matter in the choice between these two views comes down to a contest between two ideas: the idea of a plurality of causally unconnected worlds, and the idea of a succession of causally connected universes – that is to say, of universes connected across time by a causal succession that is stressed but never broken. The causal succession is stressed at moments of implosion and explosion, at which nature ceases to present a well-differentiated structure and the distinction between states of affairs and the laws governing them breaks down.

This idea deepens the conception of a history of the universe, which is the greatest achievement of the cosmology of the last century. It prevents us from having to suppose, in contradiction to our observations and experiences, that time emerges and vanishes under certain conditions. It saves us from the need to temporize with the scholastic fabrication of other universes with which, by definition, we can make no causal contact and of which, consequently, we can have no direct or indirect experience. It is in principle amenable to empirical inquiry: for previous universes, or states of the universes, must have left their mark on later universes or states.

However, the idea of a succession of causally connected universes – or of states of the universe – exacts a price. The price is the rejection of a series of metaphysical glosses on the discoveries of science. Some of these slants are so deeply entrenched in our habits of mind that rather than being recognized for the juxtaposition of science and metaphysics that they are, they are either accepted as facts of nature or embraced as requirements of reason.

First among these prejudices is the idea of an unchanging framework of the laws of nature. When the physics of the twentieth century undermined the idea of a spacetime backdrop separate from the phenomena, it nevertheless reaffirmed the notion of an unchanging framework of natural laws.

If everything changes sooner or later, so must the regularities of nature, although some of them may change more slowly or more rarely than others. If everything changes sooner or later, so must the modes of change. Indeed (as I have argued), one way to define time is to say that it is the transformation of transformation. That the laws of nature may change is a direct implication of the reality of time. Such change may be discontinuous.

This implication is feared and avoided, one can surmise, chiefly because of the threat that it is imagined to pose to the ambitions of science. However, it is not science that is threatened by the idea of the mutability of the laws of nature; it is a particular approach to science. This approach never overcame its ambivalence to recognition of the reality of time. It found reasons for this ambivalence in the use of the Newtonian paradigm as well as in belief in a privileged relation of mathematics to nature. The ambivalence persisted in the physics of the twentieth century. It achieved one of its most stunning expressions in the effort, under the aegis of the leading interpretations of general relativity, to spatialize time: that is to say, to subject it to geometrical representation and analysis as part of a spacetime manifold.

That the laws of nature may change becomes less surprising and perplexing once we allow for the variable relation between the laws and the states of affairs that they govern. The universe is so constituted that it undergoes moments of radical reformation in which the distinction between the laws and the governed phenomena diminishes or even vanishes. Causal connections may even cease to exhibit the recurrent and general regularities and symmetries that are their hallmark in the cooled-down universe. These are the same moments in which change changes more rapidly: the kinds of things that there are as well as the ways in which they turn into other things.

The phenomena may change more easily than the regularities: laws, symmetries, and supposed constants. The regularities that appear to underwrite our causal explanations but that are in fact only a codification of causal connections in their recurrent form, which we call laws of nature, change less readily and more rarely. The principles that these laws seem to obey, such as conservation of energy and least action, may change only at the limit of the most radical transformations in the history of the universe.

What changes, and how and when, are not truths that can be inferred from the logic of scientific inquiry or from the extrapolation of our local experience of nature into universal and eternal attributes of nature. They are facts of the matter. They are not, however, simply facts about local pieces of nature; they are facts about the universe and its history, including the antecedents of this universe in what may be a succession of universes or of states of the universe.

The idea that the laws of nature may change is thus quite simply the notion that these laws – the laws invoked by our causal explanations and even the basic and seemingly inviolate principles to which these laws conform, as well as the symmetries and supposed constants of nature – belong to the history of the universe. They are not outside that history and untouched by it.

* * *

Consider how the idea that the laws of nature may change, because they are inside time rather than outside it, relates to the intuitive core of what contemporary cosmology has to teach us: what it has to teach, that is to say, when considered without the blinkers of metaphysical assumptions hostile to acknowledgment of the inclusive reality of time.

Any suggestion that the laws of nature may be mutable confronts the countervailing fact that they appear to have been stable from early in the history of the present universe. Departures from this stability, such as the suggested variations in the fine-structure constant, are so uncommon and limited, as well as so disputed, that the possible exceptions seem only to confirm the permanence of the laws.

However, this fact, if it is a fact, can be read in a way that shows why the mutability of the laws of nature may be not only compatible with the findings of contemporary cosmology but also supported by them.

In any argument about the mutability of the laws and other regularities of nature, the first objection to the claim of mutability is likely to be the observation of stability. Although the stability of the regularities of nature is not, strictly, in contradiction with their mutability, it may be fairly held to support a presumption of their changelessness.

There are two classes of arguments that can be made to reconcile the observed stability of the laws of nature with the conjecture of their mutability. The first class of arguments are rationalistic: conclusions from some metaphysical principle about nature or about thought, such as the principle of sufficient reason, that we expect to hold independently of any observation, or inference from observation, of the workings of nature. In that sense, they can also be called a priori arguments (whether or not they conform to Kant's conception of the synthetic a priori). The second class are arguments that, although they may be speculative, because they cannot be supported directly by observation or experiment, nevertheless result from, and lead back to, a view of the facts of the matter about our universe and its history. Such arguments form part of a view that must be susceptible to empirical test and challenge at the periphery of its implications if not directly at the core of its most general ideas. The defense of the inclusive reality of time in Chapter 4 is an example of such speculative but nevertheless empirical reasoning.

To be faithful to the spirit and structure of these ideas, the argument for the reconciliation of stability with mutability must be of the second of these two orders of reasoning rather than of the first. It must not be tainted, so far as we can avoid it, by any metaphysical dogma, upheld independently of our struggle, in science, to develop an understanding of how nature works. Considerations about the constraints on understanding and the requirements of scientific practice (themselves open to revision in the course of the history of science) may be pertinent. However, they will be pertinent only as complements to a view that is, in a broad sense, empirical, not as substitutes for such a view.

That the laws and other regularities of nature evolve makes them more susceptible to explanation than they would otherwise be. If we cannot explain them historically, we cannot explain them: that is to say, we cannot explain why they are what they are rather than something else. The only other option that we have is to explain them as expressions or impositions of our mathematical insight, taken as a revelation of the fundamental truths of nature. As mathematical oracles or prophecies, the laws and other regularities of nature might indeed be held to be timeless, as they generally have been held in the dominant tradition of physics, given the timelessness of the relations among mathematical propositions. That we cannot justifiably explain the laws of nature in this way, and on this basis infer their immutability from their stability, is a thesis to be further developed in Chapter 6.

In my argument here, we can best understand the stability of the laws and other regularities of nature in the context of our growing insight into the history of the universe. According to this proposal, the laws, symmetries, and constants are stable in the cooled-down universe, which I described as the first or normal state of nature in my discussion of the second cosmological fallacy. The characteristics of nature in this state account for their stability.

However, by the terms of that argument against the second cosmological fallacy, nature may also be, and work, in other forms, notably those that are associated with the fiery but nevertheless finite beginnings of the present universe as well as with certain later situations in the course of the history of the universe that exhibit some of the features of those beginnings. These other forms of nature are far more hospitable to the rapid change of the kinds of things that there are as well as of the way in which they interact. Only the universal anachronism embodied in the second cosmological fallacy – the disposition to treat the workings of nature in the cooled-down universe as its sole repertory of modes of change – would justify inferring the immutability of the laws of nature from their overall stability in the observed universe.

To develop this idea, I now restate my earlier discussion of the metamorphoses of nature during the history of the universe.

* * *

Here, in pre-scientific and pre-philosophical language, is an account of the relation between the present universe and the early universe. This account accords with what we know about the history of the universe. It suggests why the mutability and the stability of the laws of nature may be a feature of this history. It shows how the same processes that give rise to the stability may also produce the mutability.

The mutability of the laws of nature ceases to contradict their stability, and begins to complement it, once we place both the stability and the mutability in the context of a historical view. It is only when we treat the stability and the mutability of the laws of nature without regard to the historical character of the universe that they appear to contradict each other.

Imagine the present, cooled-down universe, in comparison to the universe in its fiery and formative stage, as a living corpse: with limited kinetic energy, temperature, and degrees of freedom, with an established structure, and with enduring regularities – the laws of nature. Yet there was a time, of extreme density and temperature, when the distinction between states of affairs and regularities was unclear (a time that can be described alternately as one of law-giving or of lawlessness), when the present division of nature into well-defined constituents was not yet established, and when the phenomena were excited to much higher degrees of freedom than those enjoyed by the living corpse. The unexplained values of the dimensionless constants or parameters of nature may have their origin in the process by which this formative moment gave rise to the ensuing regularities and structures.

This rudimentary account – a stylized interpretation of certain central features of the history of the universe, such as contemporary cosmology represents it – distinguishes between relatively brief and formative moments of extreme density, kinetic energy, and temperature and the relatively long periods, subsequent to these moments, of lesser density, kinetic energy, and temperature, in which a universe is worked out. We have increasing reason to think that these formative moments of extreme density and temperature, resulting in a new universe, recur. The idea of their recurrence is, by another name, the idea of a succession of causally connected universes, proposed by contrast to the idea of a plurality of causally unconnected universes. To suppose that they recur is not, however, to assume that they recur with the same results: as universes with the same constituents and regularities. The recurrence of formative moments without conservation of the same constituents and regularities is the substance of the conjecture of non-cyclic succession.

The quantitative values of density, kinetic energy, and temperature in the formative moments are large but they are not infinite. Because they are large rather than infinite, they are not singularities in the present conventional, technical definition of singularity, and thus, on this account, the world does not begin in an infinite initial singularity. The importance of their being large rather than infinite is that the circumstances of the formative moment need not interrupt the causal pass-through from one universe to the next, although they may stress it. That the quantitative values of density, temperature, and kinetic energy are finite although large must be a presupposition of the thesis of succession.

For the purpose of showing how such a view of universal history may account for the combination of the stability of the laws of nature with their mutability, think of this view as a combination of two pictures of how nature works – at the fiery and formative moments and then in the long subsequent history of the formed universe. Of course, the contrast between these two faces or moments of nature amounts to a stark simplification of a historical reality in which time never ceases to flow and one state of affairs always turns into another. Moreover, the characteristics of nature in its extreme forms may be reproduced, at least to some extent, within the cooled-down, law-abiding, and well-differentiated universe: for example, in the interior of black holes.

First, in the fiery and formative moments, the values of density, kinetic energy, and temperature are extreme without, however, becoming infinite. That they are finitely large rather than infinite makes them in principle subject to scientific investigation, however indirect. This finitude will be an important consideration when we consider the translation of these ideas into an agenda of empirical and experimental inquiry. If, for example, the circumstances of the earliest universe were described by the conventional concept of a singularity, in which the parameters have infinite values, the experiments conducted with the use of particle accelerators could not even in principle mimic features of those circumstances. If, however, the values are finite rather than infinite, we may in principle be able to simulate aspects of those states of the universe. The consequences are far-reaching for a cosmology and a physics that need not take the established universe as their sole basis for insight into the workings of nature.

Second, nature does not assume the form of a sharply differentiated structure: the distinctions among the elementary constituents of nature break down. There is no longer, and not yet, an established repertory of natural kinds: of the kinds of things that exist.

Third, the contrast between laws of nature and the phenomena that they govern ceases to hold. For one thing, this contrast presupposes the transformation of nature into a discriminate structure. For another thing, the circumstance of extreme density and temperature is one in which the transformation of the phenomena is simultaneously the introduction of new ways in which nature works.

Thus, fourth, in such circumstances, new and massive degrees of freedom may be turned on. If we could witness these conditions, we would be justified in saying to one another, were it not for the unspeakable disproportion between the scales of human and of cosmic time: you have not seen anything yet. The introduction of the new, in these formative periods, is not, however, a free-for-all. It is not the spontaneous generation of uncaused effects. It takes place under the influence of what came before – of prior universes, or of the states of the universe prior to the formative events. Causality exists without laws, which is a way of saying that causal connections have not acquired, or have lost, the repetitious form, over a differentiated range of nature, that makes it possible to distinguish phenomena from laws.

The influence of the causal antecedents must leave its mark on nature during, and therefore also after, the interlude of fire, restricting the range of materials and of processes with which nature works, even when excited to high degrees of freedom and bereft of a discriminate structure. This causal continuity, shaken but never wholly interrupted, must in principle be subject to empirical investigation. Even if we cannot have direct access to any universe or state of affairs prior to the beginnings of our own universe, we may be able to study its after-effects and vestiges.

Consider now the cooled-down universe, at some time long after its incandescent origin. First, the values of density, kinetic energy, and temperature are greatly reduced. Such a reduction need not imply a contrast between the infinite (as in the traditional concept of a singularity) and the finite. It may be a contrast between orders of magnitude within the realm of the finite.

Second, nature has assumed the form of a differentiated structure. It has elementary constituents. In our universe, these constituents have the content described by particle physics and, at another level of emergent complexity, by the periodic table. The kinds of things that there are remain as real as their individual instances – at least they are until we come to the higher forms of animal life in the course of the evolution of species.

Third, together with the distinctions among natural kinds and elementary constituents of nature, there arises, or reappears, a distinction between states of affairs and the regularities to which they conform. The workings of nature are normalized in the sense that each type of thing acts in regular fashion that can be represented in effective laws – that is to say, laws applicable to certain domains, each domain being distinguished by a cast of natural kinds and of their interactions.

Such laws in turn support causal explanations, although, from a wider and deeper view, we come to understand that causal connections are a primitive feature of nature. The laws derive from the connections, not the other way around. One such style of law-based causal explanation, which presupposes both the transformation of the world into a differentiated structure and the distinction between laws of nature and the states of affairs that they govern, is the style that we call the Newtonian paradigm.

Fourth, in concert with its other characteristics, the established, cooled-down universe is one in which nature displays fewer degrees of freedom in the relation of what happens at any given moment to what may come next. This diminishment in degrees of freedom, however, need not be linear and irreversible. The expansion of the universe may, at any given time, broaden the range of the adjacent possible as well as witnessing an increase of degrees of freedom.* Whether or how the

* The idea of the adjacent possible differs from the notion of degrees of freedom by accommodating novel emergent properties. Thus, Stuart Kauffman used the idea of adjacent possible to describe the set of new species that may arise by speciation from the present set. On this view, the number of degrees of freedom possessed by the underlying atoms from which the organisms of each species are built remains the same.

expansion of the universe may allow for an enlargement of the adjacent possible is a matter for empirical investigation at the front line of contemporary cosmology.

This parable about the alternative forms of nature expresses as a duality what must in fact be a more complicated set of transitions and variations in the course of the history of the universe. It conveys, in highly simplified and stylized form, the conjecture motivating the argument against the second cosmological fallacy, set out in Chapter 1. The case for the conjecture is that it can be more fully and persuasively reconciled with the picture of universal history enshrined in the now standard cosmological model than the view that takes the workings of nature in the cooled-down universe to be the sole form of natural reality.

The resulting conception of alternative ways in which nature may work suggests how the relative stability of the laws of nature may complement rather than contradict their relative mutability. The relative stability of the laws of nature is a feature of the established universe. Their relative mutability is a characteristic of the universe in formation or, more generally, in its extreme moments. The assumption that nature always works as it does in the mature universe that we observe, with its differentiation of types of being, built out of the elementary constituents described by the periodic table and by particle physics, with its seemingly clearcut distinction between states of affairs and laws of nature, and with its severe restraint on the range of the adjacent possible, is the substance of the second cosmological fallacy.

However, the distinction between the two moments – to which I have resorted, in pre-scientific and pre-philosophical language, as a heuristic device – is only relative. In the first place, it is only relative because the historical character of the universe implies that the cooled-down universe arose from the incandescent universe. The transitions and transformations may be much more rapid at some times than at others – especially in the early life of the universe – but they are never instantaneous, just as the very large parameters of quantities at the formative moments are never infinite. There must be many states of nature that are intermediate between the two extreme situations invoked in this unscientific parable.

The distinction is also relative because the extreme forms of nature may reappear as local phenomena within the cooled-down universe (as in the interior of black holes) until they may be once again generalized at a later moment in universal history (according to the conjecture of a future contraction of the universe and subsequent "bounce"). It is relative as well because the appearance of new forms of being does not cease in the established universe: the universe in which the distinction between states of affairs and the laws governing them takes hold and in which effects seem to follow causes in each domain of reality according to the stable regularities revealed by science. Such is the case of the emergence of life on Earth and then of humanity and of our consciousness and culture.

Part of consciousness, I earlier remarked, is machine-like: that is to say, modular and formulaic. However, another part of consciousness is an anti-machine. In this second aspect of its life, consciousness can recombine structures and functions in a fashion that is prefigured and enabled, though left unexplained, by the plasticity of the brain. Consciousness exhibits a faculty of recursive infinity: the ability to recombine finite elements – of a natural language, for example – in infinite numbers of ways. It enjoys the power of negative capability: the capacity always to perceive and to think more than its presuppositions will allow. The aspect of consciousness that displays these attributes is what we call imagination.

It is because we possess imagination that we are able not always to experience the world or to act within in it according to script. We are able to tear the script up. The proponent of a weak reductionism will say that there is nothing in this subversive and transformative activity that contradicts the laws of nature, stated in the supposedly fundamental science of physics. The votary of a strong reductionism will go further and assert that even at our most subversive we act at the behest of all-determining physical forces.

Both will be making empty gestures: their claims will have little or no consequence for our actions, or even for our understanding of them. Nothing follows from the compatibility invoked by the weak reductionist other than the rehearsal of the attempt to make the strange seem natural. The strong reductionist, for his part, announces a program of explanation that no one has ever been able to carry out.

What these gestures seek to affirm is that the characteristics of nature in the fiery and formative origins of the universe have no afterlife, and leave no trace, in the established universe. If, however, the universe may become more fixed in its structure and regularities than it was in its origins, it may also witness the rise to forms of being and of experience even more open than any that existed earlier in its history.

Nothing in the recognition of this fact should encourage us to believe that the universe is on our side, or that it is open and creative, as we are, by virtue of possessing imagination. Our experience of openness ends in death. Our imaginations are incarnate in dying bodies. We are powerless to look into the beginning or the end of time, or to rest our understanding of the world on definitive and incontestable ground. The forces of the law-like universe pulse through us in the form of insatiable desire. Nothing in our experience gives support to the feel-good philosophies that fill the history of metaphysics, and abuse the prerogative of speculative thought by trading enlightenment for consolation.

The contrast between the original and the subsequent universe, in addition to suggesting how the laws of nature can be both mutable and stable, also sheds light on a source of resistance to acknowledgment of the reality of time. Our conception of science is formed on the model of the relatively cold and consolidated universe. Each of the attributes of nature in this state has a counterpart in the model of science that took hold in physics from the time of Galileo and Newton. In that practice of science, recognition of the reality of time has no sure foothold.

One of the achievements of cosmology is to have disturbed this view, both of nature and of science, through its pursuit of a research agenda founded on the idea that the universe has a history. However, it would be just as misguided to base an approach to nature and a practice of science on one of these moments of the universe, or on one of these variations of nature, as it is to base it on the other. The task is to accommodate in our ideas all these moments and variations. A feature of any conception capable of carrying out this task is that it will affirm unreservedly the reality of time.

THE CONUNDRUM OF THE META-LAWS

To affirm that the laws of nature may change is to confront a problem that is bound to gain ever greater importance in cosmology and, by the hand of cosmology, in all of physics. We call this problem the conundrum of the meta-laws.

Consider the problem first in its simplest form, as an antinomy. Whether it is a true antinomy, as I argued the antinomy of cosmogenesis to be, or a false one, as many supposed antinomies are, is not something that we can infer, in the manner of Immanuel Kant, from inquiry into the structure of rational understanding. We cannot reach a reliable conclusion without regard to any particular content of our beliefs or to what we discover, or fail to discover, about nature. It depends on whether we find in science itself, rather than in metaphysics, a path toward the overcoming of this unacceptable choice.

Suppose that the change of laws of nature is itself governed by laws: higher-order laws or meta-laws. Then the problem of the historicity of nature and of its regularities will simply recur at that higher level. We will have gained little or nothing in our effort to recognize the inclusive reality of time as well as the occurrence of a causality that may be lawless. Either we concede that the regularities of nature are themselves open to change, or we claim them to be exempt from time and change. We have simply postponed the problem, or transferred it from one level of explanation to another.

Suppose, on the other hand, that the change of the laws is not itself law-governed. Then it seems that it is uncaused, which is to say arbitrary or at least without explanation, whether deterministic or probabilistic. Then indeed the idea of a history of the universe would have driven us to explanatory nihilism. Those would have been right who feared that a full recognition of the reality of time would undermine the project of science.

Note that this statement of the conundrum elides the difference between causal explanation and the explanatory deployment of laws of nature. It is written in a way that suggests that when explanation by reference to laws of nature fails, so must causal explanation. It is an unjustified elision. To question it is to arouse the first glimmer of hope that we can begin to overcome the conundrum of the meta-laws, showing it to be a false antinomy.

In our conventional understanding of causality, the laws of nature serve as grounds for causal explanations. In the standard practice of physics, at least since the time of Galileo and Newton, no explanations are offered that fail to depend, directly and explicitly or indirectly and implicitly, on law-like regularities: symmetries and constants as well as laws of nature. The pursuit of such regularities came to be considered the chief business of science, and particular explanations of particular changes as simply instances of such regularities, especially of laws of nature.

Yet in our ordinary experience, we regularly make causal judgments that assume some measure of stability in the workings of nature but make no reference, however remote, to general connections like those that physics represents in mathematical language.

To address what is at stake in the elision from causes to laws that is implicit in my first statement of the antinomy, I now restate the conundrum in a second form: the form of an antinomy about causality. This second statement of the conundrum highlights its relation to the reality of time.

Suppose that the laws of nature are in fact exempt from time and change, as almost all physics up to now has assumed. (I have several times recalled how the physics of the twentieth century reaffirmed the idea of a timeless framework of laws of nature in the very process of overturning the ideas of a distinction between natural phenomena and their background in space and time as well as of sharp contrast between space and time.) Our causal explanations will then enjoy a firm basis in timeless laws. However, it is a basis that we shall have secured only by qualifying the reach and reality of time.

The practice of the Newtonian paradigm exemplifies how time can be accorded some limited measure of reality without being allowed to go all the way down: that is to say, to touch everything that there is so that everything can change. Under the view presupposed by the Newtonian paradigm, the laws of nature are not themselves marked by time or susceptible to change. It is because they are not so marked that they can support causal explanations concerning events in time.

Suppose, on the other hand, that time goes all the way down, or includes everything, with the result that the laws of nature may change. Then all our causal explanations are insecure. They are adrift on a sea of changing – or at least changeable – laws of nature.

From the vantage point provided by this second statement of the conundrum of the meta-laws, it seems that our conventional beliefs about causation, including beliefs that inform practices of scientific inquiry, equivocate about the reality of time. They take time to be real to some extent: to the extent necessary to allow for the reality of causal succession. However, they do not take time to be so real that no framework of natural laws can lie, immutably, outside time. Time, according to such beliefs, must be real, but not too much.

THE PROBLEM OF CAUSATION IN THE EARLY UNIVERSE REVISITED

Before considering the general character of a solution to the conundrum of the meta-laws, I revisit the problem of causation, and of the relation of causation to laws of nature, as it presents itself in the early universe. The regularities of nature are there not clearly distinct from the states of affairs, and the composite formed by the laws of nature and by the states of affairs may have been excited to higher degrees of freedom and allow for a broader range of adjacent possibles than we usually (but not always) observe in the established universe.

However, this circumstance, so disturbing to our habitual model of scientific inquiry, does not spell the suspension of causal connections. It fails to do so for two basic reasons. Each of these reasons turns out to be significant in the effort to understand the most promising way out of the conundrum of the meta-laws.

The first reason is that the convergence of the regularities of nature with the phenomena that they govern, in a composite aroused to higher degrees of freedom and with a broader periphery of adjacent possibles, need not imply the cessation of causality. It may signify, instead, that causality ceases to be lawful: it no longer works as recurrent relations among the elements of a discriminate, well-ordered structure.

That nature in this condition exhibits more degrees of freedom may be translated into the idea of a broader range of adjacent possibles around the present states of affairs: more theres are accessible from the heres. Yet as the regularities and structure of any earlier state of nature vanish into the great formative fire of the universe at its origins, there may be other more general constraints that causation continues to obey in moving toward one proximate possible rather than another.

Such more general constraints may be the ultimate regularities that we traditionally call principles rather than laws, such as the least action principle, or its twin, the principle of conservation of energy or the equally fundamental principle of reciprocated action, according to which causal influence is always reciprocal. Even these principles, however, have limited domains of application in our established scientific ideas, as well as remaining, as everything else does, within the reach of time and change. The principle of least action is satisfied only for classical physics, and fails in quantum physics. Conservation of energy is a contingent consequence of a symmetry under translation in time, and fails in general relativity.

No hard-and-fast distinction exists between the effective laws governing phenomena in certain domains and these principles, which are also described as fundamental laws. However, although everything can change, some things change more slowly or rarely than others: the phenomena more easily and often than the effective laws, and the effective laws more readily and commonly than the principles or fundamental laws.

The second reason why the circumstances of the early universe need not imply the break-up of causal connections is that in nature, as we observe it, what comes before always shapes what comes later, even if the mechanism of influence may change. That it may change, and indeed will change, results from another general feature of nature, pertaining to the character of time: that change changes.

The two presuppositions of such a causal continuity, even in the extreme conditions of the early universe, are, first, the conjecture of the succession (as opposed to the plurality) of universes, and, second, the conjecture that, in the very early universe, the values of the parameters of quantities of density, temperature, and kinetic energy are finite, although they are extreme (by comparison to the corresponding values in the established universe). These two assumptions work together to establish the possibility of causal pass-through from one universe to another. Nature can work only with the materials at hand, all of them products of transformation, including the transformation of transformation, which is the character of time.

Such a causal pass-through from one universe to another, stressed but never completely broken by the conditions characteristic of the very early universe, is compatible with the difference of one universe, or one state of the universe, from the universe or state that preceded it. Each successive universe, if there is such a succession, may have a different structure, and be made up of different materials, and display different laws. Everything in a new universe or state will nevertheless have been made with what existed in the old one, as filtered through the extreme circumstances of the formative moment.

The earlier universe, or the earlier states of the present universe, may leave vestiges or fossils. Such may be, for example, the unexplained values of the constants, in particular the so-called dimensionless parameters of nature. Set aside those unexplained parameters that are intrinsically dimensional: Newton's gravitational constant, Planck's constant, and the speed of light. We are left with the others, as candidates for the role of bearing the traces of an earlier reality. They include the masses (and the ratio of the masses) of the elementary particles, the strength of the different forces or interactions, and the cosmological constant (the energy density of space). These dimensionless parameters count among the hieroglyphs in whose forgotten language we may one day be able to read the records of a bygone world.

A premise of this approach to reasoning and empirical inquiry about the very early universe is the rejection of the idea of an infinite initial singularity. The infinite, here as elsewhere, would amount to an insurmountable obstacle to further investigation. It is an obstacle that we should not and need not willfully impose on ourselves: a mathematical conceit, announcing a breakdown in the application of a theory (in this instance, general relativity and its field equations) to some part of nature and its history rather than a description of any natural phenomenon.

THE BEST HOPE FOR RESOLVING THE CONUNDRUM OF THE META-LAWS

This exercise in relating the conundrum of the meta-laws to the picture of the history of the universe that emerges from contemporary cosmology suggests where the best hope of resolving the conundrum lies. It lies in the combination of ideas that I have already explored but have now only to bring together. In being brought together, they do more than suggest a general speculative view; they also serve as a basis for conjectures that can help inform an agenda of empirical inquiry.

One idea is that of a change, in the course of the history of the universe, in the relation between states of affairs and the laws to which they conform and, more broadly, between two major variants or moments of the workings of nature in the history of the universe. A second idea is that of a succession of causally connected universes, with the causal succession between them stressed but never broken. As a result, present nature, even at its moments of greatest upheaval and transfiguration, is always able to work with a legacy bequeathed to it by past nature. A third idea is that new laws of nature may develop coevally with the phenomena that they govern. A fourth idea is that this coeval transformation of the phenomena and of the laws may be shaped by higher-order principles or fundamental laws. Such principles may themselves be subject to evolution. A fifth idea is that causality may exist without laws. Causal connections are primitive features of nature. In certain variations of nature, causality may fail to exhibit its familiar recurrent and law-like form: such are the states associated with the very early universe in formation, or with extreme states of the cooled-down universe.

What we call laws of nature are the regular and recurrent form that causal connections take in certain states of nature or for certain periods in the history of the universe. It follows that we do better to think of the laws of nature as deriving from causal connections rather than to see the latter as deriving from the former, as we are accustomed to do. The existence of causality without laws is founded on the power of sequence: the influence of a before on an after. It is an influence exercised regardless of whether causal processes recur over a differentiated structure of reality, allowing us to state laws in the form of equations. The ability to express laws of nature as equations provides an immense benefit to scientific inquiry. By the same token, however, this practice may mislead us into thinking of causal connections as mere instances of laws of nature. It is, on the contrary, the laws that describe one of the forms, but not the only form, that causal connections take in nature: the recurrent and general form.

In previous parts of this book, I have introduced each of these five ideas. Now is the occasion to show how they can work together and establish, if not a solution to the conundrum of the meta-laws, then a class of solutions, or a marking of the conceptual space in which we can best hope to develop a solution. The first two ideas – the existence of radically different states of nature and the preference for a succession rather than a plurality of universes – set the stage for such a solution in a view of universal history. The third idea – the coeval change of phenomena and of laws – and the fifth – the priority of causation to laws – provide the key to a solution. The fourth idea – the way in which the evolution of effective laws may be shaped by fundamental laws or principles that evolve more slowly or that change only in more extreme circumstances – qualifies and refines the third idea.

* * *

That laws of nature may develop coevally with the phenomena that they govern is a notion that arouses puzzlement when stated in a physical or cosmological setting. It is nevertheless an idea long deployed in two other domains: the life sciences and the field of social and historical studies.

The regularities by which we have come to understand the evolution of life, with respect to Darwinian natural selection and, even more remarkably, with regard to genetic recombination, arose together with life itself. They do not suspend the laws of chemistry and of physics; they work through them. The physical constraints on body shapes and on the pairing of structures with functions are, for example, a traditionally underemphasized element in the evolution of life forms. However, life, whenever or wherever it arose, represents something new in the world, with respect both to the phenomena that constitute it and to the regular causal connections – the laws – that characterize it. In any attempt to reduce these regularities to the supposedly more basic laws of physics and chemistry, the strong reductionist has always found that whatever victories he wins are Pyrrhic. The interesting information, the information about the kinds of things that there are and the way in which they function, is left out of such reductionist accounts.

In the study of society and history, the idea that the phenomena and the regular relations by which we explain them develop together is a familiar view. It takes many different forms, according to the conceptions of different social, economic, and political theories. For example, those who believe that there are types of social and economic organization such as "capitalism" or the "market economy" also often hold that each of these types have certain "laws": a logic of reproduction and transformation that cannot be defied with impunity. A Marxist may claim that there are higher-order laws, commanding the historical succession of "modes of production" and therefore as well of the sets of laws governing each of them. Yet those who have ceased to credit the more heroic claims of dialectical materialism, with its idea of meta-laws, have not believed themselves required to abandon the more limited claim that each new system of social and economic organization emerges together with its own laws. In this confidence they are matched by their conservative counterparts, who think that the laws of a market economy are, with minor variations, always and everywhere the same.

The belief that the laws and the phenomena develop in tandem, which we find widely expressed in the earth, life, and social sciences, may be puzzling, but it is not nonsensical. In fact, physics may be the only major science in which it is not a standard explanatory move. (It is not because they make this move that the neo-Marxists and the practical, conservative economists are mistaken, but for other reasons, including the idea that capitalism or the market economy represent indivisible, law-like systems. These other reasons concern the substance of their understanding of history and humanity.)

Part of the solution to the conundrum of the meta-laws lies in the introduction into physics and cosmology of the idea that the laws of nature may change and evolve in concert with change and evolution of the phenomena that they govern. This coeval development of the laws and of the phenomena may occur more readily and rapidly in those moments in the history of the universe when the distinction between laws and states of affairs has broken down, or in which a differentiated structure has not yet emerged. However, it may also recur later, in the formed and cooled universe, whenever and wherever something genuinely new happens.

Of things new, the most astonishing to us, its products, is life, and then, later, the work of the imagination in society and in culture. Not only is life an emergent phenomenon but it also becomes a means for the generation of more emergent phenomena: that is to say, more novelty in the world. It does so first through natural evolution and then through the conscious work of humanity.

The joint transformation of the laws and of the phenomena is forceful and quickened in those states of the universe that witness a collapse of the distinction between the laws and the states of affairs that they govern. Causality, as I have argued in Chapter 1, may exist without laws, or not in law-like form. It is discontinuous or episodic in those conditions of the universe in which the laws have become distinct from the states of affairs. In either situation, however, the coeval change of the phenomena and the laws always works with the materials produced by sequence. It is, as the evolutionary biologists like to say, path-dependent. These path-dependent materials include both the structure of nature and its regularities as they result from the prior history of nature.

Moreover, some regularities are more fundamental than others. That need not mean that the more fundamental regularities are immutable. It means only, for a view refusing to allow exceptions to the sway of time, that they change more rarely and marginally. On this conjecture, they can help shape the content and the evolution of the lower-order or effective laws without being themselves timeless and changeless. They are therefore meta-laws in one sense (the sense of having as their object the lower-order or effective laws) but not in another (the sense of being the eternal framework of the universe). They can move without being themselves immovable. They are nevertheless the most stable part of nature, although our understanding of them is subject to revision in the light of the advance of science.

Our contemporary view of the content of these principles is a summation of the history of physics from the seventeenth century to today. Here is a non-exhaustive list of them, stated in loose conceptual order rather than in historical sequence.

First among them is the principle of least action (the principle of Maupertius), which all by itself can support an alternative account of the phenomena explained by Newton's laws of motion.

A second principle is the principle of conservation of energy (the principle of Mayer), intimately related to the principle of least action. The conservation of energy, however, can be derived from the principle of least action only when there is a translation symmetry in time, as described by Noether's Theorem below.

A third principle is the principle of the equality of action and reaction or, more inclusively, of the conservation of momentum (the principle of Newton). It can be further generalized in the form of the core idea in Noether's Theorem: that on the basis of any theory using a Lagrangian or a Hamiltonian to describe a symmetry, we can, once we specify the values of the variables, apply abstract algebra to infer a conserved quantity, be it a quantity of energy, linear momentum, angular momentum, or something else. In this generalized form, we might call it the principle of Noether.

A fourth principle is the principle of the degradation of energy (the principle of Carnot).

A fifth principle is the principle of the invariance of the laws of nature for fixed observers or observers in uniform movement – relativity (the principle of Galileo or of Einstein, commonly known as Galilean relativity).

These principles, whether taken one by one or in concert, are not effective laws (although they may become such laws in the context of general relativity): they do not suffice to support particular causal explanations of particular phenomena. They have as their proximate subject matter the effective or domain-specific laws of nature, rather than the phenomena. Many – but not all – of the states of affairs that we observe in the established universe and many – but not all – of the effective laws that science has thus far propounded conform to all of them. In tandem, these principles or fundamental laws describe the most stable and universal aspects of the workings of nature. It does not follow, however, from their pre-eminent stability, and it is not required for their power to shape the effective laws, that they be themselves either immutable or universal.

We are faced, in the present condition of our knowledge, with a choice between two ways to understand them. Both of these accounts are speculative, in the sense that we cannot directly subject either of them to direct empirical confirmation or challenge. They can nevertheless play a part in agendas of scientific research that are open, at their ample periphery of implications, to empirical confirmation or invalidation. In this sense, the two speculative accounts can also face empirical test: indirectly, by virtue of their share in such research agendas.

The first understanding is that these principles are indeed an immutable framework of reality. This view gains force from the idea – against which we argue in this book – that the universe of ours is only one of many universes, in all of which the same principles are realized. It also wins support and meaning from the notion that time is not all-powerful and all-inclusive: something lies beyond its reach, if only the fundamental laws of nature.

The second understanding is that these principles are only relatively distinct from the effective laws and thus as well from the phenomena. They form part of time-drenched and changing nature, although they represent the part that changes least or less often. The fire must be yet greater for them to burn. Nevertheless, they too can change.

This second view makes sense in light of the two combined ideas that have been central to the argument of this book: the solitary existence of the universe and the inclusive reality of time. The most important attribute of the world is that it is what it is and not something else. It happens to be one way rather than another. The "it" that happens to be one way rather than another includes the effective laws of nature. The principles, or more fundamental laws, describe the way that the effective laws happen to be: that is to say, their family resemblance (to use Wittgenstein's phrase). We do not say, like the God of the rationalist philosophers, that the scandalous particularity of nature and of its history is the disguise of rational necessity. The universe happens to be one way rather than another in time, and it keeps happening and changing – all of it, not just part of it.

The coeval transformation, or co-evolution, of the laws and of the phenomena, hastened and extended in those states of the universe that are characterized by breakdown of the distinction between them, is shaped both from below and from above. It is shaped from above by the relative recalcitrance of the higher-order laws or principles to change. It is shaped from below by the restrictive force of sequence, even at the extreme and testing limit of a non-cyclic succession of universes.

The conception of co-evolution of the laws and of the phenomena, informed by sequence or path dependence and constrained, at any given time, by the family resemblance of the effective laws (a resemblance codified as principles or fundamental laws), against the background of the conceptions of alternative states of nature and of a succession of universes, suggests the beginnings of a response to the conundrum of the meta-laws. These ideas fail to provide a definitive solution to the conundrum, if only because they have yet to touch the ground of an empirical and experimental research agenda. However, they mark out a space in which to look for a solution.

An aspect of this view requiring special attention is the relation between the idea of the co-evolution of laws and phenomena and the idea of causality without laws. The way in which these two ideas operate together becomes clear only when both of them are developed and understood with acknowledgment of the different ways in which nature may work over the course of universal history. Such an acknowledgment begins in the rejection of the second cosmological fallacy.

When nature may not yet be, or has ceased to be, organized into distinct and differentiated entities or elements (as described by particle physics and by chemistry), the distinction between laws and states of affairs may fail to apply, and the range of the adjacent possible may be large, the *before* may nevertheless continue to influence the *after* in time. Causality survives laws. Even the principles or fundamental laws may be subject to change, although less readily and rapidly than the effective laws. Causal connections, however, do not require that any part of the workings of nature, not even the fundamental laws or principles, be unchanging. Their ultimate basis is not the laws, whether effective or fundamental. It is time. On the argument of this book, time, and only time, is what always remains. Everything else changes, including the ways change happens. These propositions – the residual and inclusive reality of time and the mutability of everything else – are not distinct propositions; they are equivalent formulations of the same idea.

In guiding the co-evolution of the phenomena and the laws, the influence from above – the effect of slowly or rarely changing fundamental laws or principles – and the influence from below – of sequence or primitive causal connection – are, in this view, not at the same level. The former is variable and relative. The latter is radical and permanent. It borrows its power from time.

This approach to the solution of the conundrum of the meta-laws has as one of its assumptions the primacy of historical explanation over structural explanation, in all science, not in cosmology alone. Structure results from history, rather than the other way around. It then both constrains and enables later historical development. No simple conflict holds between explanation from history and explanation from structure.

FROM SPECULATIVE CONCEPTION TO EMPIRICAL INQUIRY

These are speculative ideas. They can, however, help inform an agenda of cosmological theory and research. Equipped with them, we can cross the frontier between the philosophical representation and the empirical study of nature. Natural philosophy, as we understand it and here seek to practice it, outreaches what empirical and experimental science is yet able to establish. It does so, dangerously, with the intention of engaging the work of science.

The conceptions presented in these pages – about the singular existence of the universe and the inclusive reality of time – do not lend themselves to instantaneous translation into scientific work. However, they are worth little if they cannot help inspire a new and better way of thinking about problems of contemporary cosmology and, through cosmology, of physics as a whole. To deal with their bearing on these problems is an ask that Lee Smolin takes up in Part II of the book. For the moment, it is enough to suggest that these views can inform a way of thinking about riddles that have become central to the scientific understanding of the universe and of its history.

It is by their contribution to a more comprehensive and successful way of thinking about these problems that our speculative argument about the solitary existence of the universe and the inclusive reality of time can reach the front line of empirical and experimental science. The speculative leaps may be forgiven if they can be shown in the end to bring us closer to nature as she so strangely happens to be.

Consider, briefly, first, the feature of the history of the universe that our argument must principally be concerned to elucidate if it is to show its empirical value; second, the range of practices that can test these conceptions along their periphery of empirical consequence; and third, the chief conceptual obstacles that must be overcome to advance in the development of such a research agenda.

The aspect of universal history offering the most promising terrain for such a translation of speculative conception into empirical inquiry is the explanation of the initial conditions of the universe. (I use the term initial conditions in this context to denote simply the facts of the matter about the very early universe, without the baggage that the term carries in what we describe and criticize as the Newtonian paradigm.) These conditions are so peculiar, and yet so improbable by the standard of the range of variation observed in the subsequent universe, that they baffle and cry out for explanation.

The placement of our universe among an indefinitely large or infinite multitude of other universes, motivated by the attempt to erase the massive underdetermination of prevailing theories in contemporary particle physics, cannot dispel this sense of strangeness. On the basis of such a placement, we can imagine the cosmic dice rolled many times, until a universe with the initial conditions and other characteristics of ours results from the many rolls. Even, however, if we agree to enter into the spirit of such a quasi-scientific fantasy, we are not entitled to expect it to support a calculus of probabilities.

In such a circumstance, there is no closed set of alternative states on which the calculus can operate; the dice have an infinite or indefinitely large number of sides. The legitimate domain of probability is inside the universe, not outside it, at least so long as we have no grounds to assume the existence of a closed set of universes. If we concede, for argument, the existence of a plurality of universes, we would need further reason to believe that it is a limited plurality. No such reason exists. The consequence is to undermine the application, in that setting, of any calculus of probability. (In the technical literature of contemporary cosmology, this problem goes under the name of the cosmological measure problem. Many have attempted to reestablish the basis for a well-formed judgment of probability when the cosmic dice have an infinite or indefinitely large number of sides. They have worked in vain.)

A number of conundrums result from this many-sided problem of the initial conditions of the universe. I have alluded to several of them in earlier parts of this work; others are explored, in depth and in detail, in Part II of this book. In each instance, our argument suggests an approach to the elucidation of the issue at hand that offers an alternative to now prevailing views. It is an approach in which structural analysis becomes ancillary to historical explanation.

The major practices by which, given the centrality of the problem of initial conditions, we can give an empirical character to such an agenda are of two main kinds. The first class of procedures is the evocation of conditions similar to those that we believe may have prevailed in the early universe. Today we can best hope to produce such an evocation through the use of particle accelerators as well as computer simulations. Whatever their present physical limits, they are, at least in principle, capable of shedding light on the initial conditions, or on its many ramifications. They can inform us so long as an indispensable requirement is met: that we represent the initial conditions as remaining within the realm of finite values rather than as hiding behind the screen of the infinite.

The second class of methods is the observation and the interpretation of events or phenomena, in the consolidated universe, such as the interior of black holes, that may share some of the traits of the very early universe: supercondensation resulting in the breakdown of discriminate structure. Once again, the threshold requirement of openness to empirical inquiry is that the values of these phenomena stay this side of the boundary between the finite and the infinite.

The conceptual obstacles to the development of such an agenda of empirical research are numerous and formidable. To overcome them, by suggesting different points of departure for our thinking about fundamental problems in cosmology and physics is a major aim of this work. Two such obstacles deserve special attention in this context because they have a direct bearing on the elucidation of the initial conditions of the universe.

A first obstacle is the dangerous appeal of the runaway multiplication of universes under the thesis of plurality. Such a multiplication disposes of the mystery of the initial conditions by the false conversion of an explanatory failure into an explanatory achievement. Earlier and later parts of this book explore reasons to resist this temptation.

A second obstacle is the prestige lent by the dominant understandings of general relativity to the idea that a time-eviscerating and finitude-breaking singularity must lie at the beginning of the universe. The Hawking–Penrose theorem, for example, states that any cosmological solution to the field equations of general relativity will have timelike geodesics that cannot be extended, without limit, into the past so long as certain conditions are fulfilled. Among these conditions are that the field equations of general relativity be universally applicable, that the energy density of matter be universally positive, that there be a spacelike surface at which the universe expands, and that the solution be untainted by special symmetries restricting its range of application.

At least since the work of deWitt and Wheeler, it has been argued that the consideration of quantum effects would eliminate the cosmological singularity. Quantum theory helps suggest the incompleteness of the singularity theorems, or of the view from which they result. However, it cannot itself supply a basis for a comprehensive view capable of elucidating the problem of the initial conditions of the universe: it is, by its nature and design, an account of patches of the universe, lacking a cosmological vocation.

If we have many other reasons to affirm the inclusive reality of time and to avoid the jump into the infinite, it may be wiser to regard the singularity theorems that have been inferred from general relativity as outcomes of a process often repeated in the history of science: the extension of a powerful theory beyond its proper realm of application.

It is the fate of science that what at first appears to be a theory of unrestricted generality later turns out to be a special case: a piece, an approximation, a variation of another understanding. So long as we do not allow ourselves to be bewitched by our momentary and fragmentary insights, we can continue to turn enigma and failure into discovery.

IMPLICATIONS OF THE INCLUSIVE REALITY OF TIME FOR SOME FUNDAMENTAL IDEAS

The reality of time, far from being an empty platitude, is (I earlier argued) a revolutionary proposition. Radical acknowledgment of the reality of time undermines many of our conventional beliefs about nature as well as some of the conceptual categories central to the work of scientists. A common characteristic of many of the ideas thus subverted is to equivocate about the reality of time, affirming it in some respects but denying it in others. To explore some of these implications is further to elaborate the meaning of the reality of time.

Causes and laws

The second statement of the conundrum of the meta-laws suggests a sense in which our conventional ideas about causality are confused. Causal judgments presuppose the reality of time. The relation among logical or mathematical propositions does not. Laws of nature have been commonly understood to justify causal explanations. If time goes all the way down – or, to use a metaphor less spatial, is all-inclusive – the laws of nature should not be understood to be outside time. Laws of nature codify causal connections when such connections recur over a discriminate structure, as they generally do in the cooled-down universe. The language of laws of nature thus has, among its defining assumptions, that nature appears as discriminate structure, with a relatively stable but nevertheless mutable repertory of natural kinds and that causal connections among pieces of this structure display the features of generality and recurrence.

Nature often satisfies these conditions, but not always. In this view, laws can change, as can symmetries and supposed constants, although some of them (the fundamental laws or principles) may change less readily than others (the effective laws). The stability and the mutability of the laws need not contradict each other once we place both in the setting of a natural history of the universe.

It follows that we cannot hope to ground causality in a timeless and changeless foundation. Our causal explanations may then seem to be vitiated by the timeliness and the mutability of the laws of nature.

Our conventional beliefs, received into the inner sanctum of the dominant interpretations of the discoveries of science, fudge the difference between the two horns of this dilemma. They grant the reality of time, but not to the point of allowing that the laws of nature are within time rather than outside it and that they may therefore change.

To accept this criticism of those conventional beliefs is to recognize the need to revise our view of causal explanation. Causal explanation is ordinarily both a hypothesis about the temporal connections among phenomena in the world and a conjecture about the regularities of nature. A view of the causal connections among phenomena or states of affairs, as they usually occur in the cooled-down universe, implies as well a view of regularities of nature that are time-bound and mutable. The regularities as well as the phenomena remain enveloped within a world of time and change.

However, on the argument developed here, these two implications are not of the same order or on the same level. Causation always involves the force of sequence: the shaping of a before on an after. It need not always require that this shaping by sequence assume regular and recurrent form. Causality may exist without laws, in what I have called the second state of nature (in the discussion of the second cosmological fallacy) as well as in extreme events in the colder, differentiated universe that exhibit features of that second state. And insofar as causality is law-like, the laws may change together with the phenomena. The limit to generalizing explanation on the basis of laws and symmetries may come from nature rather than from our ideas. The conundrum of the meta-laws, at least in the form of its second and causality-related statement, is, therefore, not a true antinomy.

In the practice of much ordinary scientific explanation, we may disregard these constraints on law-like explanation resulting from both the mutability of the laws and the existence of causality without laws. We may disregard them because the regularities of the cooled-down universe in which we find ourselves appear to be remarkably stable. However, the broader the scope and ambition of our theories, the more dangerous it becomes to disregard the historical character of causation and the priority of causal connections to the regularities that they exhibit in the established universe.

Necessity and contingency

The argument about the reality of time has implications as well for our ideas about necessity and contingency. Such modal categories have no fixed meaning; their referents are the workings of nature or of society. Among our ideas about the workings of nature, our cosmological ideas take pride of place because they deal with the universe as a whole. In so doing, they condition the character and the measure of the necessity that we are entitled to attribute to the most necessary relations in nature.

It is sometimes said that a necessary relation is one that is necessary in every possible world. What, however, if there is only one universe at a time – the view we argue to be more in accord with what science has so far discovered about nature? If the universe is indeed solitary, the concept of possible worlds can at best serve as a heuristic device by which to explore the accessible transformations of the universe in which we find ourselves: the changes that the universe might suffer given what it has become. It makes no sense, under such a view, to search for the attributes shared by imaginary worlds and to interpret them as the touchstone of necessity.

Different views of the structure and history of the universe have distinct implications for our understanding of why nature works as it does, or might work in other ways. Cosmology sets the outer horizon of our beliefs about natural necessity. We look in vain in mathematics for the bedrock of universal necessity, more absolute and transparent than what we can discover in nature. In mathematics (I later argue) we can be sure of finding only ourselves. We can count on no pre-established harmony between our mathematical inventions and the facts of nature.

Consider, for example, the implications of a steady-state cosmology, according to which the average properties of the universe never change with time. As the universe expands, if it does expand, new matter is continuously created and keeps the density of the world constant. Such a cosmology requires that features of the universe be self-propagating. At any given time, the universe can have only those properties that make constant density possible.

A view of this type establishes a secure basis for the idea of timeless and immutable laws of nature. However, it says nothing that goes to the question of why the universe must be so constituted as to possess the quality of self-propagation or even why a very different kind of stuff from the stuff that in fact exists in the universe might not also possess this quality. The only useful way to expound the necessity of the most necessary relations under such a view is to describe the detailed content of the view. As always, the modal categories of necessity and contingency turn out to be nothing more than shorthand, incorporating by reference what we otherwise believe about the workings of nature.

The trouble is that what twentieth-century science discovered about the world suggests that such a cosmology is false. The present standard cosmological model represents the universe to have begun in an explosive formative moment and to undergo a ceaseless transformation. Such an account gives us even more reason to qualify strong claims of necessity about the order and history of the world. From its vantage point, we can describe the structure and evolution of the universe as necessary in some senses and as contingent in others. The sense of their necessity and of their contingency is just the sense implied by the details of our cosmological understanding.

There is no issue of the necessity or the contingency of natural processes that is separate from the explanatory direction taken by cosmology, with one important qualification. The qualification is that cosmology can cast no light on the presuppositions that are codified in what I earlier described as a proto-ontology: that there be something rather than nothing (the postulate of reality), that there be more than one uniform thing or state of affairs (the postulate of plurality), and that the plurality of things or states or affairs be constituted by a web of relations among them defining what they are (the postulate of connection). The primary form of such connection is causality.

The assumptions of plurality and of connection entail that reality not take the form of a single uniform thing, without internal differentiation, and related only to itself. We cannot imagine nature working in a way that fails to conform to these presuppositions without, by that very act, surrendering our ability to investigate and to explain natural processes. However, it is not only our natural constitution that requires us to accept these minimalist assumptions about nature: both science and pre-scientific experience confirm them. If indeed the world is so constituted that it always conforms to the postulates of reality, plurality, and connection, regardless of our powers or infirmities, it is, for all we know, factitious that nature is so constituted. It might have been arranged differently, although, given our constitution, we might then be unable to understand it, even if we were able to exist.

By describing conformity to these postulates as factitious rather than as contingent, I mean to avoid implicit reference to the conception of contingency. Any conception of contingency or necessity is parasitic on a particular understanding of how nature works, or of the universe and its history. What we call necessary relations are just the relations that are most entrenched, and least susceptible to change, in such an understanding. The universe just happens to be what it is.

Once, however, we accept these presuppositions as unavoidable, all issues about necessity and contingency turn into questions concerning the content of our views, especially our cosmological views, about the ways nature works. A version of the now standard cosmology that traces the origins of the universe to a singularity in which temperature, density, and energy achieved infinite dimensions will be one in which the causal sequence reaches a point zero, beyond which we cannot look. It will be able to invoke processes of causal and statistical determination reaching back close to the origins of the universe. However, it will be powerless to represent either the antecedents of the singularity or the transformations that brought nature from infinite to finite values, and thus shaped a universe in which the laws of nature can begin to operate. No matter how aggressive it may be in its affirmation of the conformity of the laws of nature to timeless laws, it will find these laws tainted by a species of contingency resulting from this barrier to the operation of causality.

Causality, however, is much more fundamental to science than is the conceit of timeless laws. Physics can dispense with the latter, although at the cost of abandoning some of the metaphysical gloss to which it has become attached. It cannot dispense with the former.

Consider a cosmology that represents the very early universe as never having left the realm of finite values and as therefore open to a causal connection with a preceding universe or with an earlier state of the present universe. The development of such a cosmology requires a reinterpretation of what we already know about the history of the universe and motivates a distinctive research agenda. It demands that we reject the second cosmological fallacy: we must expand our vision of how nature can work by admitting that some of the ways in which it works allow neither for fixed distinctions among elementary constituents of nature nor for a contrast between laws and phenomena.

It radicalizes and generalizes the idea of natural history to the point of compelling us to admit that the laws of nature form part of this history and may therefore be susceptible to change. It leads to a view of mathematical reasoning that denies to mathematics the privilege of ensuring an inside track to the representation of reality. In all these ways, it requires radical deconstruction and reconstruction of many of our cherished beliefs about both nature and science. However, it need not deny or disregard the chain, backward in time, of empirical inquiry and causal reasoning.

Such a cosmology gives a central role to discontinuous change in a universe that is better called factitious than radically contingent. The advantage of the preferred term is its relative neutrality. The idea of radical contingency is tainted by the suggestion of a disappointment: the failure to live up to some predefined conception of necessity. To affirm radical contingency seems to be to make a broad claim about the world. In fact, it is to make a claim about the gap between two ways of understanding the world: the one that we can defend and the one that some admired and influential approach to the world (such as the approach committed to the Newtonian paradigm and to Laplacean determinism or the block-universe view today) would have us adopt. The concept of necessity is context-bound and theory-related. So, at least derivatively, is the concept of contingency.

To say that the universe is factitious means that it just happens to be the way that it is, for reasons only historical explanation, supplemented by structural explanation, can partly explain. Structure is ancillary to history because it arises out of history rather than the opposite. Historical explanation is ineradicably incomplete because it extends only to a variable but limited portion of both the past and the future.

Subordinating structure to history makes room for a species of facticity: the species that is inherent in the conception of coeval and discontinuous transformation of the laws and of the phenomena, constrained from below by the path dependence of the history of the universe and from above by the relatively greater recalcitrance to change of the fundamental laws or principles of nature. However, it no longer needs to rely on the mystery implied by the notion of the emergence of the universe, and of time itself, out of a singularity that defies understanding by taking refuge in the infinite.

The contrast between a conception of universal history relying on the ideas of a permanent repertory of elementary constituents of nature – particles and fields – as well as out of an immutable framework of natural laws and a view dispensing with these ideas is analogous to the contrast between the idea of a law-governed world, whose basic constituents and laws remain immutable until the end of time and whose rigid conformity to unchanging laws is made all the more mysterious by the factitious quality of its regularities and initial conditions, and the idea of a natural history. In a natural history, time never stops, and the forms and agents of change also change, together with everything else.

Possibility

There is a mistaken notion of the possible that is deeply entrenched in our conventional beliefs. Many philosophical doctrines have embraced and elaborated it. It has long formed part of the background to the interpretation of the discoveries of natural science.

Call the dominant conception, which we need to reject and replace, the spectral idea of possibility. A possible state of affairs is, according to this view, a ghost stalking the world. It is waiting for its cue to come onto the stage of actuality. There is only a relatively short step from this idea to the notion that a possible state of affairs, unrealized in our universe, may be actualized in another universe, with which we have no hope of causal contact.

Whether or not there are many universes, there is, according to this spectral idea of possibility, an outer horizon of all the possibles. In this view or in any view, to understand a state of affairs is to grasp what it can become. A premise of the spectral idea of possibility is that the possible in the light of which we understand the actual is forever fixed and forms part of reality. The more penetrating our insight becomes, the better are we able to understand the accessible transformations of a state of affairs in the light of our insight into the ultimate horizon of the possibilities, drawn, like a wide concentric circle, around actual nature. Laplacean determinism, with its assumption that a mind with godlike insight into the world would see future events in present ones, and the block-universe conception, with its use of the concept of spacetime to conflate past, present, and future, are simply extreme variants of such a view, their limiting cases.

The idea of multiple or possible worlds, both in philosophy and in science, draws on this notion of possibility and in turn lends it support. It also converges with the thesis that mathematics provides us with a favored representation of reality. Under one variant of this conception, mathematics explores the most general structures in all possible worlds and states of affairs. The theses of the singular existence of the universe and of the inclusive reality of time are incompatible with all such approaches. They suggest along which lines we need to reshape our understanding of the possible.

A corollary of the spectral idea of possibility is the denial or diminishment of the reality of the new. The new can be new in only a limited sense if it consists simply in actualizing an already established possibility. It is as if the possible had every attribute of reality except the property of being enacted in this world.

The argument for the solitary existence of the universe and for the encompassing reality of time lays the basis for a different conception of the possible. If there is only this one real world (the present universe and the earlier universes, or states of the universe, from which it may have sprung), it makes little sense to speak of possible worlds. The idea of possible worlds can at best serve as a device – and a dangerously misleading one at that – by which to suggest directions that, at any given time, change in the one real world might take.

The only possibles about which we can speak with some measure of realism and confidence are the proximate possibles: what can happen next; where we can get to from here. The argument for the reality of time and for the historical reconciliation of the stability and the mutability of the laws of nature gives us reason to expect that, the further we move away from the proximate possible, the less reliable our speculative causal conjectures become. We may still make predictions about, for example, the ultimate fate of the universe, based on our understanding of its prior history and of the stable laws now governing the phenomena. However, we have as yet no knowledge of previous universes, or previous states of the universe, or of how they ended. If the now-stable laws of nature may again change, we must contextualize and qualify the force of our predictions.

The possible understood as this variable penumbra of transformative opportunity around all phenomena and states of affairs is the possible about which we can hope to have well-grounded ideas. Our view of possible transformation makes explicit our understanding of present and past actuality. We revise our beliefs about the possible, as we develop our understanding of the actual. We do so either because we discover something that is new to us or because something new in fact happens in the world.

For this reason, in the order of discovery, the possible comes after the actual rather than before it. It is the cumulative product of our encounters with the changes of nature. (Among philosophers, only Henri Bergson had such an understanding of the nature of possibility. It is telling that he developed it as a byproduct of his defense of the reality of time.)

The sense in which the new is new depends on the way in which causality operates. A world that changes in time may nevertheless be one that remains largely in the grip of deterministic or probabilistic forces. The affirmation of the singular existence of the universe and of the inclusive reality of time gives little reason to believe, in the spirit of the feel-good metaphysical systems, that the universe favors our concerns and that its evolution is as open to surprise as are life, culture, and consciousness. Nevertheless, the rejection of the spectral idea of possibility, and of the assumptions on which it relies, overthrows an impediment to the recognition of novelty in nature. It enables us to imagine a world that has room for what never existed before.

6 The selective realism of mathematics

THE PROBLEM

A view of mathematics and of its relation to nature and to science complements the two ideas that are central to this argument: the idea that there is one real universe (as opposed to a multiplicity of universes, of which our universe would represent one) and the idea that time is real and touches everything (with the consequence that everything, including the laws of nature, changes sooner or later). These ideas make trouble for received accounts of mathematics and of its applicability in the scientific study of nature: they cannot be adequately accommodated within any familiar interpretation of the nature of mathematics. They require a different approach.

The core of this conception is that mathematics is an understanding of nature emptying the world out of all particularity and temporality: that is, a view of a world without either individual phenomena or time. It empties the world out of them the better to focus on one aspect of reality: the recurrence of certain ways in which pieces of nature relate to other pieces. Its subject matter are the structured wholes and bundles of relations that outside mathematics we see embodied only in the time-bound particulars of the manifest world.

The distinctiveness of the mathematical perspective – its evisceration of particularity and its suppression of time – helps explain the power of mathematics to illuminate a world in which time holds sway and particularity abounds. This power, nevertheless, perpetually subjects us to a twofold risk.

The first risk is to mistake the mathematical representation of a slice of the one real world – the slice that has to do with bundles of relations and with structured wholes – for privileged, indubitable insight into a separate, nature-transcending realm of mathematical truths. There is no such realm, any more than there is – so far as we have reason to believe – a panoply of possible worlds, in wait for their cue to become actual.

The second risk is that we allow ourselves to be lulled by the effectiveness and beauty of our mathematical propositions into the belief that nature shares in their timelessness. It would do so, most convincingly, by operating under the force of eternal laws and symmetries. Such regularities achieve adequate expression only when they can be represented mathematically. Their susceptibility to mathematical representation confirms, according to this illusion, their claim to participate in the freedom of mathematics from time. It does not. To believe that it does is to spoil the gift of mathematics to physics.

MATHEMATICS AS DISCOVERY AND MATHEMATICS AS INVENTION

If we put aside the technical disputes of the contemporary philosophy of mathematics, we can see that almost all the known options in our philosophical understanding of mathematics fall into two families of ideas.

According to the first family of ideas, mathematics is discovery. It is the progressive (or recollected) discovery of the truths that exist in a domain of mathematical facts uncompromised by the vicissitudes and the variations of the manifest world. Vulgar Platonism (denied in the writings of the real Plato by a variety of approaches to the relation between natural complication and mathematical simplicity) is the limiting case. ("To be a mathematician," wrote David Mumford, "is to be an out-and-out Platonist.") Many, more qualified lines of reasoning move in the same direction.

According to the second family of ideas, mathematics is invention: the free development of a series of conventions of quantitative and spatial reasoning. This conventional practice of analysis may be rule-guided or even rule-bound, but the rules are themselves inventions. There is no closed list of motives for this inventive practice. Some have little or nothing to do with the deployment of mathematical analysis in natural science. Others take this deployment as their goal.

For the discovery theories, there is a fact of the matter in every true mathematical question, although it is not a fact about or in the natural world. For the invention theories, there can be no fact of the matter. We who invented mathematical propositions remain their arbiters.

* * *

Both families of ideas about the nature and applicability of mathematics suffer from a common defect. They fail to account for why mathematics has been as useful to science as it has. They fail to make adequate sense of what Eugene Wigner called "the unreasonable effectiveness of mathematics" in the scientific study of nature.

The discovery views fail to do so by setting up a discontinuity between natural fact and mathematical truth for which they then neglect to provide any bridge. A metaphysics like Plato's would, if it could be believed, build such a bridge. It would do so in the form of an account of how natural phenomena come to share, though dimly and imperfectly, in the fuller, deeper reality of mathematical truths. Ever since Plato, however, most who have been attracted to the discovery approach have balked at any such ontology. In its absence, they must contend with the metaphysical dualism – natural history and particularity on one side, timeless and faceless truth on the other – that their position implies.

The invention views fail to account for the applicability of mathematics in natural science by making its applicability appear to be either a happy accident or an abstract engineering. It would be happy accident if it just turned out – mysteriously – that so many of our mathematical inventions, undertaken for other reasons, supplied what each succeeding wave of theories in physics happened to need. It would be abstract engineering if we could always rely on making up, after the fact, the mathematical theories that our physical conjectures require. If mathematics were so elastic that we could be sure of getting from it, by invention or construction, whatever we need formally to represent our causal ideas about the workings of nature, it would fall under suspicion. The ease with which we could get from it whatever we want would rob it, by the same token, of the power to provide any independent check on our theory-making in natural science.

Neither happy accident nor abstract engineering do justice to the historical interactions between mathematics and physics: as now mathematics, and now science, advance ahead of the other. Neither thesis takes adequate account of the interplay between the two great forces that have influenced the history of mathematics: the internal development of mathematical reasoning (as each breakthrough generates a new set of problems) and the provocation offered by the development of natural science (when the available mathematical tools prove inadequate to the advancement of novel physical theories or to the refinement of disconcerting physical intuitions).

THE ATTRIBUTES OF MATHEMATICS

Here is a view intended to redress the defects of these contrasting conceptions of mathematical discovery and invention. It is also designed to reconcile our understanding of mathematics to the ideas of the singular existence of the world and of the inclusive reality of time.

I proceed by offering an account of the attributes of mathematics, beginning with those that are relatively shallow and uncontroversial and advancing to those that have been much less appreciated. The superficial and familiar features of mathematics turn out to be fully intelligible only in the light of the deeper and more elusive ones.

Once this picture of attributes is complete, it will yield the rudiments of an account of mathematics as a distinctive understanding of the one real world. The power of mathematical reasoning, including its usefulness to natural science, is directly connected with its unavoidable disregard for some of the most basic and pervasive characteristics of nature.

* * *

Among the familiar characteristics of mathematics are explication, recursive reasoning, and fertility in the making of equivalent propositions. They are connected, and they overlap.

Explication is the working out of what is implied in a particular conception of a structured whole or of a bundle of relations: not just any structured whole or bundle of relations, as we shall soon see, but one foreign to the natural experience of time-bound particulars. Once the world is robbed of its flesh – the flesh of the particulars that begin, move, and end in time – and reduced to the skeletal form of its most general characteristics, it continues to have, in this skeletal mode, structure or content.

This content is described by mathematical conceptions. Each mathematical idea refers to a piece of the residual structure; each serves as a summary reference. Explication is the progressive development of the mathematical propositions that are implied in the summary reference and that depict the relevant piece of the world without flesh.

A comparison with what in legal thought has been called conceptualism helps make the point. In the nineteenth century, jurists used to believe that concepts like contract or property have a built-in legal content: they are summary references to entire systems of legal rule and doctrine. This view has long been derided as a primitive scholasticism: the concepts can supposedly yield only what we put into them. It nevertheless made sense in the light of an idea that has exercised vast influence on legal and social thought over the last two centuries.

According to this belief, which we can call the typological idea, there are types, or indivisible systems of social and economic organization, expressed in institutional detail as law. Even if a society can choose which type to establish, the legal-institutional content of each type is predetermined. We reveal the content when we analyze concepts like contract or property. It is much harder to come to terms with the intellectual and political consequences of rejecting the typological idea, as we should, than to dismiss the biases of nineteenth-century legal analysis as a crude fetishism of concepts.

The commitment to explication as a method of legal analysis is misguided. The basic reason is that it treats a social construct as if it were a natural phenomenon. The analogous procedure of explication in mathematics is neither misguided nor dispensable. It does address the natural world. However, it does so with the forceful selectivity of a form of reasoning that has no truck with the particularity and the temporality marking all nature.

A second attribute of mathematics is its reliance on recursive reasoning. Reasoning is recursive when it takes itself for a subject, or, more precisely, when it applies to the procedures that it deploys. Recursive reasoning enables us to pass from enumerations to generalizations; we jump off from the particular to the general by suggesting the general rule implicit in what, up till then, had seemed to be a mere enumeration of particulars. (This is the aspect of recursive reasoning that Pierce called abduction, the better to emphasize its contrast with induction, for which it is commonly mistaken.) It allows us to reach strong and rich conclusions on the basis of weak and parsimonious assumptions.

Recursive mathematical reasoning enables us to develop our insight into structured wholes and bundles of relations indirectly. It does not do so through direct study of nature or even of the presuppositions or implications of theories in natural science. Instead, it proceeds through a generalization of mathematical ideas used to explore more particular bundles and wholes. It is bootstrapping. Nevertheless, it is bootstrapping of an activity that is turned outward to nature, viewed under a particular aspect. Its proximate subject matter is mathematical reasoning itself. Its ulterior subject matter is the eviscerated natural world – eviscerated of time-bound particulars – that mathematics addresses.

A third attribute of mathematics is its fecundity in the statement of equivalent propositions. A major part of mathematical reasoning consists in showing how one line of analysis can be restated in terms of another. The practical importance of this feature of mathematics for natural science is manifest in the vital role that gauge symmetries play in physics and cosmology.

Just as explication can be mistaken for a superstitious conceptualism and recursive reasoning for induction, so can the multiplication of equivalent propositions be mistaken for the marking of synonymy. It is as if we already understood the truth about the aspect of the world under consideration and wanted only to organize better the mathematical language in which to represent this achieved understanding. We would organize the language of mathematics better, according to this misinterpretation of the facility for equivalence, by clarifying which combinations of symbols are and are not synonymous.

The basis and nature of this third trait of mathematics lie in another direction. The abstraction of mathematics exposes it to a danger to which natural science has other antidotes. It is the danger of failing to distinguish the ordered wholes and sets of relations it studies from their conventional expressions. (Those who embrace the view of mathematics as simple invention take succumbing to this danger as their program.) Precisely because in mathematics we lack the manifest, time-bound world to surprise, baffle, and correct us, we must try, at every turn, to distinguish our ideas about nature, abstracted, eviscerated of particulars, from their conventional expressions. The best way to do so is to insist on restating the ideas in equivalent forms – in alternative conventions.

* * *

These three attributes fail adequately to account for the distinctive nature and the unique power of mathematics. Not only are they insufficient to account for mathematics; they are also insufficient to account for themselves.

Our mathematical and logical reasoning has a characteristic that places it in sharp contrast to our causal explanations. A cause comes before an effect. Causal explanations make no sense outside time; causal connections can exist only in time. However, the moves in a mathematical or logical chain of argument do occur outside time. To take a simple example, there is no temporal succession – no real-world before and after – in the relation between the conclusion of a syllogism and its major and minor premises. They are not so much simultaneous, as they are outside the realm of time altogether.

In nature, temporality and particularity are tightly linked. Every particular in the world exists in time. Everything that exists in time is a particular, although the nature of particularity – that is to say, how particulars are distinguished from one another – is itself subject to change in the course of time. The manifest world is a world of particulars as well as a world of time. Even the angels, to be able to intervene in historical time, require distinct personality, despite being said to lack bodies.

That the relations among mathematical propositions exist outside time is a fact consistent with two other truths that might incorrectly be thought to contradict this claim. The first such truth is that mathematical and logical reasoning takes place in time. Its temporal enactment in the minds of time-bound individuals says nothing about its content. The propositional content is one thing; the psychological phenomenon, another.

The second such truth, more remarkable and puzzling than the first, is that a form of discourse that is not temporal can be used to describe movement and change in time. The most striking example is the single most important instance of the applicability of mathematics in natural science: the work that the calculus does in mechanics. The calculus is used – indeed in part it was invented by Newton and Leibniz – to furnish a mathematical representation of movement and change in a configuration space limited by initial conditions (and therefore representing a part of the world rather than the world as a whole) and governed by supposedly immutable laws. The seemingly paradoxical use of non-temporal connections to represent changes in time has ceased to mark physics only insofar as physics despairs of the attempt to explain change and puts structural analysis in the place of causal explanation.

The use of statements about connections outside time to represent phenomena in time is not just an aspect among others of the problem of the "unreasonable effectiveness of mathematics." It is the kernel of that enigma. The enigma has an explanation, in fact two explanations: one, psychological and evolutionary; the other, methodological and metaphysical. Before exploring them, however, I must go further in depicting the war that mathematics wages against time.

Mathematics deals with nature as well as with itself. However, it addresses a nature from which time and, together with time, all phenomenal distinction have been sucked out. The world that it represents is neither the real one nor another one – a domain of timeless mathematical objects. It is the real world – the only one that exists – robbed of time. The subject matter of mathematics is a visionary simulacrum of the one real world. Unlike the real thing, the simulacrum is shadowy and timeless. It is preserved against corruption and change only because it is removed from nature, in which time and particularity rule.

How can the analysis of such a proxy for the world prove so useful to the representation of causal relations in the real thing? I later suggest an evolutionary conjecture about why the mind as a problem-solving device could come to be equipped with a faculty for relating structured wholes and bundles of relations to one another outside of time. It requires no such account, however, to understand the explanatory advantages of such a disregard for natural time and natural distinctions.

We cannot set aside the particularity of phenomena without also depriving them of their temporality: everything in the natural world is sunk in time. However, it is only by disregarding both time and natural distinction that we can deal with relations and combinations in their most general form. We can then more easily form ideas and inferences about these general connections, free from the spell of the embodied, particular, and temporal forms in which we encountered them. We can use such inferences and ideas to explain change and movement in time. They provide natural science with what it could never achieve if its imagination of the possibilities of connection were limited to the forms in which we meet them through our senses or our instruments.

But what exactly is it that we address when we deal with the simulacrum – the world without time or particularity? This is the fourth attribute of mathematics: the subject matter distinctive to a way of thinking that sucks the world dry of time.

A simple but incomplete answer is that we are left with space and with number, or more precisely with the connection between number and space. A more adequate and comprehensive answer is that we are left with the most general relations among parts of the world: structured wholes and bundles of relations. In mathematics, we deal with them in their incorporeal and therefore timeless form. It is then only by exercising a self-critical vigilance that we avoid the pseudo-Platonism that has always been rife among mathematicians: the unwarranted supposition that if we are able to conceive and to reason about them in this abstracted form, and if in reasoning about them we experience a constraint that our axioms and rules of inference seem insufficient to explain, they must be objects of a special type, inhabiting a distinct part of reality. We turn a practice into an ontology.

In so doing, we fail to consider a possibility for which there is a much better basis: that the referents of our mathematical notions are not a distinct realm of objects. They are the very same objects of the one real world, viewed from the vantage point of a special way of thinking, one that is blind to the phenomenal distinctions and the temporal variations of nature.

The temptation to use mathematical analysis and logical reasoning as pretexts for an ontology of mathematical objects is nowhere stronger than in number theory. That the prime numbers are distributed in the manner described by the Riemann hypothesis or that, according to Hardy's result, an infinitude of zeros of the Riemann zeta function lie on the critical line may seem to be facts of the matter as much as any fact about nature. Indeed, they are: they are as much brute features of the world as that water freezes at 32 rather than at 31 or 33 degrees Fahrenheit. It does not follow, however, that the truths that they announce apply to a special category of entities rather than to the same natural world in which 32 degrees Fahrenheit is the freezing point of water.

It turns out that in a world characterized by the postulates of reality, plurality, and connection (enumerated in the proto-ontology of Chapter 4), that is to say a world of discrete but connected phenomena, the relations among the numbers by which we count these phenomena have certain definite and surprising relations. We could, by convention, invent a symbolism and a way of connecting these symbols in which other relations would hold. However, this symbolism would not be the powerful and dangerously seductive mathematics that has made the fortune of physics. It would not refer, even by the starkly selective mechanisms of this mathematics, to the world that we encounter in science as well as in perception. It would have rules and constraints, as games that we invent do, but it would be no more than a game of our invention.

To understand what is at stake in the distinction between this conception and the ideas that it superficially resembles, it is necessary to grasp the relation between number and space, arithmetic and geometry, the two perennial starting points of mathematics, as Leibniz saw. Space and number are intimately related. Our ideas about number are ideas about the one real world, in just the same sense as are our ideas about space. Moreover, each of the two sets of ideas is parasitic upon the other.

The world as a differentiated structure exists in spatial extension. Its spatial extension is the condition of its variation. There is not just one thing; there is a plurality, although the nature of this plurality may change: not just the range of difference that exists but also the ways in which different parts of the world differ from one another as well as the ways in which they turn into something different from what they were before.

The variable complexity of the world turns out to be decisive for the reality and the nature of time. One way to understand time – I have argued – is to think of it as the transformation of transformation. Time is real only insofar as change is uneven. It cannot be uneven if there is just one thing – at least with respect to change – rather than there being many things. It is space, or spatial extension, that allows for plurality of being and consequently for variability of change.

As soon as there are space and complexity, there is also reason to count or to quantify. That is the ultimate ground of number and the substance of its relation to space.

We do not count objects in another world. We count them in the only world that there is. In this world, their numerical relations and their spatial disposition are two sides of the same thing. The correspondence between geometry and algebra is not an accidental invention of early modern mathematics; it reveals an important truth about nature.

The numerical relations that we then discover, such as the facts about the distribution of prime numbers, are not facts about a realm of mathematical objects. They are facts about the only world there is, although they are, so to speak, second-order facts. They are not facts about the interactions among the natural phenomena that make up nature. They are facts about the second-order or higher-level relations among the ways of numbering, counting, or quantifying that are suggested to us by such interactions.

They are neither facts about a world other than ours, nor made-up, free constructions. They are facts about this one real world, although facts about what this world looks like to us at one order of remove: the order of remove established by the special perspective of mathematics. This order of remove does not save them from having the same startling, it-just-happens-to-be-that-way character of ordinary natural facts.

An implication of this view is that there is a basic asymmetry in the relation of mathematics to space and to time. Its relation to space is intimate and internally connected with number. Its relation to time is distant and external, even when movement and change in time are (as in the original applications of the calculus) the very subject that the mathematical reasoning is used to represent.

The history of twentieth-century physics shows the influence of this asymmetry. Even when it proposed to connect space and time and to deny that spacetime could be understood as an absolute and invariant background to physical phenomena, twentieth-century physics spatialized time rather than temporalizing space. The use of a spatial metaphor to describe time as a "fourth dimension" is a crude manifestation of this bias. Its most significant expression, however, is persistent equivocation about the reality of time (unaccompanied by any such wavering about space). One of the results was and is the reaffirmation, mostly unthinking but sometimes considered, of the idea of a timeless framework of natural laws underwriting our causal explanations of nature.

* * *

Mathematics is deficient in the imagination of time and of temporal continuity as well as of phenomenal particularity, realities central to nature and therefore to natural science. It has, however, excelled in representing the infinite. The taming of the infinite was one of the greatest achievements of nineteenth-century mathematics. Ever since, a view seeing mathematics as the oracle of nature and the prophet of science, and seeing science, including cosmology, as applied mathematics, looks for the mathematical infinite in nature. It cannot find there the infinite – not at least the infinite for which it is looking.

The absence of any counterpart to the mathematical infinite in nature signals the unyielding limits to any parallel between mathematics and the natural world. Nature is prodigal in what mathematics has trouble representing: singular events taking place in the unbroken flow of time. Mathematics offers, as one of its supreme accomplishments, the representation of what nature fails to allow: the infinite.

The infinite is a mathematical idea, the mathematical representation of the unlimited or the unbounded, when it is not a theological or metaphysical conception, or a physical notion with no clear counterpart in mathematics. As many have argued (including thinkers as different as Aquinas, Hume, and Hilbert), the mathematical notion of the infinite has no presence in nature. The exclusion of the mathematical infinite from natural science gains additional significance in the context of the cosmological view presented in this book.

Criticism of the accommodation of the infinite in natural science traditionally begins with the remark that no one has ever observed in nature an example of the infinite, as mathematics understands it. The infinite is invoked by loose and illegitimate analogy to the indefinitely large in number, in space, or in time. This invocation often serves the aim of representing, or concealing, a limit to the applicability of a physical theory. Such is the case with the inference of an infinite initial singularity from the field equations of general relativity. It is the significance, as well, of the use of the infinite in quantum mechanics. There, it is used to make up for the failure to explain what actually happens, as opposed to what might happen, at the Planck scale.

This conventional criticism fails to reach the root objection to admitting the infinite into science. Everything that exists in nature, including the universe and all of its phenomena and events, results from other events and phenomena in time. Everything, as Anaximander wrote, turns into everything else, under the dominion of time.

How then could the infinite come to exist, given what we see and know of the workings of nature? The universe may be indefinitely large, and some of its rudiments indefinitely small. Its history may extend indefinitely back into the past and far into the future. There is, nevertheless, an infinite difference between indefinite largeness or smallness and infinity, or between indefinite longevity and eternity, which is infinity in time.

No natural event analogous to any process that we observe in nature could jump the gap between indefinite largeness or longevity and infinity or eternity. The universe can no more become infinitely large or infinitely small by becoming larger or smaller than it can become eternal by growing older. Consequently, the infinite could exist only if it always existed.

The world could not have become spatially infinite just as it could not have become eternal. It would have to have been spatially infinite or eternal or both, always. Nothing within the world would be infinite or eternal. However, the world itself would be eternal and infinite.

We fall into confusion when we equivocate about the relation between the idea of the universe, of the succession of universes or of states of the one real universe, and the concept of being or reality. Where or when the universe stops, we imagine that being, or some potential for it, must continue to exist. We struggle to conceive nothing, and nothing easily becomes for us a shadow or a signature of being.

If, for example, the universe is reduced, at certain times in its history, to an unstable vacuum field, it remains not nothing. It continues to be something. This something has a history, extending further into the past and into the future than we can grasp: before the 13.8 billion years to which we now date the present universe or its present state and beyond its conjectured future heat death and imagined contraction and rebirth.

For science, the universe and nature are one and the same thing (if we allow for the complication introduced by the existence of bubble or domain universes). For science, the whole of the universe, observed and unobserved, and of its history, known and unknown, is the world. Nothing can be outside them. We may nevertheless think that something is outside them, at least a potential for being and even for a particular kind of being. This way of thinking may seduce us into believing that the world must be spatially and temporally infinite and that the infinite must be conceived in mathematics only because it is also present in nature.

Science, however, is powerless to solve the enigma of being. It cannot legitimately dispel the factitiousness of the universe and of its history – their happening to be one way rather than another – by inferring their attributes from supra-empirical constraints as rationalist metaphysicians, such as Spinoza and Leibniz, claimed to have done. It cannot, by its own devices, pursue either being or becoming beyond manifest nature. It cannot represent manifest nature to form part of a larger narrative or a more fundamental regime, to which, inspired by mathematics, it then assigns the attributes of infinity and eternity. Neither, however, does science have any basis on which to trump or prohibit other channels of experience and insight by which, outside science, we may try to push our luck against the mystery of being.

The self-denying ordinance that is the source of its power provides no license to impose that ordinance on the whole of experience. Scientism is not science. It should have no place in the temporal naturalism espoused in this book.

The postulates of being (that there is something rather than nothing), of plurality (that there is more than one thing), and of connection (that these plural things interact) – which figure in the proto-ontology or the anti-ontology sketched earlier – describe minimalist working assumptions of science. These working assumptions say nothing, one way or another, about the infinity or eternity of the universe.

Once we have overcome these confusions, we can acknowledge with greater clarity the objections to claims of eternity and infinity in nature. The problem in supposing the world to be infinite or eternal, or both, is not just that we could never know that the world is infinite or eternal, given the infinite difference between indefinite largeness or longevity and infinity or eternity. The problem is also that the overall character of nature would be at odds with nature as we encounter it piecemeal, through science as well as through perception. We may, unjustifiably, feel driven to this extremity by the difficulty of accepting, or even conceiving, the apparent alternative to it: that the succession of universes, or of states of the universe, is temporally finite even if geometry now allows more readily to understand how it may be spatially finite. (The unacceptability of these contrasting and seemingly exhaustive options was the quandary that Kant sought to capture in his "first cosmological antinomy" in an epoch in which the spatial finitude of the universe was as hard to grasp as its temporal finitude.)

There is, however, another, unassuming alternative, which we reject only when we rebel within science against the limits of science and try to turn it into something that it cannot become. According to this alternative, the one real universe is indefinitely large and the succession of universes, or of states of the universe, is indefinitely old. Nothing is gained for science, and much is lost, by importing the mathematical idea of the infinite and by using it to disguise ignorance as pseudo-science.

The preceding argument about the finitude of natural processes deals with only one conception of the infinite: the view of the infinite as the unlimited or the unbounded that is suggested by mathematics. This view is far from being the only one. On other understandings of the infinite, the infinite may exist in nature. The natural presence of these other versions of infinity, however, rather than reinforcing the parallels between mathematics and nature, highlights the distance between them.

Consider, for example, Aristotle's notion of the infinite as that which cannot be traversed, like a "treacherous river" or a "uniform circular racecourse." The infinite, on this view, is that which always has some part outside it, as distinguished from that which has no part outside it. It is that which is never completely present because it exists in time and goes on. In Aristotle's vocabulary, inseparable from recognition of the reality of time and informed by his imagination as a metaphysician who was also a biologist, the infinite always exists potentially, never actually, and over time. As applied to temporal events, it refers to their incompleteness and open-endedness, and therefore requires an idea of indefinite continuance rather than of eternity.

If by the infinite we mean some such feature of nature, then the infinite does indeed exist in nature, at least in nature as it is represented in a view like the one that we develop in this book. The universe may be infinite in this sense because, like Aristotle's treacherous river, it cannot be crossed. This infinite, however, far from being an import from mathematics, is incapable of full representation, if only because of its temporality, in mathematics. It is a physical idea that can be translated only in part into mathematical language.

* * *

On the view defined by the preceding discussion of its attributes, mathematics is neither invention nor discovery, not at least in the senses in which mathematical invention and discovery have traditionally been understood. The conceptions of mathematics as discovery and as invention each contain elements of truth. Nevertheless, an adequate account of mathematics cannot be developed as a combination of these elements. It requires a different point of view: one that supports, and receives support from, the ideas of the singular existence of the universe and of the inclusive reality of time.

Consider first the aspects of truth in the notion of mathematics as invention. There is no realm of entities outside nature that mathematics could have for a subject matter. Nor can its subject matter be simply our faculties and practices of reasoning and inference; they are too large and indistinct to mark out a specific approach to reality.

We can narrow the field further by saying that mathematics is reasoning and inference with respect to number and space and, more generally, to structured wholes and bundles of relations. Even then, however, we have left much too much room to distinguish the mathematical from the non-mathematical. If, on the other hand, we insist that the reasoning with respect to number and space be deductive and axiomatic, we have gone too far; there is much in mathematics that is not deductive and an even greater part that is not axiomatic.

By exclusion and in desperation, we may be driven to conclude that mathematics is defined by its history as well as by its focus on only the most general and abstract parts of reality. Taken together, the focus and the history leave a vast margin for maneuver and innovation. The availability of that margin is the truth in the view of mathematics as invention.

Consider now the pieces of truth in the notion of mathematics as discovery. Mathematics does have a subject matter. Its subject matter is not simply itself, despite the preponderant role that what I later call internal development has played in the history of mathematics. Its subject matter consists in the most general and abstract features of the world: the ways that the relations among discrete entities are bundled into larger wholes or unbundled into discrete and therefore countable fragments. Mathematics achieves this focus by abstracting, in the formulation of its concepts and in the way of connecting them to one another through chains of inference, from both phenomenal distinction and time. Such abstraction may make mathematics more rather than less useful to the representation of events that are both temporal and phenomenally distinct.

The second element of truth in the notion of mathematics as discovery is that branches of mathematics do not develop conjecture by conjecture, theorem by theorem, or inference by inference. They develop discontinuously; each new field or approach, like a new style in the history of music, comes laden with a certain structure: not just a distinct set of conceptual tools but also a range of combinations and applications. The content of this structure does not become evident instantaneously; it takes time to be grasped and described. To explicate and develop it either through its application to its own problems and propositions, or through its application to problems in natural science, is the work of the mathematician when he is not inaugurating a new area of mathematics.

That a mathematician must submit to the discipline of an established branch of mathematics with which he engages may seem no more than a truism. That the creators of a branch of mathematics find that their creation comes with its own intrinsic properties, potential, and constraints, and even its own repertory of inferential moves, is much more surprising but no less true. When this fact is then combined with the distinction between time-bound causal explanations and timeless mathematical connections, it is easy to fool ourselves into thinking that mathematics is the prospecting of a realm of being outside tangible and visible nature. Recognition of the rigid and intrinsic properties of each field of mathematics then becomes an invitation to vulgar Platonism.

In this respect, mathematics is no different from natural science: any large theory is distinguished by its research agenda and structural possibilities and limitations, all of which become manifest only over time, in the course of inquiry. It is a mistake to attribute this feature of mathematics to the role of axiom-based thinking. Mathematics can never be fully rendered as an axiomatic system, and axiomatic reasoning plays only a modest and marginal role in its endeavors. The subordination of mathematics to axiomatic reasoning would, on the contrary, result in the effacement of the distinctiveness of different mathematical disciplines and in their reduction to standard logic and deductive inference.

A third fragment of truth in the view of mathematics as discovery qualifies and complicates the second fragment. The differences among parts of mathematics are real, and the content of mathematics is rich, precisely because mathematics is always more than a formal axiomatic system in the making. It is a way of thinking about reality, and of representing it, from a unique perspective: the perspective of forgetting about the particularity of the phenomena and their placement in the flow of time. It is not deduction (much less axiomatic reasoning); it is the use to which we put all our forms of inference – including deduction, abduction, and induction – when we are disposed to accept the limits and the promise of such a singular austerity of vision. In this sense, mathematics is discovery of the world, when we view the world in a certain way. This way is useful to the development of our causal explanations precisely because it is not, and cannot be, causal. It cannot be causal because it abjures time and phenomenal particularity alike.

It may seem at first that this account of the elements of truth in the ideas of mathematics as invention and as discovery moves in the direction of a synthesis, or a half-way house, between the two ideas. It does so, however, only by relying on a conception that is alien to the traditions of thought from which each of those approaches emerged. This conception so radically alters the meaning of both invention and discovery that there is as much reason to view the outcome as a repudiation of both the ideas as to see it as their marriage.

If mathematics exhibited only the first three of the four attributes that I earlier enumerated – its pervasive practice of explication, its devotion to recursive reasoning, and its fertility in equivalent propositions – we might be justified in taking something from the view of mathematics as discovery and something from the view of mathematics as invention, and in using each to make up for the inadequacy of the other. In this undertaking, however, we would have missed the most decisive and unique feature of mathematics, and the one by virtue of which it can be neither invention nor discovery, as they are conventionally understood. This feature – the fourth attribute of mathematics – is the study of a simulacrum of the world, of the only world that there is: a version from which the flesh and flow of things have been wiped away.

* * *

The view invoking these four attributes of mathematics helps explain and resist the two temptations to which our mathematical capabilities subject us. They form part of the price that we pay for these powers.

The first temptation is the temptation to imagine that our faculties of mathematical abstraction give us access to a doubly privileged form of insight. It is privileged by virtue of addressing a realm of timeless truth, distinct from the natural world in which we move. It is also privileged as the result of enjoying a species of certainty for which we dare not hope in the practice of natural science. In this light, mathematics begins to seem not just like higher insight into this our world but also like insight into a higher world.

The second temptation is the temptation to deny or to discount the reality of time, given that the simulacrum of the world with which mathematics presents us is timeless. Our mathematical and logical reasoning is a fifth column within the mind, working against recognition of the inclusive reality of time. In the conflict between what nature seems to be – steeped in time as well as endlessly varied – and what mathematics appears to say about reality, we may be seduced into siding with mathematics against nature.

The two temptations threaten us with different degrees of illusion: the second more dangerous than the first. The relation between them is asymmetrical. We can succumb to the first without giving in to the second. However, it is difficult to surrender to the latter without having accepted some of the former, for reasons that are conceptual as well as psychological. We are led to diminish or even to reject the reality of time by the conviction that the most reliable truths, the truths of mathematics, are timeless.

The preceding argument gives reasons to reject both temptations. The refusal of the two temptations helps create a basis for a deflationary and naturalistic approach to mathematics.

A NATURAL-EVOLUTIONARY CONJECTURE

The removal from time and from phenomenal distinction that lies at the center of mathematical reasoning need not be viewed as a miraculous flight from reality. It may be capable of explanation in straightforward naturalistic terms. One such account appeals to natural evolution.

I do not hope to show that this account is true. I present it only to show how in principle one might make sense of this exceptional and decisive feature of mathematics without contradicting our present understanding of ourselves.

The mind is a problem-solving device, but it is more than a machine. One of the most basic ways in which we distinguish ourselves from other animals lies in the manner in which we can use our minds to deliberate on courses of action in circumstances requiring initiative. We can compare different ways of intervening in a circumstance and of changing it through our intervention or at least of escaping its dangers.

The extent of our power to solve problems, in the service of transformative or evasive action, depends in part on the range of our ideas about the ways in which pieces of the world can combine and recombine. It is a power that may be strengthened when such ideas cease closely to track the apparent structure of the circumstance. This power may be enhanced even more if these views of how parts of nature can connect outreach any conclusion at which we might arrive by induction from past experience.

At one limit, of constraint on our ideas of connections, lies whatever, by way of tacit thinking, is implicit in the behavior that used to be called instinctive. At the opposite extreme, of the broadening of our ideas of connection, is our capacity for mathematical and logical reasoning, with its radical disengagement of the relations among parts of the world from their material and temporal embodiment.

The apparent exception to the acknowledgment of time and phenomenal variation vastly increases our problem-solving capability. The power of mathematical and logical abstraction is useful to our abilities for circumstantial problem-solving. It enables us to form, to develop, and to deploy our most general ideas about structured wholes and bundles of relations beyond the confines of any particular circumstance.

The link between situated, action-oriented problem-solving and our abstract relational thinking recurs in science. For science, the counterpart to the practical imperatives and restraints of action is the central role of the scientist's instruments: the tools by which he surpasses the limits of the senses to investigate nature and to subject it to his experiments. The tools extend the reach of the senses, but only up to a point. All they can deliver is a fragment of the world, within a circumscribed horizon.

Mathematical invention and inference enhance and expand the stock of our ideas about how pieces of the world connect, and do so in ways susceptible to numerical and spatial representation. They rescue the instrument-conditioned practice of science from its narrowness of view.

The precondition for such rescue is the existence of a two-sided mind. To excel at problem-solving, the mind cannot be merely a modular contraption, operating formulaically. Two sides of the mind must coexist. In one side, the mind is modular: it is made up of discrete components, with specialized functions. It is also formulaic: it operates repetitively, as if according to formula and therefore as a machine. In another side, however, the mind is non-formulaic. It can conceive or discover more than can be prospectively generated or countenanced by any formula or system of formulas, an advantage enabled but not explained by the plasticity of the brain. It exercises a faculty of recursive reasoning: an ability to combine everything with everything else. It enjoys a power of negative capability: a power to defy and outreach its own settled methods and presuppositions. This other side of the mind is what we call imagination.

Our problem-solving abilities are at their strongest when we are able to think more than any pre-existing formula can accommodate, and then retrospectively come up with formulas to make sense of our formula-breaking discoveries. To this end, we have to be able to develop new ways of understanding, of explaining, of seeing what stands before us, in the scene of imminent action, as an ordered whole or as a set of relations. When, in natural science, we extend the reach of our senses by using our scientific instruments, we expand the setting of this dialectic between problem-solving and mathematical or logical abstraction. We do not, however, change the character of the dialectic.

THE HISTORY OF MATHEMATICS RECONSIDERED: SOARING ABOVE THE WORLD WITHOUT ESCAPING IT

The history of mathematics over the last century and a half inspires directions for developing this view of the attributes of mathematical reasoning and of its relation to natural science. At the same time, the view suggests an understanding of this history that is at odds with entrenched preconceptions about the role of mathematics in natural science.

Two overriding tendencies have characterized the advance of modern mathematics. The first tendency has been movement progressively away from perceptual experience and common-sense reasoning. This movement has occurred in both the main registers of mathematical insight: space and number. With respect to space, its chief manifestation has been the development of non-Euclidean geometries. With regard to number, its expressions have been countless. None, however, has been more subversive of common-sense prejudice than the mathematical techniques for taming the idea of the infinite, or more far-reaching in its scope and implications than the theory of sets. Antinomies in set theory led to the crisis, which came to a head in the 1920s, in thinking about the foundations of mathematics.

To understand the significance of this movement beyond the limits of the manifest world is to deepen the point of the natural-evolutionary conjecture previously stated. The relational ideas that form the subject matter of mathematics are initially inspired by our experience of the reality that we encounter through the senses. Those ideas present the spatial and numerical realities of that world abstracted from their embodied, individualized, and time-bound expression in nature. Once abstracted, they then become the starting points for the internal development of mathematics.

In principle, our ideas about space and number, and more generally about structured wholes and bundles of relations, can expand in two distinct ways. One way is by the direct analogical amplification of our interpreted perceptions. Around our perceptual experience, we develop, by the combination of analogy and abduction, a periphery of broader ideas about the connected realities and possibilities of quantification and spatial disposition. The instruments around which we organize our practices of natural science extend the reach of our senses, and thus push forward the point at which the work of abstraction and abduction begins.

The second way in which the ideas develop accounts for a far greater part of modern mathematics: the indirect analogical amplification of our perceptual experience. The proximate source of analogy and abduction becomes mathematics itself rather than the perceptual world lying behind it. The raw material of the amplifying activity lies, in the first instance, in the established body of mathematical ideas and procedures and only at a second remove in the reality that we perceive through the unaided senses or the senses enhanced by our instruments. (The truth in Brouwer's "intuitionism" is to recognize that mathematical insight ultimately must be insight into givens that just happen to be one way rather than another. The mistake is to fail to acknowledge that these givens are those of nature, grasped either directly or indirectly, rather than simply of our mental experience.)

As mathematics progresses, the indirect reference to the natural world comes to overpower the direct one. Mathematical reasoning leaves perceptual experience, even as augmented by the technology with which we equip it, back in the dust. (This overtaking of perception misled Kant into defining mathematics as the form of inquiry in which rational concepts are "constructed.")

The divergence of mathematics from perceptual experience, which has become the most striking feature of its history, suggests an objection to the view of mathematics presented here. According to this objection, we cannot plausibly believe mathematics to concern the one real, time-drenched world, albeit from a perspective foreign to both time and phenomenal particularity. We cannot because the preponderant element in the history of mathematics has been internal to its own work: one set of mathematical ideas and problems has generated another. One branch of mathematics inspires the creation of another, either as the result of questions that it poses to itself but is unable to answer or by the force of analogies that it arouses in the imagination of its students. If mathematics is not about an independent realm of mathematical objects, it must therefore, according to this line of argument, be chiefly about itself.

The ordinary activity of the mathematician seems to confirm this interpretation of the history of mathematics. Although he may often find inspiration in the inquiries of natural science, his most fundamental and persistent experience is likely to be that of an autonomous development of mathematical ideas: independent of any engagement with the natural world. The incitement to mathematical innovation provided by science may, on this view, serve only to enhance a force of development that comes chiefly from within mathematics.

The lessons of the history of mathematics do not, however, undermine the thesis that mathematics explores a counterfeit version of the world, seen in its most general and abstract features, and robbed, to this end, of time and phenomenal particularity. Those lessons qualify and deepen this thesis. They teach us how the power of mathematics arises from its paradoxical combination of connection and disconnection to nature. Once understood, they begin to dispel the air of mystery surrounding the uses of mathematics in science. Rather than suggesting that the truths explored by science are prefigured in mathematics, whether the mathematics that we already know or the one that we may develop, a reading of the history of mathematics suggests that there is no such guaranteed convergence between scientific discoveries and mathematical insights. We are not entitled to expect either that any given piece of mathematics will be scientifically useful or that any given scientific discovery will be best represented in mathematical language.

The two decisive forces in the history of mathematics have long been its internal development and its engagement with natural science. Consider them first separately and then in concert.

By internal development I mean the power of mathematical reasoning to inspire an agenda of problems and innovations on the basis of its own resources, without further regard to the representation of any part of nature. The key point in this expression lies in the "further." Mathematics begins, historically and psychologically, as an account of the most general and abstract features of nature. The spatial disposition of distinct things in the world becomes geometry. The plurality of countable things in the world becomes arithmetic. The faceless and timeless world of early or primitive mathematics reveals the one real world under an austere dispensation. It forms a core repertory of ways of representing structured wholes and bundles of relations. The distinctiveness of this way of considering the world is not limited to its selectivity. It consists also and above all in a feature that mathematics shares with logic: the timeless character of the relations among its propositions.

The elements of this repertory can then be varied and extended. Each successive variation or extension can take place without any one-to-one relation to any branch of science. Having begun in a selective representation of nature, mathematics takes off; it develops its toolbox as if nature were no longer the object. Its original abstraction and selectivity make it all the more useful to the investigation of the world, by providing us with ideas about structured wholes and bundles of relations that are not confined to any particular setting of perception or of action. The subsequent history of mathematics, under the impulse of internal development, reaffirms the principle by which mathematics increases its power to inform our understanding of nature by distancing itself from any particular understanding of how nature works or of what it is.

Sometimes the internal development of mathematics is no more than recursive reasoning at work: the assumptions or axioms of one branch of mathematics become the subject matter of another. The methods of reasoning deployed in one branch are redefined as a special case of a more general domain of inquiry. At other times a field of mathematics generates enigmas that cannot be solved without recourse to new conceptions and methods. There is then a more radical break with established ideas and methods rather than a continuous, progressive movement to greater depth or generality of vision. At yet other times, neither the conversion of presuppositions into problems nor the struggle to reckon with a riddle is the driving force of internal development. One set of mathematical ideas and methods suggests, by analogy, another. Seemingly disinterested in the world, mathematics then finds sufficient inspiration in itself.

The condition of the internal development of mathematics is that mathematics rob experience of some of its most basic and pervasive features: its temporality and particularity. However, the condition is also that mathematical reasoning retain a connection to the world. Without such a connection, mathematics would become a fantasy and a hallucination. The thesis that it explores an independent realm of timeless objects would provide this hallucination with a specious metaphysical basis. The thesis that mathematics has itself for a subject would describe the hallucination, reassuringly, as self-reference.

Mathematics, however, is not a hallucination. It remains connected to the world – the only world – through engagement with natural science as well as through its origins in abstraction and selection from perceptual experience. Science now comes, increasingly, to occupy the place of perception as the bridge between mathematics and nature.

Sometimes science offers mathematics the challenge of finding a way to represent a discovery that will make patent certain symmetries and connections latent in nature. Mathematics then responds to science. At other times, the mathematical ideas that can serve such a purpose are already available, although unapplied or applied to other problems.

This two-way relation between natural science and mathematics is so central a trait of the histories of both mathematics and science that it elicits an impression of mystery: the mystery of the "unreasonable effectiveness of mathematics" in science. This mystery in turn strengthens the attractions of vulgar Platonism. It reinforces the conviction that mathematics must be about an independent realm of being, if it is not simply the explication of a game that, once invented, is found to have more content than its inventors supposed it to possess.

In every instance, however, what appears to be an astonishing and strange example of the convergence of a physical phenomenon or process with a mathematical equation turns out to have an independent physical explanation. That the process or phenomenon lends itself to mathematical representation at all is the result of two circumstances, which operate as unspoken presuppositions of its mathematical portrayal.

The first circumstance is the character of the mature, cooled-down universe, invested with the attributes that I earlier described: a defined structure of division into natural kinds (all the way down to the subject matter of particle physics and to its extension in the periodic table); a distinction between recurrent states of affairs and the laws of nature governing them (so that causal connections appear to be the consequences of the laws when, from a wider perspective, they are in fact primitive features of natural reality); and a diminished susceptibility to transformation, or a restraint on the range of the next steps (which physics has sometimes described in the language of "degrees of freedom"). This singular universe happens to be extraordinarily homogeneous and isotropic. As a result, it creates a setting hospitable to repeated phenomena and regular connections.

The second circumstance is the limited importance to the phenomenon subject to mathematical representation of the features of reality from which mathematics, by its very nature, abstracts: time and phenomenal particularity. This abstraction exacts a greater price in some branches of science than it does in others. (I later discuss the implications of this variance for an understanding of the relation of mathematics to nature and to science.)

That mathematics can be useful to scientific inquiry at all depends on the combination of these two circumstances. That one mathematical connection rather than another strikes one of nature's chords, however, always has physical reasons: reasons that can be formulated in non-mathematical language. Mathematics may help suggest the physical picture. It can never, on its own, establish or even imagine such truth. We do not overcome the limitations of a scientific insight simply by giving it mathematical expression.

In the history of physics, no example of the supposedly preternatural power of mathematics to lift the veil of nature is more striking than Newton's inverse square law of gravitation, according to which the gravitational force connecting two bodies varies in direct proportion to the product of their inertial masses and in inverse proportion to the square of the distance between then. Why the square of the distance rather than some other, less simple and pleasing measure? Why the neat and disconcerting symmetry? And why does an inverse square rule apply to a number of physical phenomena other than gravitation?

Newton's inverse square law accords with a visual representation that we have independent, physical reasons to take as accurate, appealing to conserved lines of force, or flux lines. Imagine the gravitational field of the Sun to be represented by lines fanning outward from the center of the Sun, always in the radial direction. Suppose further the existence of a planet moving in a circular orbit at a distance d from the Sun's center.

Let us assume that the force felt by the planet is proportional to the density of the field lines at the distance d. The density falls off as $1/d^2$. It falls off at this rate because the number of field lines intersecting a sphere of any radius around the Sun remains constant. The density, which is the number divided by the area of the sphere, diminishes according to this area, which is d^2. Consequently, the force weakens at the measure of $1/d^2$. Only if the force falls off according to the rule 1/distance-squared will the picture of lines of force that I earlier invoked hold.

An instance of the same way of thinking, not directly related to gravity, suggests the generality of the spatial reasons for its wide application. Compute the dot product of the force field with the unit normal to the surface and integrate the resulting function over the surface. The result is proportional to the mass of the interior bounded by the surface. This relation recurs no matter how the mass is distributed.

The same mathematics works for the electric force, with electric charge in the place of mass. It works as well in a number of other contexts, for reasons that have to do with the regular and recurrent spatial disposition of natural forces in the established universe.

The less we grasp the non-mathematical reasons for the application of mathematics (and in each of these examples we understand them only very incompletely), the more enigmatic and disconcerting the application of mathematics will appear to be. We will be tempted to bow down to mathematics as the custodian of nature's secrets. That the laws of parts of nature are written as mathematical equations then bewitches us into believing that all the workings of all of nature are foreshadowed in the truths of mathematics.

A study of the history of mathematics, properly understood, can help save us from these illusions. Having departed initially from our direct experience of the world through its extirpation of time and phenomenal particularity, mathematics moves yet farther away from it thanks to the overriding influence of internal development. Only its engagement with science qualifies this influence, as science comes to play for mathematics the role that our sensible experience of nature once performed. Nevertheless, by the paradox that characterizes the entire history of mathematics, this removal from the world of the senses only increases the power of mathematics to suggest new ways of thinking about how parts of reality may connect.

As a result of the sovereignty of internal development, the relation of mathematics to nature, now mediated through science, becomes so indirect that whole branches of mathematics may shift, or even reverse, their use in the mathematical representation of natural reality. For example, algebra, culminating in set theory, was traditionally understood as the mathematical investigation of structure; analysis, beginning in calculus, as the methodological instrument for the depiction of change. Yet, in the dealings between mathematics and the science of the recent past, this allocation of roles has been largely inverted. Analysis has been married to geometry, as in harmonic analysis. Quantum theory has depicted change by linear transformations on Hilbert space.

It follows from this interpretation of the history of mathematics that any given mathematical construction will have no assured application to the one real world. The price that mathematics pays for the enhancement of its power through internal development is the loss of any guarantee that its ideas will find application in the study of nature. Some will, and some will not. Having fallen short of the world (through its exclusion of time and particularity), its ideas will also overshoot the world. They will be both too little and too much to hold a mirror to reality. We shall have to dispense with the hope of a pre-established harmony between nature and mathematics.

* * *

Consider now the reverse question. Will every part of nature be susceptible to mathematical representation? Once again, the answer is negative. The greater the salience of time and phenomenal particularity, the more change changes, and the more richly defined and individuated the natural kinds and their individual embodiments populating the aspect of the world that we study, the more limited the use of mathematics will be.

Imagine a spectrum from the domains in which we should expect mathematics to play the largest role to those in which we can expect it to play the smallest. Each place in the spectrum is defined by the marriage of a part of nature to a practice of explanation. This marriage may be dissolved from time to time. The substitution of one explanatory practice for another, in the study of a certain part of nature, is likely to result in a displacement on this spectrum.

The easiest case for the widest application of mathematics is the tradition of physics that Galileo and Newton began: in particular, this tradition insofar as it conforms to what in this book we call the Newtonian paradigm. The objects of this science are recurrent phenomena governed by time-reversible laws. The particularity of the explained phenomena is reduced to a minimum: they are a small number of forces operating among entities whose distinctive characteristics are disregarded except with respect to variations of mass, energy, and charges. This science finds a way to deal with change locally that leaves both the observer and the laws governing change safely outside the domain of the phenomena to be explained. Movement along a trajectory, under the aegis of timeless laws, is then all that is left of time. The stipulated initial conditions defining the configuration space of the phenomena to be explained are deprived of a history. They wait to be promoted from stipulation to explained subject matter in some other application of the same practice.

That this tradition of science should have been considered, for much of the history of modern science, the gold standard of the scientific method helps account for the overpowering authority of mathematics not just as the language in which the laws of nature are written but also as the form of thought that is alone capable of prophesying what they are. Even the minor tradition within physics that leads from early thermodynamics and hydrodynamics to kinetic theory, and then from this theory to a revision of thermodynamic and hydrodynamic ideas, limits the sovereignty of mathematics. It does it only by invoking path-dependent and irreversible (or imperfectly reversible) processes such as those of entropy. Yet this restraint on mathematics also has a mathematical expression, given that, at the microscopic level, entropy can be defined as the logarithm of the number of accessible states.

Even in this its strongest field of application – the representation of local interactions among elementary constituents of nature – mathematics exhibits a revealing weakness. It struggles, and fails, to do justice to the continuum: the aspect of nature that has to do with continuous flow and that resists division into discrete elements or "betweens." Here I allude to the continuum as a feature of nature, characterized by undivided and indivisible flow, rather than as a mathematical concept, traditionally represented as the real number line. Thus, Brouwer (in his inaugural lecture at the University of Amsterdam in 1912) referred to "... the intuition of the linear continuum, which is not exhaustible by the interposition of new units and which can therefore never be thought of as mere collection of units." To do justice to this pre-mathematical intuition would require, however, what mathematics is unable to give. Its origin in counting stands as an insuperable obstacle.

The part of nature that displays most completely the attributes of the continuum is the passage of time. Movement in space exhibits them as well, if less starkly, given the different ways in which such movement may be understood and represented. The powerlessness of mathematics adequately to represent undivided flow is yet another aspect of its antipathy to time. It is, moreover, a limitation revealing the limitations of mathematics as a window on the central subject matter of physics.

The difficulty that mathematics experiences in faithfully representing continuous flow or movement helped provoke, in the late nineteenth century, Dedekind's redefinition of the continuum as numerical completeness. By reimagining the continuum as the completeness of the real number series, Dedekind and his successors assured the triumph of a program of discrete mathematics. The point of the "Dedekind cut" was to show how discrete mathematics, with its reduction of mathematical reasoning to numerical operations, could make sense of the "between." In so doing, his approach did more than offer a way of thinking about numbers; it also disengaged the concept of the continuum from any conception of unpunctuated flow in time and even of unbroken movement in space. Continuity in time and in space became close numerical succession.

This was the direction that Weierstrass, having taken his cue from Cauchy, generalized and radicalized. In his treatment of limits and functions, he reduced calculus to arithmetic and provided a basis for analysis free at last from geometrical metaphors. Because that basis was non-geometrical, it was also untainted by the vulgar – that is to say, by the true – idea of the continuum as uninterrupted flow or movement, which can never be completely expunged from geometry.

Uninterrupted movement in space is the most similar thing in the world to uninterrupted flow in time. Their shared quality of non-interruption is what the vulgar idea of the continuum connotes and what the program of an intransigently discrete mathematics cannot brook; thus, the insistence of this program on identifying the continuum with numerical completeness and, in particular, with the real number line. It is no adequate vindication of the use of the real numbers as a proxy for the continuum to argue that the real numbers are uncountably rather than countably infinite. Discrete steps fail to turn, by virtue of being indescribably numerous, into unpunctuated flow.

It may be objected that, despite its enormous influence and its claim to have shaped part of contemporary mathematical orthodoxy, this line of thinking fails to exhaust the powers of mathematics to represent uninterrupted flow in time or movement in space. There have been, in the history of mathematics, other points of view about this topic. Of these alternatives, the most significant was Leibniz's idea of the infinitesimal, later renewed by Abraham Robinson and others.

The defenders of the orthodoxy championed by Dedekind and Weierstrass were not mistaken, however, in seeing the infinitesimal as a desperate attempt to outreach the boundaries of the mathematical imagination. It was desperate because it tried to split the difference between a world in which discrete things can be counted by discrete numbers and a world in which very small differences (such as the infinitesimals were designed to mark) fail to do justice to unbroken continuity.

Counting separate things with separate numbers, and relating configurations in space, is what mathematics does easily and well. Taking account of the idea or the reality of unbroken continuity is what it can try to do only by approximation and in defiance of its congenital biases and intrinsic limitations. The history of mathematics shows that these limits can be stretched. It does not suggest that they are non-existent.

Consider now the opposite pole of this spectrum of openness to mathematics, before confronting the additional problems presented by the study of humanity and of society and culture. This pole is defined by the attributes of natural history: path dependence, the mutability of types, and the co-evolution of regularities and structures – the change of change. In this realm, mathematics may remain a powerful instrument for the representation and analysis of large-scale phenomena such as population dynamics, just as it may prove indispensable to the investigation of the microscopic realm of genetic recombination. Nevertheless, the more decisive the role of a unique and irreversible history, resulting from many loosely connected causal sequences; the more significant the differences among natural kinds (such as animal species) as well as among the distinctive individuals who exemplify them; and the more rapid or radical the transformation of transformation, the greater the explanatory price that mathematics must pay for its denial of time and phenomenal particularity.

When we turn from natural to human history, restraints on the usefulness of mathematics in the development of causal explanation acquire a new and stronger force. All the traits of natural history that cause trouble for mathematical simplicity continue to apply. To them, however, we must now add the distinctive role and origin of the formative arrangements and assumptions of a society. These institutional and ideological structures, which shape the terms of people's relations to one another and represent the most decisive element in the history of mankind, differ from natural phenomena. They are constructions or inventions, although they are always built within the constraints of a historical circumstance, with the limited practical and conceptual materials that history and imagination make available. We can know them from within, as we cannot know natural phenomenon, because we are their half-conscious authors. By the same token, however, we can change them.

The institutional and ideological structures of society do not exist univocally. They exist with greater or lesser force: with greater force to the extent that they are organized in a way that inhibits their revision, and with lesser force to the extent that they are set up in a fashion that makes it easy to challenge and change them. By inhibiting their own revision, they acquire a false appearance of naturalness, necessity, and authority. It is always in our interest to deprive them of this delusive patina, which some of the classic social theorists called reification. Our stake in the revisable character of institutional and ideological regimes is closely connected to our material interests in the development of our practical capability, on the basis of an enhanced freedom to recombine people and resources in new ways. It is also bound up with our moral interests in the subversion of entrenched schemes of social division and hierarchy. Such always depend on the relative immunity to challenge of the arrangements and assumptions sustaining them.

The source of such variation in the force of the structures is their origin. They do not form part of the furniture of the universe. They result from the containment or interruption of conflict over the organization of society. They are frozen politics, if by politics we understand not simply the contest over the mastery and uses of governmental power but also the struggle over the terms of our access to one another in every area of social life.

Nothing in the resources of mathematics allows it to capture the qualitative features that are central to these realities. Its instruments are of greatest use in informing the analysis of some of the aggregate phenomena that may be shaped by the structures of society and culture and that have a quantitative or spatial presence. In understanding and representing the structures themselves, however, it may be of little use.

Is not economics, the most influential social science – it may be objected – a standing refutation of this claim of the limited use of mathematics in the study of society and history? Does it not owe its pre-eminence among the social sciences to its devotion to mathematics? The tradition of economics that resulted from the "marginalist revolution" of the late nineteenth century, the economics inaugurated by Walras and his contemporaries, adopted the strategy of viewing the economy as a series of interlocking markets. Its explicit aim was to develop a theory of relative prices that would be free at last from the confusions of the classical theory of value. Its implicit goal was to immunize economics against causal and normative controversy. It sought to advance this goal by transforming economics into an analytic apparatus – at the limit, a species of logic – generating explanations or proposals only insofar as it is informed by causal theories and normative commitments supplied to it from outside.

The post-marginalist economics that culminated in the general equilibrium theories and in the "neoclassical synthesis" of the mid-twentieth century achieved its rigor and generality at the cost of four connected defects. Its first and most important flaw is its separation of formal analysis from causal explanation. There is no basis in such an economics for causal theories of any kind; they must be imported from somewhere else – for example, from some variant of psychology, as in the contemporary practices of behavioral economics or neuro-economics. Its empirical turn is therefore misleading because wholly parasitic on the disciplines from which it must obtain if not actual conjectures then their theoretical foundation. Its mathematical devices are predictably trivial, focused as they are on a stringently reduced understanding of its subject matter. There can then be a proliferation of idle models in violation of Newton's warning not to feign hypotheses. Such a pseudo-science has trouble learning from its mistakes; it has condemned itself to eternal infancy.

A second vice of this tradition of economic theory is its deficiency of institutional imagination. As soon as it moves beyond analytic purity and tautology, it is tempted to identify maximizing behavior with behavior in a market and, more momentously, to identify the abstract concept of a market economy with a particular set of market institutions that happens to have become predominant in the recent history of the West. Such an economics is thus either pure and empty or potent but compromised.

A third complaint against post-marginalist economics is that, unlike the economics of Smith or of Marx, it has no substantial account of production. Its ideas about production are ideas about exchange, under disguise. It views production under the lens of exchange, as implemented through a flow of funds or through a complex of contractual relations. The persistence of wage labor as the predominant form of free labor – to the detriment of the superior forms of free work: self-employment and cooperation – makes it easier than it would otherwise be for economics to view production under the aspect of relative prices and, consequently, of a theory of exchange.

A fourth failure of this tradition of economic theory is that it amounts to a theory of competitive selection without a corresponding account of the genesis of the diverse stuff on which the mechanism of competitive selection can operate. However, the fecundity of any mechanism of competitive selection depends on the range of the material from which it selects. Such an economics resembles half of the neo-Darwinian synthesis: the half about natural selection, unaccompanied by the half about genetic variation.

The centrality of mathematics to this tradition of economic theory is inseparable from this fourfold taint, as a result of which it cannot serve as a model worthy of imitation. The very same features that make it deficient in the imagination of the structure of the economy, and therefore as well of alternative economic arrangements, are the ones that reinforce its mathematical devotions. The development of an economics free from the fourfold taint will have as one of its byproducts a reconsideration of what economics can expect from mathematics.

This account of the part of the spectrum at which the pertinence of mathematics diminishes and its limitations become paramount is not yet complete. There is one other topic in nature and in science, besides natural and human history, that defies and exceeds the resources of mathematics. That topic is the universe as a whole, and its history: the subject matter of cosmology. It is a subject matter that does not lend itself to the explanatory practices of the Newtonian paradigm. What may succeed as local explanation must fail, for the reasons that I earlier explored, as universal explanation.

To make this claim is not to deny that mathematics is indispensable to the work of cosmology: the science of the universe as a whole cannot develop in isolation from the science of features or of parts of nature, reliant on mathematics. It is to affirm that mathematics meets its match, and reveals the limitations intrinsic to its procedures, when the phenomenon to be elucidated is the whole of the universe and of its history. The reason for this limitation is that an adequate view of the universe as a whole must emphasize the two attributes of reality to which mathematical reasoning can never give adequate expression.

What for cosmology, as natural science rather than just as applied mathematics, stands in the place of phenomenal particularity is the singular existence of the universe. The universe cannot be counted. The most important fact about it is that it is what it is and not something else. What makes time – real, inclusive, global, irreversible, and continuous – of decisive importance to cosmology is that every feature of the universe, including the laws of nature, has to do with the history of the universe and forms part of this history. Cosmology must be both historical and a science. Nothing in the toolbox of mathematics shows how a historical science is to be conceived and practiced.

It follows from these considerations that the effectiveness of mathematics in science is reasonable because it is relative. In the representation of the workings of nature, mathematics is good for some things and bad for others. Its power is the reverse side of its infirmities. Its power as well as its infirmities result from a vision of the most general and abstract features of the one real, time-soaked universe, from a vantage point that denies both time and phenomenal particularity.

Both the reach and the limitations of mathematics increase by virtue of the process of internal development that has become the commanding force in its history. As a result of internal development, mathematics becomes ever more prodigal in the wealth of its ways of grasping how parts of reality may connect. By the same token, however, it loses any assurance that its ideas of connection will find application in the analysis of nature. Its growing stock of speculative conceptions never brings it any closer to overcoming its fundamental limitations. These restraints arise from its character and remain inseparable from its capabilities.

THE HISTORY OF MATHEMATICS RECONSIDERED: RIGHT AND WRONG IN HILBERT'S PROGRAM

A misreading of the history of mathematics helps inflame two contrasting sets of vanities and illusions. We may mistake mathematics for a godlike way of jumping out of mortal bodies, with their limited vision, and of seeing the world with the eyes of God. Alternatively, we may understand mathematics as a preternatural faculty of invention, providentially and mysteriously useful in the scientific investigation of the world.

As a further corrective to these illusions, consider another development in the history of mathematics: the one most famously exemplified by Gödel's proof and Turing's thought experiment with his "machine." The most far-reaching implication of this development is the rebuff to any attempt to transform mathematics into a closed system, all of the insights of which could be attributed to the combination of agreed procedures of inference with stipulated axioms. If such an attempt were successful, mathematics would indeed resemble the workings of a machine rather than expressing, as its does, the power of the mind to act, in its other nature, as an anti-machine.

Gödel did more than show that arithmetic cannot be fully axiomatized. He demonstrated both that a large consistent system (such as the propositional calculus presented by Russell and Whitehead) can make statements that cannot be proved or disproved within that system and that a large consistent system cannot be proved to be consistent within the system itself.

The bearing of these results on the preceding argument about the nature and applicability of mathematics can best be made clear by correcting a common mistaken view of their relation to the history of twentieth-century mathematics. According to this view, mathematics before Gödel and Turing was on a track defined by Hilbert's ambition to reduce it to a closed system under axioms. Then Gödel and others appeared, and drove mathematics out of a paradise to which it has never since been able to return.

The half-truth contained in this reading of the recent history of mathematics prevents us from grasping the strange message of this history for physical science. Hilbert, and many of his contemporaries and successors, had in fact three programmatic goals, rather than one alone. The frustration of the first, rather than jeopardizing the other two, has made it possible to realize them more fully than Hilbert and his party of mathematical orthodoxy could have imagined feasible.

The first aim was to organize and to vindicate all of mathematics under a system of finitistic axioms and of rigidly defined methods of inference. This aspiration proved to be a dead end despite the enormous influence it exerted thanks to Hilbert, Peano, and others as well as to the undying example of Euclid's geometry. The second objective was to affirm the unity of mathematics against the specialties into which it threatened to fall apart, as every science had, under pressure from the conventions of the university system. The third purpose was to move the focus of attention from mathematical objects (whether spatial or numerical) to mathematical methods.

Hilbert and many of his contemporaries thought that the second and third goals could not be separated from the first. They were mistaken. The first was not only based on an illusion – the illusion that Gödel and others worked to dispel – it also operated as an obstacle to the attainment of the other two. It prevented us from seeing what we can now recognize: that, freed from the illusory pursuit of all-inclusive axiomatic systematization, insistence on the unity of mathematics and on the ascendancy of its methods over its objects enables us to understand and to practice mathematics as a comprehensive expression of the two-sided human mind. The salvageable part of Hilbert's program helps undermine the view of mathematical reasoning as a machine the workings and results of which follow ineluctably from the instructions that we have programmed into it.

Once rid of the delusive effort to reduce mathematics to the implications of a closed set of axioms, we can reinterpret and reorient it as an instance of the mind at work simultaneously in its two registers. The first is its modular and formulaic register. Here the mind, made of discrete parts with specialized functions, works as if it were a machine. A machine is a device for reenacting the activities that we have learned how to repeat. The formula codifies the routine of repetition, which is then embodied in the machine. If the attempt fully to axiomatize mathematics had been successful, a machine is all mathematics would be, in everything save for physical embodiment.

The second register of the mind is the respect in which mind work does what no machine can accomplish. The mind is surprising, and transcending. It profits from the ability of the brain to alter, according to need and circumstance, the relation of structure to function (plasticity). It tries out what it has not yet learned how to repeat, or therefore to reduce to formulaic expression (surprise). It discovers what the established axioms and the canonical methods do not allow but cannot prevent, and then establishes retrospectively the assumptions and procedures that enable it to make sense of them (transgression or transcendence).

Mathematics does not work in machine-like fashion. For all its war against time and its indifference to phenomenal particularity, it expresses the two-sided nature of the mind too fully to resemble a machine. It helps us understand a world – the one real world, in which time is for real – that is also not a machine. The results of mathematics are not forever foretold in its assumptions and methods, just as the present laws of nature are not necessarily for keeps and consequently cannot be counted on to foretell the future of the one real universe. The laws cannot be counted on to foretell the future not because they are indeterminate (although sometimes they may be) but because they are susceptible to evolution and change.

A DEFLATIONARY AND NATURALISTIC VIEW OF MATHEMATICS

What results is a view that recognizes the unmatched powers and the unique perspective of mathematics. It nevertheless repudiates the Pythagorean claim, made on behalf of mathematics for at least twenty-six hundred years, that mathematical insight represents a shortcut to eternal truth about incorruptible objects. It sees mathematical reasoning as inquiry into the world – the only world that there is, the world of time and fuzzy distinction – only at one step of remove.

By an apparent paradox that goes to the heart of what is most interesting about mathematics, its removal from time and phenomenal variation helps explain the power of mathematics to assist science in the investigation of a world from which phenomenal variation and time can never be expunged. The denial of these intrinsic features of nature turns out to be the condition for the development, in geometrical or numerical language, of our most general ideas about the ways pieces of reality connect.

The apparent paradox is so disconcerting in its content and so far-reaching in its implications that it constantly tempts us to mistake its significance. It induces a dream: that mathematical insight is a way out from the limitation of the senses, even as that limitation is loosened by our scientific instruments.

However, the exceptionalism of mathematics – its lack of regard for certain pervasive aspects of the world, for the sake of the enhancement of our powers to understand and to act – justifies no such privilege. On the contrary, mathematics can best or only be understood as part of our natural constitution. It can be explained in straightforward evolutionary terms and justified by the epistemological advantages that, at a price, it secures us.

Such a view cuts mathematics down to size only in the sense that it presents it as part of our embodied and action-oriented humanity. It does not, however, deny its distinctive vantage point or its irreplaceable service to natural science. On the contrary, it helps explain them.

Once interpreted along these lines, the "unreasonable effectiveness of mathematics" is subject to a natural explanation that dissolves the appearance of unreasonableness in two convergent ways. It does so in one way by showing the cognitive and evolutionary advantages that the timeless abstractions of mathematics can have for inquiry into the temporal world. It does so in another way by suggesting that insofar as these abstractions are only obliquely connected with the phenomenal world studied by science – extending as they do, by analogy and abduction, what our senses and instruments allow us to see – there is no guarantee that they will be applicable in natural science. They may or may not be. There is no pre-established harmony between physical intuition, or experimental discovery, and mathematical representation.

Many aspects of this view have precedents in the history of twentieth-century thinking about mathematics as well as in the larger history of Western philosophy (for example, in Hermann Weyl's essay of 1928 on the "Philosophy of Mathematics and Natural Science" in the *Handbuch der Philosophie*, or in Leibniz's treatment of number and space). Nevertheless, their significance can be fully appreciated only when they are taken in conjunction with the ideas of the singular existence of the world and of the inclusive reality of time.

If time were illusory or even just emergent, mathematics could belong to the study of what lies behind emergent or illusory time. If our universe were only one of many, inaccessible universes, mathematics could form part of the science of the timeless totality of these remote and hypothetical worlds. However, if all we have is this world in which we awake, and nothing remains outside of time, we can give ourselves – or mathematics – no such excuse. What many have mistaken for an escape from ourselves turns out to be a road back into time, nature, and humanity.

Part II **Lee Smolin**

I Cosmology in crisis

THE CRISIS INTRODUCED

For those who want to know the answers to the big questions about the natural world, the fundamental sciences whose role is to discover the answers to those questions are in a most puzzling state. There is certainly reason to celebrate the greatest triumphs in the history of the physical sciences. Powerful theories based on great principles have given us an unprecedented understanding of nature on a vast array of scales of space, time, and energy. The great theories invented through the twentieth century – general relativity, quantum mechanics, quantum field theory, and the standard model of particle physics – have yet to fail experimental test. These theories have given us detailed and precise predictions for phenomena as diverse as gravitational waves emitted by orbiting neutron stars, the scattering of elementary particles, the configurations of large molecules, and the patterns of radiation produced in the big bang – and in all these phenomena, which were not even dreamed of a century ago, those predictions have proven correct.

But just as the physical sciences celebrate their greatest triumphs, they face their greatest crisis. This crisis lies in our failure, despite what has been now many decades of effort by increasing numbers of scientists, to complete the scientific revolution that was initiated by Einstein's discovery of the quantum nature of radiation and matter and his simultaneous invention of relativity theory. These two revolutions are each successful in its own domain, but remain unjoined.

Adding to this disappointment is our failure to go beyond the partial unification of the standard model of particle physics. Compelling hypotheses for further unification have been developed, but none of the novel phenomena implied by these theories have been found by experiments purpose-built to discover them. Among the beautiful ideas that experiment has not smiled upon are grand unifications tying together all the forces of nature save gravity and supersymmetry, which promised to unite matter and forces.

This crisis is, ironically, a measure of the success of models written down in the 1970s to explain all the experimental data gained since. In elementary particle physics that model is called the "standard model," and it has been confirmed by all experiments including the recent observations at the Large Hadron Collider. In cosmology there is also a standard model, and its predictions have likewise been confirmed by the most recent data.

The crisis is due to our inability to go deeper than these models to a further unification of physics or to explain the features of the models themselves. They reveal a universe that on rational or aesthetic grounds appears preposterous, and each has a long list of parameters which must be tuned very finely to agree with experiment. Many ideas have been proposed to explain why these parameters have the values they have; none has definitely succeeded.

The standard model of particle physics develops a few simple ideas: unification via increased symmetry and the spontaneous breaking of that symmetry, the gauge principle. These ideas have failed to take us further. The few additions to our knowledge of nature gained since the 1970s fulfilled no theoretical need and confirmed no deeper idea: neutrinos have mass, there is dark matter and dark energy. These are good to know but they just deepen the mystery.

In cosmology, all observations support a very simple picture of a universe expanding from a hot big bang, governed by Einstein's general theory of relativity. But it is an extraordinarily special universe, formed with extremely special initial conditions that tempt us to say we live in an improbable universe – tempt us until good sense holds us back from asserting that our deepest understanding of the unique, single universe is that it ought never to have happened at all. One way the universe appears to be improbable is that it is highly asymmetric in time, so that a strong directionality to time is apparent in a large spectrum of phenomena over a wide range of scales. This appears inexplicable given the fact that all the known laws of physics are reversible in time; this is an aspect of the present crisis that has been with us since Boltzmann's invention of statistical thermodynamics in the late nineteenth century. So one way to measure the depth of the current crisis is by the fact that basic features of our universe, such as its dominance by irreversible processes, appears to be inexplicable based on current knowledge.

In this book Roberto Mangabeira Unger and I propose a radical solution to this crisis of fundamental physics and cosmology. The road back to reality, we suggest, begins by making two affirmations about nature: the uniqueness of the universe and the reality of time. These together have an immediate consequence which is the central hypothesis of our program: that the laws of nature evolve, and they do so through mechanisms that can be discovered and probed experimentally because they concern the past.

There are several arguments that lead to this program. In this Part II of the book I develop three of them. The first goes directly from the data itself, because this preposterous universe with its seemingly inexplicable fine tunings of parameters, both elementary and cosmological, does make sense if it is seen as the result not of a-priori principles but of a historical and evolutionary process, which acted over time and continues to act.

The second way into this program comes from an analysis of the basic principle that governs the nature of the fundamental forces: the gauge principle. Both the local gauge invariances that form the weak, strong and electromagnetic interactions and the diffeomorphism invariance of general relativity reflect a single idea. Physicists call this idea local gauge invariance, which philosophers of physics understand to be the application of the philosophy of relationalism, the view of nature introduced by Leibniz and developed by Mach, Einstein, Weyl, and others, according to which properties of elementary particles rest fundamentally on their participation in a dynamical network of relationships that form the universe.

The current crisis in theoretical physics and cosmology can also be understood as a crisis of relationalism, which has two causes. The first is that the program of relationalism, while essential and successful – in that it led to general relativity and Yang–Mills theory – faces its own limits. Not every property can be a relation – there must be intrinsic properties which the relations relate.

Relationalism is also in crisis because its expression within the usual framework of timeless, immutable laws embodies a tension, if not a contradiction, because the essence of relationalism lies in the relations being dynamical. If the laws are timeless, they cannot themselves be aspects of the developing networks of relations. Yet a basic tenet of relationalism is the idea that everything that acts must also be acted upon – and this should apply to everything that serves as a cause of change or motion, including the laws themselves.

The path to resolve the crisis of relationalism is then to make the laws themselves subject to change and dynamics, that is to embrace the reality of time in the strong sense that everything changes, sooner or later; everything is in the throes of dynamics and history – even the laws of nature.

Behind the crisis of relationalism is a larger crisis – a crisis in the naturalist philosophy which underlies the progress of science and technology. The heady idea that all that exists is natural, physical stuff is more plausible now than ever, due partly to progress of physics and digital technologies, but even more to the triumph of reductionist strategies in biology and medicine. Yet it is in crisis because of an embrace of the old metaphor that the world is a machine. In its modern incarnation the mechanical philosophy becomes the computational philosophy – that everything, including us, are, or are isomorphic to, digital computers carrying out fixed algorithms. This leads to the failed but – to its proponents – inevitable program of strong artificial intelligence and also to the identity theorists in the philosophy of mind who proclaim that conscious experience, agency, will, and intentions are all illusions. This third crisis of naturalism is explored elsewhere [1].

But it is becoming increasingly apparent that we are not gadgets [2] and neither is the universe. We would like to have a naturalism that does not reduce human experience and aspirations to illusion. At the very least we need a naturalism that does not straitjacket our understanding of complex systems such as the human brain to failed metaphors coming from an early twentieth-century formulation of what it is to make a computation. Why can't the brain be a physical system that does not happen to be a programmable digital computer? Are we sure there are not still new principles to be discovered in complex systems, biology, and neuroscience?

The root of the crisis in naturalism is its being wedded to the picture that the universe is a machine. This in turn is a consequence of the idea that nature is governed by laws which are timeless, immutable and mathematical. The path out of crisis is to embrace a new form of naturalism based on the reality of time and the evolution of laws. I call this temporal naturalism, a term I introduced elsewhere [3].

Another indication of the crisis that afflicts all three manifestations is the growing fascination, if not embrace, of multiverse cosmologies [4, 5], according to which our universe is just one of a vast or infinite collection of other universes, within which the properties we have failed to explain – like the parameters of the standard models of physics and cosmology – are distributed randomly. This surrender of the hope for sufficient reason – the hope to satisfy our curiosity as to the root of things – is nothing but an indication that a philosophy wrongly assumed to be an essential part of science has failed.[1] The remedy is not to throw away our hopes of understanding but, rather, to discard our excess metaphysical baggage and to let science progress.

To focus us on this task, the first principle we adopt is that the universe is unique – single and singular.

This uniqueness of the universe means there is just one causally connected domain, but we take it to mean more than this, we mean

[1] The case that multiverse cosmologies cannot yield falsifiable predictions is made in [8, 7, 10], where several claims to the contrary are shown to be fallacious.

that the causally closed universe contains all that exists. The single, unique universe must contain all of its causes, and there is nothing outside of it. This assertion, together with the reality of time, has a further implication, which is that there are no immutable laws, timeless and external to the universe, which somehow act as if from the outside to cause things to happen inside the universe. Instead, laws of nature must be fully part of the phenomena of nature. The distinction, absolute in physics till this point, between laws which act and states which are acted on, must break down. Laws evolve. Our most important point is that taking laws to be mutable and subject to evolution rather than timeless and immutable brings questions as to the choices nature has had to make about which laws govern the single universe within the domain of scientific explanation. We will show that this increases rather than decreases the empirical reach of our theories because the hypotheses as to mechanisms which acted in the past to select the laws have consequences which are testable, not just in principle, but in real experiments. We will give several examples of such theories.

When worked out in detail these ideas lead to the denial of the Pythagorean dogma that the aim of physics is the discovery of a timeless mathematical object isomorphic in every respect to the history of the universe. This leads to a third affirmation, about mathematics rather than nature, which is that mathematics is an adjunct to scientific description of nature and is not a description of a separate or parallel reality or mode of existence. Nor can any mathematical object serve as a complete mirror of the universe or its history, in the sense that every property of the universe is mapped to a property of that mathematical object.

A key observation which is central to our argument is that it is fallacious to take methods and formal frameworks which have proved successful when applied to small subsystems of the universe and apply them to the universe as a whole. We demonstrate this by describing a precise schema for physics of subsystems, which we call the Newtonian paradigm, which includes the major theories of physics including Newtonian dynamics, field theory, quantum mechanics, and general relativity, and show in detail how it breaks down when applied to the universe as a whole. This leads to paradoxes, fallacies and dilemmas that plague the literature on theoretical cosmology. If we aspire to be scientific cosmologists, we must invent new paradigms of explanation.

On the basis of this critique, we propose a direction for research to resolve the crisis facing our understanding of the whole universe. The place to begin is to assert that we do not now have anything that could suffice as an adequate theory of the whole universe. A substantial part of this essay is devoted to explaining that a theory of the whole universe cannot just be like the theories we know in physics, just scaled up to the whole universe. The universe as a whole is a very different kind of system from those usually studied and modeled in physics, and its comprehension will require a new paradigm, and one that is novel, not just at the level of the content of the theory.

One reason we need a new paradigm – apart from the fact that the usual paradigms break down when applied to cosmological questions – is that when we attempt to understand the whole universe we face novel kinds of questions. These include the components of the "why this universe" question, particularly "why these laws" and "why these initial conditions." The standard methodology of physics cannot address these questions because, as I will describe, it takes laws and initial conditions as inputs.

A new kind of methodology and framework is needed to answer these questions. Therefore the crisis that cosmology faces is not a crisis of a theory in progress, it is a birth crisis, which accompanies our efforts to invent a new scientific methodology. The goal is the invention of a truly cosmological theory, which is to say a theory that could apply to the whole universe and explain its features to us, including the choices of laws and initial conditions.

It is often said that cosmology is in the midst of a golden age in which the quality and quantity of observations have increased dramatically.

But anyone who attends a conference in theoretical cosmology may, hearing the range of speculation about multiverses, conclude that cosmology faces the greatest crisis of its short history, because those observations severely challenge our abilities to make sense of them. There are reasons behind both assertions. Is there a paradox, or are the celebrants and worriers talking about different things?

Indeed, there is no crisis if our attention is just restricted to the data itself. Looking back into our past we can reconstruct a story of ceaseless change which goes under the name of the standard big bang cosmology. We can model the observations using standard general relativity and quantum field theory. So the crisis is not located in the data nor in any difficulty modeling the data in a way consistent with established principles of physics. The crisis is rather in attempts to go beyond modeling, to explain the data. It arises when we turn from describing the universe in which we live and understanding how it evolved, according to the known laws of physics, and turn to a new question: Why this universe? We confront crisis when we expand our ambitions from describing the part of the universe we can see to having a theory of the whole universe.

To make this clear, let us distinguish between astronomy on the scale of the time since the big bang, which we can call large-scale astronomy, and the notion of the universe, by which we mean all that exists. The former we can also call our observable universe, to distinguish it from the universe (which we will sometimes call the whole universe), which may be very much larger. The great progress in observational astronomy concerns the description of our observable universe; the crisis concerns our knowledge of the whole universe. To make things clear, when I use the noun cosmology or the adjective cosmological I will be referring to a theory of the whole universe. Unfortunately some of what are usually called cosmological models are not cosmological in this sense. This kind of confusion is unavoidable, but to make the distinction clear, when I need to I will call these models of large-scale astronomy.

TEMPORAL NATURALISM

One way to frame the project this book advances is to emphasize how fundamentally our conception of nature is shaped by our understanding of time. The notion of a law of nature is much changed if one thinks that the present moment and its passage are real or are illusions hiding a timeless reality. If one holds the latter view, then laws are part of the timeless substance of nature; whereas on the former view this is impossible, as nothing can exist outside of time. Even the creed of naturalism, i.e. that the natural world is all that exists, can mean two very different things, depending on whether you think existence is only real in each moment or only applies to timeless entities such as the history of the universe taken as one.

To make this distinction clear I propose to call the view we set out in this book *temporal naturalism*, and to distinguish it from its opposite, *timeless naturalism*. Temporal naturalism holds that all that is real (i.e. the natural world) is real at a moment of time, which is one of a succession of moments. The future is not real and there are no facts of the matter about it. The past consists of events or moments which have been real, and there is evidence of past moments in presently observable facts such as fossils, structures, records, etc. Hence there are statements about the past that can have truth values, even if they refer to nothing presently real.

Timeless naturalism, on the other hand, holds that the experience of moments of time and their passage or flow are illusions. What really exists is the entire history of the universe taken as a timeless whole. *Now* is as subjective as *here* and both are descriptions of the perspective of an individual observer. There are, similarly, no objective facts of the matter corresponding to distinctions between past, present, and future.

Timeless naturalism is similar, but not identical, to the view philosophers call "eternalism" and temporal naturalism has elements in common with the philosophers' "presentism," but my categories differ from the older ones because of an emphasis on the nature of law with regard to time.

Timeless naturalism holds that the fundamental laws of nature are timeless and immutable. Temporal naturalism holds that the laws of nature can and do evolve in time and that, while there may be principles which guide their evolution, the future may not be completely determined. This is consistent with the claim that there are no facts of the matter about the future.

NATURALISM IS AN ETHICAL STANCE

To appreciate the import of such a change in our conception of nature, it is helpful to contemplate for a few moments how little we know about the substance of the natural world. When we say that as naturalists we believe in the existence of only the natural world, we are making first of all a negative statement. We don't believe in ghosts, spirits, heaven or hell. We don't believe in a separate mental world or the Platonic world of mathematics. But on the positive side, what exactly is it that we believe exists? We say matter, and indeed we know a lot about matter, for example that the material world is constructed from atoms and radiation, which are in turn made of ... But what we know of elementary particles is the laws they satisfy, which determine how they move and interact. We do not know substantially what an electron is. By that I mean we do not have any conception of the intrinsic nature of being an electron except that its motion and interactions are governed by the Dirac equation and the standard model.

But if we have no conception of the intrinsic existence of an electron, we have no such conception also of a rock. Thus we are open as naturalists to conceiving wildly divergent conceptions of the natural. Some of us will say that there is no such thing in the actual world as the rock I am holding in my hand, as it is, this instant, in the moment; that is an illusion – what exists is only the whole history of the rock taken as one. Others will assert the opposite: that all that ever exists is the rock in the moment.

How are we to decide between two such different conceptions of nature, especially given that as naturalists we are committed to the use of evidence defendable within science? I would suggest that the most reliable test we have is the pragmatic one: whichever view contributes more to the progress of our knowledge of the world is more reliable to take as a hypothesis for further explorations of nature.

Given the fact that such divergent conceptions of nature are possible, it is helpful to give a definition of naturalism:

> Naturalism is the view that all that exists is the natural world that is perceived with, but exists independently of, our senses or tools which extend them; naturalists also hold that science is the most reliable route to knowledge about nature.

Part of my definition of naturalism refers to science as the most reliable path to knowledge about nature. This is unavoidable, as, without reference to a conception of what it means to argue in good faith to a conclusion, it will not be possible to resolve disagreements between competing conceptions of nature.

Note that my definition doesn't claim that science is the only path to knowledge, nor does it call on nor require that there be a scientific method. So I have to flesh out the definition by explaining what I mean by science as a route to knowledge about nature. Most importantly, I need to emphasize that while, as Feyerabend convincingly argued [11], there is no scientific method, science is most fundamentally defined as a collection of ethical communities, each organized around a particular subject. An ethical community is a community, membership in which is defined by the holding and following of certain ethical principles. In [7, 10] I argued that the scientific community is defined by two ethical principles. To quote from [10]:

> *Scientific communities, and the larger democratic societies from which they evolved, progress because their work is governed by two basic principles.*
>
> 1. *When rational argument from public evidence suffices to decide a question, it must be considered to be so decided.*
> 2. *When rational argument from public evidence does not suffice to decide a question, the community must encourage a diverse*

range of viewpoints and hypotheses consistent with a good-faith attempt to develop convincing public evidence.

I call these the principles of the open future.

Naturalism is then in part an ethical commitment. To quote from [8]:[2]

> *Science is that activity by means of which we display the same respect for nature that we aspire to show to each other in a democratic society.*

If naturalism is an ethical commitment, temporal naturalism is a deepening of it because it rejects the subversion of the naturalist impulse which occurs when a scientist substitutes nature, in all its immediacy and primacy, with an imagined world, believed to be transcendent but, in reality, just a construction of our imaginations.

The naturalist stance is vulnerable to this kind of subversion because we know about nature second-hand, through our sense impressions. Unless we are idealists we do not believe that all that exists are our perceptions. What we believe is that our senses give us evidence for the existence of a natural world, which can be learned about through our sensations but which exists independently of them.

However, our senses, and the experiments and observations which we carry out to extend them, only give us direct acquaintance with the qualia which are the sensory elements of our experience. They do not give us immediate acquaintance with, or direct knowledge of, the rest of the natural world. They can then only provide evidence for hypotheses which we make concerning the natural world. Thus, as naturalists we are constrained to deal in indirect knowledge of the object of our study and we must be always conscious that this knowledge is incomplete and never completely certain. But since we believe all that exists is the natural world we must admit that incomplete and tentative knowledge is the best that can be had concerning what exists.

[2] A quote I mistakenly attributed to Richard Dawkins.

Because of this, naturalists can hold quite strikingly different views about nature – and still be naturalists. For example, many naturalists believe that everything that happens in nature is governed by universal and unchanging laws. But one doesn't have to believe this to be a naturalist – because we must admit the possibility that experiment could provide evidence for phenomena that are governed by no definite law. For example, if we believe that no hidden-variables theory determines the precise outcomes of measurements on quantum systems for which quantum mechanics only gives probabilistic predictions, then we believe there are phenomena that are not law-governed at all. Indeed, if we follow Conway and Kochen [12], then quantum phenomena are in a well-defined sense free [13]. Or, if we believe the standard big bang cosmology expressed in the context of classical general relativity, then we implicitly believe that no law picks the initial conditions of the universe. Or to put it another way, no law governs which solution to the equations of general relativity is somehow uniquely blessed with describing the actual history of the universe.

Another thing that some, but not all, naturalists believe is that everything that exists in the natural world can be completely described by the language of physics. There are varieties of positions held with respect to emergence and reduction; but it is quite reasonable to believe that matter is made out of elementary particles which obey general laws, but that complex systems made out of many atoms can have emergent properties not expressible in or derivable from the properties of elementary particles.

Many naturalists hold beliefs about the natural world that are more firmly held and expressed than the tentative nature of scientific hypotheses allows. These are often beliefs of the form:

> *Our sense impressions are illusions, and behind them is a natural world which is really X.*

Such a view can either be an ordinary scientific hypothesis or a metaphysical delusion, depending on what X is asserted to be. When X is a statement like *made of atoms*, this is an innocuous scientific hypothesis which carries little metaphysical baggage and is, in fact, very well confirmed by diverse kinds of experiments. (But this was of course not always the case.) But statements of this form can be traps when X is a big metaphysical assertion which goes way beyond the actual evidence.

A common and widely believed example is the claim that: *X (the universe) is really is a timeless mathematical object* [5]. Whether that mathematical object is a solution to an appropriate extension of general relativity or a vector in an infinite-dimensional space of solutions to the Wheeler–deWitt equation of quantum cosmology, there is a big stretch from a statement of the form

> *Some experimental evidence concerning a specified range of phenomena is well modeled by a mathematical object, O,*

which is a statement that might or might not be supported by evidence, and a metaphysical assertion that *The universe is really a mathematical object,* which is not by any reach of the imagination a hypothesis that could be tested and confirmed or falsified.

What is troubling is that statements of the form *Experience is an illusion, the universe is really X* are common in religion. When naturalists make statements of this kind, they are falling for what might be called the transcendental folly. They are replacing the concrete natural world by an invented conception, which they take to be "more real" than nature itself. Thinking like this turns naturalism into its opposite.

Much that passes for naturalism and physicalism these days are instances of transcendental folly.

2 Principles for a cosmological theory

The idea that an acceptable cosmological theory needs to be formulated in a framework different from that of the so far successful theories of physics is not new. There is a tradition of critique of Newtonian physics, which leads to what is often called the relational position on the nature of space and time. Relationalism is associated with Leibniz [15], Mach [16], and Einstein [17] and, in the present period, Barbour [18], Rovelli [19], and others. Much of the critique concerns issues that only arise if you aspire to a theory of the whole universe rather than a part of it. General relativity is partly – but only partly, as I will explain – a response to that critique.

THE ROOTS OF RELATIONALISM

We can draw criteria for a truly cosmological theory from that tradition of critique. The starting point is Leibniz's great principle:

> *The principle of sufficient reason (PSR).* For our purposes we state it thus: for every question of the form *Why does the universe have property X?* there must be a rational explanation. This implies that there should be rational explanations for the selection of the effective laws we see acting in our universe, as well as for any choices of initial conditions needed for that universe. This application of the PSR was echoed in Peirce's insistence that "nothing is in so need of explanation as a law" [14].

My view is that we should take the PSR as an aspiration and a goal, perhaps never to be completely reached but, nonetheless, a beacon illuminating the direction in which we are to search for the answer to cosmological questions. We can state this as follows:

> *The principle of differential sufficient reason (PDSR).* Given a choice between two competing theories or research programs, the one which decreases the number of questions of the form *Why does*

the universe have property X? for which we cannot give a rational explanation is more likely to be the basis for continued progress of our fundamental understanding of nature.

In this form the PDSR is best used as a means to evaluate progress (has sufficient reason increased?) or to judge the likelihood of success of competing research programs. It is especially useful in judging the promise of novel research programs.

The PDSR is especially helpful in cases in which the PSR acts to resolve a question by removing it from the list of questions that a theory must answer. The paradigmatic example of this is in the debate between Newton's absolute conception of space and time and the competing relational notions put forward by Leibniz and Mach and instantiated in a dynamical theory by Einstein. Here the question to be removed was *Why did the universe not start five minutes later?* On a relational account this question does not make sense, whereas in an absolute account it both makes sense and is unanswerable rationally. Thus, by removing the question for which sufficient reason cannot be given, a relational account increases sufficient reason. This standard argument implies that the PDSR implies that space and time must be relational rather than absolute. In Einstein's hands this led to general relativity, which definitely was progress.

In this book when I refer to the PSR I will mean the PDSR.

More generally, the PSR/PDSR insists that there be no ideal elements or background structures in the formulation of a truly cosmological theory. These are structures or mathematical objects which are specified for all time, have no dynamics, participate in no interactions, but are necessary to give meaning to the degrees of freedom that are dynamical. Examples of such ideal elements are Newton's absolute space and time and the Hilbert space of quantum mechanics.

Einstein formulated this as the demand that there be no unreciprocated actions [17]. These are instances where an object A acts on an object B, which, however, does not act back on A. We can call this *Einstein's principle of reciprocal action (PRA).*

In contemporary physics we distinguish between background-dependent and background-independent approaches to quantum gravity, and prefer the latter for being closer to the principles of relationalism [7, 10, 21]. Examples of the former are perturbative approaches to quantum gravity, based on fields or strings moving on a fixed background geometry. By this time several background-independent approaches are proving more fruitful; these include loop quantum gravity, group field theory, causal dynamical triangulations, causal sets and quantum graphity.

Another consequence of the PSR was stated by Leibniz as the *Principle of the identity of the indiscernible (PII)*. This is that if two elements of the world have the same set of relations to the rest of the world, or identical properties, they are in fact a single object. There cannot be two distinct objects in the universe with the same properties. This rules out symmetries, in the sense of global symmetries, i.e. transformations which take a physical system between two physically distinct states which have the same values of all conserved quantities.

Many systems studied in classical and quantum mechanics have symmetries, including the standard symmetries of Euclidean space or Minkowski spacetime. In every instance these can be taken to be instances where an isolated system is moved relative to an external frame of reference. These symmetries thus make sense in the description of an isolated subsystem of the universe. But the PII insists that the universe as a whole has no symmetries. This implies it can have no non-vanishing conserved quantities, because, by Noether's theorem, the basic conserved quantities, energy, momentum, and angular momentum, are consequences of the corresponding symmetries of spacetime.

General relativity gets this right. A system can only have conserved quantities or symmetries in general relativity if it has a finite or infinite boundary, where external boundary conditions are imposed. This is how an isolated subsystem of the universe is treated in general relativity. When general relativity is applied to a spatially compact universe, there are no global symmetries and no conserved quantities [22].

It is then proper to regard the great conservation laws of physics – of energy, momentum and angular momentum – as emergent and approximate. So one way a cosmological theory must differ from the standard dynamical theories of physics is that it must have no global symmetries or conservation laws.

This does not rule out gauge symmetries, which are entirely different, as these take the description of a single physical system to a mathematically different description of the same system.

The realization that a cosmological theory has no symmetries runs directly counter to a methodological slogan that, the more fundamental a theory is, the more symmetry it must have. This imperative governed the progress of field theory for a bit more than a century, from Maxwell through the standard model of particle physics. But it has since failed us as a good guide, and it is important to emphasize that what went wrong is that the imperative of more symmetry pointed to the construction of unified models of physics beyond the standard model, which were falsified by experiments. Two major versions of this were grand unification and supersymmetric extensions of the standard model. Each of these theories implied new phenomena as well as a natural energy scale at which those phenomena would be expected to be seen. The experiments built to discover those phenomena have now pushed the limits beyond the scales naturally indicated.[1]

One problem with the use of symmetry as a route to unification is that we must then explain why the symmetry is broken. This can make the effective laws which appear to govern our universe contingent – as they are consequences of a particular solution which breaks the symmetry spontaneously. The problem of explaining our observed laws is then pushed back to a question of initial conditions, because different solutions may lead to different symmetry-breaking patterns. The limit of this line of thought is string theory, which comes in an infinite number of versions depending, in part, on the symmetry-

[1] A phenomenon may still exist and be discovered but to explain it would require unnatural fine tunings of parameters.

breaking pattern coded into the geometry of the extra, compactified dimensions.

On the contrary, I will argue below, on the basis of the PII, in a truly fundamental theory each elementary event will be unique [23, 24]. Our universe should not be seen as a vast collection of elementary events, each simple and identical to the others, but the opposite, a vast set of elementary processes, no two of which are alike in all details. At this level fundamental principles may be discerned but there are no general laws in the usual sense. General laws apply to large classes of phenomena, which emerge from the fundamental theory only when details which distinguish the elementary events from each other are forgotten in a process of coarse graining.

This leads to the understanding that symmetries are always consequences of willed ignorance, which is the result of ignoring small differences between states. For example, in reality no physical system is translationally invariant because the universe is complex enough that each and every event has a unique curvature tensor reflecting the influences of distant masses as well as gravitational waves, neither of which can be screened by any physically real material. Another example is the global symmetries of the standard model, which are only approximate when the effects of the fermion masses are ignored. To put it in simple terms, the proton is slightly lighter than the neutron. Of course these and other symmetries hold to very good approximations – the gravitational field is very weak and the consequences of assuming the proton and neutron are identical up to their charges are extremely useful for understanding the spectra of nuclei. But there are no unbroken, exact global symmetries.

We also require that a theory of the whole universe be explanatorily closed. This means that chains of explanation and causation do not point back to entities outside the universe. This is of course another way to state the requirement that there be no ideal or background elements.

Finally, we require that our theory be successful as a scientific theory. To be scientific, a theory must reproduce what is known in its domain. Since the domain is all that exists, this means it must reproduce and explain the successes of the standard models of particle physics and cosmology. But to be judged to have expanded our knowledge, a theory cannot just retrodict, it must predict, and these predictions must be checkable in the near term. The usual and best case is that the predictions be falsifiable by doable experiments or observations. Not as good as, but acceptable in the short term, are theories which are strongly confirmable. This means that such a theory makes predictions that can be confirmed by doable experiments such that, were they confirmed, there would be no other plausible explanation other than that this theory is correct.

Confirmability is not as good as falsifiable because there is always a good chance that someone will invent an alternative theory that also plausibly explains the data.

One may object that it would be utopian to imagine that we live in the period where cosmology can be completed in agreement with these principles. Especially since the study of scientific cosmology is in its infancy, we can expect many steps of progress on a long road to a complete understanding of the universe. It is important then to say that the principles stated here do not require that everything is explained. Chains of explanation and causation may be followed back to a point where we are ignorant of what preceded them. This is no problem, so long as the principle of explanatory closure is respected, so where chains of causation end due to our ignorance, they end inside the universe. Then there can always be hope that they can be picked up again when observations improve.

The most important thing is then not to explain things in a way that precludes further progress. This happens when explanations are built on uncheckable claims about unobservable phenomena or when chains of explanations end in background or ideal elements, outside of dynamical influence. So the principles we have proposed do not insist or imply that science can within a finite time come to answer every question. Rather they keep the future of science open, by keeping high standards for what has been explained.

THE NEWTONIAN PARADIGM

We now turn from the goals of a cosmological science to the theories we have available to us.

The standard methodology of physics was invented by Newton, and frames all the successful physical theories since, including quantum mechanics, quantum field theory and general relativity. It can be described as follows. The system to be studied is always a subsystem of the universe, idealized as an isolated system. The theory appropriate to that subsystem is defined by giving separately the kinematics and the dynamics. The kinematics comes first and is described by giving a state space, C, of possible states the system may have at any moment of time.

The dynamics is then given by specifying a law which, given a point of C, gives a unique evolution from that point. The evolution takes place in time, measured in many cases by a clock outside the system. The state space and the law are timeless, while the law evolves the state in time.

We will call this the Newtonian paradigm for a dynamical theory.

To use this paradigm, one inputs the space of states, the law, and the initial state, and gets as output the state at any later time. This method is extremely powerful and general, as can be seen from the fact that it characterizes not just Newtonian mechanics, but general relativity, quantum mechanics and field theories, both classical and quantum. It is also the basic framework of computer science and has been used to model biological and social systems.

In classical theories the state space is the phase space, given by coordinates and momenta. In quantum theories it is the Hilbert space.

THE FAILURE OF THE NEWTONIAN PARADIGM WHEN APPLIED TO COSMOLOGY

Successful as it is, this method cannot be used as the basis for a truly cosmological theory. There are two strong reasons for this. The first is that we want a cosmological theory to answer the questions of "why these laws" and also to account for the initial conditions of the universe. Since laws and initial conditions are inputs for the method, they cannot be outputs. Much inconclusive work has resulted from attempts to explain the choices of laws and initial conditions making use implicitly of the Newtonian paradigm. These failures are due to the reliance on a paradigm of explanation that cannot answer the questions we ask of it.

Another reason the Newtonian paradigm cannot be extended to the whole universe is that its success relies on our ability to cleanly separate the roles of the initial conditions from the laws in explanations of physical phenomena. But this separation in turn relies on our ability to do an experiment many times while varying the initial conditions. Only by making use of the freedom to run an experiment over and over again with different initial conditions can we determine what the laws are – for the laws code regularities that are invariant under variation of the initial conditions.

That is to say, the Newtonian paradigm is ideally structured to be applied to small subsystems of the universe, which can be prepared in many copies. In these cases the configuration space corresponds to the operational fact that the experimenter has the freedom to prepare the system in any initial state in C.

In cosmology we do not have this freedom, both because there is only one system, with one history, and because we were not there at its origin to choose the initial conditions. So when we attempt to extend the Newtonian paradigm to the universe as a whole, we take it outside of the domain where its logic and structure tightly fit the experimental methodology.

In usual applications of the Newtonian paradigm we are seeking to verify a hypothesis about the dynamical law which applies to the subsystem. To do so we prepare the subsystem many times, varying the initial conditions, and we seek to verify whether the law applies in all these cases. What we mean by a general law is a feature or invariant of the motion which is unaffected by varying the initial conditions.

When we take this method out of its domain of validity and attempt to apply it to cosmology, several things go wrong. One is that we only get one try, so it is not clear what meaning a general law has in this context. A second is that we have to test *simultaneously* hypotheses as to the choice of initial conditions *and* hypotheses as to the choice of laws. This can lead to degeneracies in which the same data set can be explained by different choices of laws and initial conditions that cannot be resolved by doing experiments with more cases because there is only one case. This lessens the predictive (or postdictive) power of the scheme.[2]

On the other hand, to limit the application of a theory expressed in the Newtonian paradigm to a subsystem of the universe is to make a necessary approximation because interactions between the subsystem and degrees of freedom outside of its boundaries are neglected. Any model of a subsystem by means of the Newtonian paradigm is then necessarily a truncation of a more exact description. This is captured by the notion of an effective theory, which is a version of the Newtonian paradigm with limits on its regime of validity spelled out explicitly. Newtonian mechanics, ordinary quantum mechanics and all known quantum field theories, including the standard model, are all understood to be sensible only when thought of as effective theories. There is also good reason to suspect that general relativity is also an effective theory.

An exact theory can only be obtained by extending the description to include all the degrees of freedom a subsystem interacts with, which means to the universe as a whole. The question is then whether that more exact cosmological description is to be formulated within the Newtonian paradigm or within a new, presently unknown paradigm. If one presumes the cosmological theory falls within the Newtonian paradigm, then, however, we run into all the failures of explanation just discussed – the failure to answer queries about the choices of state space, laws and initial conditions as well as the

[2] An example of this is discussed in [10].

degeneracies arising from the inability to probe experimentally the separation in the role of initial conditions and laws. Thus, the result of extending a theory formulated within the Newtonian paradigm to the universe as a whole is to strongly decrease rather than increase the empirical adequacy of the theory. This can be called the *cosmological dilemma*.

This dilemma can be formulated also as follows. A general law gets its empirical confirmation from its applicability to many cases. But by definition, if there are many cases a law applies to, each is a small subsystem of the universe. Each application is then necessarily approximate because of the truncation which removes the influence of interactions between degrees of freedom in the subsystem and degrees of freedom in the rest of the universe. To make the application of the law more exact, one can seek to expand the subsystem to include interactions with and dynamics of an increasingly larger set of degrees of freedom. The limit of this is to include all the degrees of freedom in the universe. But at that point there is only one case and one run of each measurement so the operational context which defined the notion of a general law no longer applies.

This dilemma has teeth because there are questions about phenomena on the level of the universe as a whole which cannot be formulated as a question about a great many subsystems. These include questions about background structures such as the geometry of space, spacetime or the state space, or the adequacy of grounding observables on fixed reference systems. This also includes questions about the choice of laws and initial conditions.

To resolve the dilemma we need a new methodology which transcends the limitations of the Newtonian paradigm and will be uniquely suited to address questions about the unique universe. If we instead ignore the cosmological dilemma and seek to apply the Newtonian paradigm to the universe as a whole, we find ourselves trapped in fallacious reasoning. We can call the taking of the methodology of the Newtonian paradigm outside of its domain of validity, where it corresponds to experimental procedure, the *cosmological fallacy*. The consequences of committing this fallacy are easily observed in the many puzzles and paradoxes which fill the literature of contemporary cosmological theory.

THE FAILURE OF THE NEWTONIAN PARADIGM TO SATISFY THE PRINCIPLES FOR A COSMOLOGICAL THEORY

It should be immediately clear that any theory formulated within the Newtonian paradigm fails to satisfy several of the principles stated above. Let us begin with the PSR. It demands that there be a rational reason for every property of the universe, and this certainly includes the choice of laws and initial conditions. Could there then be, within the Newtonian paradigm, a rational reason for such a choice?

There have been several proposals for how the sufficient reason might be supplied. The most important of these, which was very influential during the twentieth century, is that there is but a single mathematically consistent theory that unifies the fundamental particles and interactions. Stated this way, it fails right away because you can ask what is the sufficient reason for a universe to have the four forces we observe. Let's start with gravity. We know that there are consistent mathematical descriptions of worlds without gravity; these are given by conventional classical and quantum field theories. Or let us ask "why the quantum?" We know there are consistent theories of possible worlds, with and without quantum mechanics. But perhaps what is meant is that there is a unique mathematically consistent quantum theory that includes gravity, and some mix of gauge fields and fermions. This was initially the hope of string theory, and on present evidence it is simply not the case.

Nor do other approaches to quantum gravity give any reason to hope that there is only one consistent mathematical description of a world where gravity and quantum physics are unified. Loop quantum gravity seems equally consistent coupled to any set of gauge fields, fermions and scalars.

One might posit that the list of consistent theories is reduced down to a unique case by the anthropic principle. But there is no reason to suppose this is the case. Indeed, many of the parameters of the standard model of particle physics can be varied within large ranges without affecting the possibility that life exists. These include the masses and couplings of the two more massive generations of fermions.

The bottom line is that mathematical consistency is simply too weak a requirement to supply sufficient reason for the choices of the known laws of nature, and this remains the case even when supplemented by the anthropic principle.

Almost the opposite proposal has been made as well: there are many consistent mathematical descriptions of possible universes and they all exist [5]. We happen to live in one of them, picked out – but far from uniquely – by the anthropic principle. This fails the criteria of good science because the main hypothesis – a vast or infinite set of other universes, disconnected causally from our own – is in principle not subject to any kind of check or confirmation. This also fails the principle of sufficient reason because when every property or choice is manifested in some universe there is nothing to explain.

There is then no possibility of a sufficient reason for the choice of theory within the Newtonian paradigm. The fact is there are a vast or infinite number of possible configuration spaces, and on most of them an infinite set of possible dynamics. Restricting to quantum theories or use of the anthropic principle does not change that.

We reach the same conclusion when we ask whether there could be a sufficient reason for the choice of the initial conditions. It suffices to restrict attention to the initial conditions of a single theory: general relativity, coupled to the matter fields which are observed. If we neglect quantum effects to begin with, this is a completely well defined theory and it has an infinite-dimensional space of solutions. Each of them is a possible universe, but at most one of them describes our universe. Even if we restrict by the anthropic principle that leaves an infinite number of solutions that are roughly like our universe. So there is a choice, and within the theory there can be no rational reason for one solution to be picked over another.

If we ignore this, and believe in the application of this theory to the whole universe, we end up asking an apparently silly question: Why does the theory of the whole universe allow an infinite number of solutions when a single one would suffice, because there is only a single case for the theory to apply to? This question just restates the failure of any theory within the Newtonian paradigm to satisfy the PSR when applied to the whole universe.

THE FAILURE OF THE NEWTONIAN PARADIGM FOR ELEMENTARY EVENTS

Remarkably, the Newtonian paradigm doesn't only fail when applied to the universe as a whole; there is an argument that it must also fail when applied to the most elementary events [23, 24]. This argument begins with a discussion of the limits of reductionism.

Reductionism and its limits

Reductionism is the good advice that if you want to understand a composite system, which means a system composed of parts, you will do well by explaining the properties of the composite system in terms of the properties of the parts. The same holds for compound processes, whose properties can often be usefully explained in terms of the interactions of the properties of the subprocesses that make it up.

There are many cases where the properties of the composite are not of the same kind as those of its component parts or processes. This occurs when the properties of the composite would not make sense when applied to the parts. In these cases we say that the property of the composite is emergent, by which we mean that it must be invented and added to the list of properties. For example, a liter of gasoline can have a mass, a momentum, a temperature, and a density. Its component molecules have mass and momentum, but it makes no sense to talk about the temperature or density of a molecule (ignoring its internal degrees of freedom). So we say that temperature and density are emergent properties of the gas.

This common circumstance does not represent a limit of the method of reductionism; instead it represents its intensification. For emergent properties, in the sense I've defined them here, can often be elucidated in terms of the properties of the parts. This is the case for both temperature and density; the temperature of a gas, for example, was discovered to be the average kinetic energy of its constituent molecules.

But there is a limit to reductionism. The method of reductionism can be iterated, as the parts are broken up into smaller parts. But it fails the moment we get to a level of constituents that are deemed to be elementary, meaning that they have no parts. But they still have properties and reductionism is moot as to how we are to explain them.

Let us suppose that the quarks and electrons of the standard model are truly elementary; that is, they have no constituents so their description in terms of local quantum fields is exact down to all distance scales. Then their masses and charges would be in need of explanation and this, by assumption, cannot come out of further reduction.

How are we to proceed?

Relationalism offers a strategy that can take over at the point that reductionism fails. The properties of the elementary particles can be understood as arising from the dynamical network of interactions with other particles and fields. A property of a particle or event that is defined or explained only by reference to the network of relations it is imbedded in can be called a relational property; its opposite, a property that is defined without such reference to other events or particles, is called intrinsic. The ambition of a purist relational approach would be satisfied if all properties of elementary particles and events are relational.

An early attempt to realize such a relational explanation of the properties of the elementary particles was the bootstrap approach of

Chew and collaborators, developed in the 1960s. They situated the observed hadrons within a complex network of interactions in which the properties of composite and constituents were mixed, i.e. a proton was composed of a neutron and a pion, but the neutron was also a composite of a proton and a pion, while the pion was a composite of the proton with the anti-neutron. These relations gave a coupled non-linear set of equations for the amplitudes of these processes of compositeness. They conjectured that mathematical consistency plus a few basic properties would constrain these amplitudes sufficiently to give a unique set of properties to the elementary particles.

The bootstrap program failed at the time to produce results that supported this conjecture, but it recently has been revived by combining it with another seemingly failed program, twistor theory, and shown to work in the case of quantum gauge theories with maximal supersymmetry [25]. But the original bootstrap program was superseded in the early 1970s by the standard model of particle physics, which is a conventional quantum field theory in which the protons, neutrons and pions emerge as composites of elementary quarks and gluons. So this was a further victory of reductionism.

However, the standard model of particle physics is only partly reductionist, for it is partly relational. The quark and lepton masses all come from their coupling to the Higgs field and, through those couplings, depend on the phase of the vacuum. In a symmetric phase the masses all vanish; in a spontaneously broken phase they are all proportional to the resulting vacuum expectation value of the Higgs field.

In more highly unified models the Higgs potential can be quite a bit more complex, with many local minima, each of which is characterized by a pattern of symmetry breaking, which, in turn, defines a different phase of the quantum field theory. In each of these the properties of the elementary fermions will be different. It seems like the larger the symmetry and the greater the unification, the more potentially complex the Higgs potentials are and hence the larger the possibility that the elementary particles have available to them a potentially large number of arrays of properties to choose from.

Of course this doesn't explain everything. The quark and lepton masses are all different because each is proportional also to a coupling constant by which they couple to the Higgs field, and these are among the unexplained parameters of the model. These are not so far explained relationally.

We will come back to the question of whether the pure relationalist ambition can be satisfied or whether some properties of elementary events or particles all turn out to be intrinsic. We need to first consider a consequence of the principle of the identity of the indiscernible for the description of elementary events.

The uniqueness of fundamental events

For a naturalist, the universe, being the totality of all that exists, must be unique. For a relationalist, this unique universe must contain all of its causes. But a little known consequence of Leibniz's principle of the identity of the indiscernible is that every elementary event must be unique, in the sense of being distinguishable from every other event in the history of the universe by its location in the network of relations.[3]

It is most straightforward to make this argument within an ontology in which the history of the universe consists of discrete events whose relational properties are fundamentally causal relations [26], but the argument can be made in other ontologies as well. These events may, as we have just discussed, also have intrinsic properties. But they have no intrinsic labels. If we want to refer to a particular event, we cannot just give it a name, such as "event A"; we must give it a unique description in terms of its relation to other events, which uniquely distinguishes it from all the others. And in that description you cannot just say event A is the one whose immediate causal past is events B and C – for B and C must be specified relationally as well. So the description must be based on the past causal set, going far enough into the past that event A's causal past is distinct from the causal past of any other.

[3] The ideas of this subsection are developed in [23, 24].

So the point is that elementary events are not simple to name, because a complete specification in terms of relational properties must contain enough information to single each out from the vast number of other events in the history of the universe.

Given this, let's consider what goes into spelling out the most elementary causal relation, i.e. events B and C are the direct causal progenitors of event A. Developed in enough detail to relationally describe or address the three events, this is a highly complex statement that carries a vast amount of information. So elementary events are not simple and neither are elementary causal relations.

It follows that at the fundamental level nature cannot be governed precisely by laws that are both general and simple. A complete law of nature would have to explain why event A occurred, which means why the events B and C gave rise to a new event. To give a definite answer to this question would be to pick out which pairs (or small sets) of events give rise together to new events. As there are vastly more pairs or subsets that do not have common immediate future events, the question is what makes B and C different from the vast numbers of pairs that do not give rise to new events. Any explanation for this must be based on what makes those common progenitors different from the other pairs, i.e. it must involve enough information to pick out by means of their relational properties what makes those that are progenitors distinct from those that are not.

So any complete explanation for the elementary causal process "B and C together cause A" must make use of information that distinguishes B and C from all the others. So it cannot be a general and simple law of the kind we usually envision governs elementary events. Candidates for such laws are discussed in [23].

In particular, a definite law that deterministically explains and predicts the causal relations among unique elementary events will not be of the form of the Newtonian paradigm. As each event and each causal link is unique in the history of the universe, they make up a vast set, when fully described. There is no general dynamical law that acts on a configuration space of possible states of a small system.

One may ask how the approaches to physics which have an events ontology deal with this issue. They do it by summing or averaging over an ensemble of all possible histories that connect given "in" states to given "out" states. Two examples are Feynman diagram expansions in quantum field theories and stochastic or quantum approaches to the dynamics of causal sets [26]. This ducks the issue of explaining which events take place – and also weakens the events ontology, as the elementary events appear only in terms that are averaged or summed over. But even if we buy this as a necessary expression of the freedom of the quantum world, it is still the case that this methodology departs from the Newtonian paradigm in that there is no continuous evolution on a fixed configuration space.

Indeed, nowhere in physics do we have a theory that explains why individual events occur. Most theories with an events ontology are quantum theories, such as Feynman diagram approaches to quantum field theory, according to which every causal history, with every possible set of events and causal relations, has an amplitude it contributes to the sum over histories. There are, to my knowledge, no deterministic causal set models; instead those causal set models that have been studied take a stochastic or quantum approach to dynamics. They therefore do not attempt to answer the question of why particular causal sets may occur. The only exception I am aware of is the model developed in [23].

On the other hand, dynamics fitting the Newtonian paradigm can be derived for coarse-grained descriptions of large sets of events. So, if we are willing to have coarse-grained, rather than exact, laws, then we can speak of classes of events. For example, A is in the class of events that has two immediate progenitors. These classes require much less information to describe than the individual events, hence we can have simple laws applying to large classes of events. These may be quite general, applying, as they do, to large classes of events. So we see that general and simple laws of the usual kind can emerge from a more detailed description by coarse graining, so that they apply to large classes of events.

RELATIONALISM AND ITS LIMITS: RELATIONAL VERSUS INTRINSIC PROPERTIES

Relationalism is not just a philosophical position, it is a methodological imperative: *Progress in physics can often be made by identifying non-dynamical background structures in the description of a subsystem of the universe and replacing it with a real dynamical physical interaction with degrees of freedom outside of that subsystem.*

The paradigmatic example of this is Einstein's use of what he called "Mach's Principle," by means of which the role of absolute space in defining the distinction between inertial and accelerated motion in Newton's physics is replaced by the action of the distant stars and galaxies, acting through the dynamical gravitational field of general relativity to influence the selection of local inertial frames. A key step in Einstein's discovery of general relativity was his "hole argument" which pointed to the role of active diffeomorphisms in wiping out the background structure of the differential manifold, rendering bare manifold points physically meaningless. The result is that spacetime is NOT identified with metric and other fields on a manifold. It is identified with equivalence classes of such fields under diffeomorphisms. As already discussed above, in general relativity physical events are defined relationally and contingently in terms of physical effects perceptible there.

The use of diffeomorphisms in general relativity serves as an example of a general method for eliminating background structure, which is to introduce a first, kinematical, level of description which is encumbered by background structure and then to reduce the description to those engendered by a system of relations by defining physical observables to be those invariant under the action of some group acting on that background structure. This is called gauging away the background structure. Three major examples of this are local gauge transformations in Maxwell and Yang–Mills theory, spacetime diffeomorphisms in general relativity and reparameterizations of the string world sheet in string theory.

This may seem like an awkward and indirect way to proceed, but it allows us to write simple, local equations of motion for fields; this would be much more difficult once the naive local fields are eliminated because they are not gauge-invariant. Indeed, in all three cases all the physically meaningful observables are non-local.

Two relational paths to general relativity: Einstein and shape dynamics

Since the important work of Stachel [27], Barbour [18], and others, the interpretation of general relativity as a relational theory, and the key role of spacetime diffeomorphisms in wiping out background structure are well known and appreciated. But recently there is a new development in the interpretation of general relativity which is highly relevant for the nature of time, which is *shape dynamics* [28]. This gives a different way of defining general relativity by gauging away background structure.

In the old way developed by Einstein, space and time are treated on an equal footing. Spatial temporal coordinates provide a background structure which is gauged away by imposing spacetime diffeomorphism invariance. Indeed, one of the things that is gauged away in this story is any distinction between space and time, because there are spacetime diffeomorphisms that will turn any slicing of spacetime based on a sequence of spaces into any other. This means there is no meaning to simultaneity.

Barbour has emphasized for years that there is a nagging flaw in the beauty of this story. This resides in the fact that there is a big piece of background structure that is preserved in general relativity, which is an absolute scale for the size of objects. We must assume the existence of fixed scales of distance and time which can be compared with each other across the universe. In general relativity two clocks traveling different paths through spacetime will not stay synchronized. But their sizes will be preserved, so it makes absolute sense to say whether two objects far from each other in spacetime are the same size or not.

You can gauge away this background structure on top of the spacetime diffeomorphism invariance of general relativity, but the result will not be general relativity.[4] The reason is that imposing another gauge invariance changes the number of physical degrees of freedom. But the amazing thing is you can get to general relativity by trading the relativity of time of that theory for a relativity of spatial scale, so that the number of gauge transformations, and hence the counting of physical degrees of freedom, are unchanged. The resulting theory is called shape dynamics [28].

Shape dynamics lacks the freedom to change the slicing of spacetime into space and time. Consequently there is a preferred slicing, i.e. a preferred choice of time coordinate that has physical meaning. This means that there is now a physical meaning to the simultaneity of distant events. But physics on these fixed slices is invariant under local changes of distance scale.

Shape dynamics is not actually a new theory – it is for the most part just a reformulation of general relativity. Its preferred slices are expressible in the language of general relativity and, indeed, were already known to specialists of classical general relativity. They are called *constant mean curvature* (CMC) slices because certain components of curvature are constant on each slice. The technical statement is that shape dynamics is equivalent to general relativity so long as the spacetime has such slices – and most of them do. (This is modulo spacetimes with black hole horizons, the interiors of which may be different in shape dynamics than in general relativity [29].)

You might object that these preferred slicings represent a return to a Newtonian conception of absolute time. But they do not because the CMC condition is a dynamical condition so that which slices satisfy it depend on the distribution of matter, energy, and curvature throughout the universe. Moreover, because the predictions of shape dynamics match those of general relativity, these preferred slices cannot be detected by any local measurements. The slices nonetheless

[4] I know of two ways to combine spacetime diffeomorphism invariance with local scale invariance: one leads to a theory full of instabilities, the other was invented by Dirac and requires extra physical fields to implement.

play a role, which can be seen in how the Einstein equations enjoy an impressive simplification when expressed in terms of them.

Shape dynamics, just by its existence, has two important implications for the present argument. First, the impressive empirical success of general relativity cannot be taken as evidence for claims that the universe is fundamentally timeless, or even that there is no preferred simultaneity of distant events. These common claims are nullified by the fact that there is an alternative formulation of general relativity that does feature a preferred simultaneity.

Second, the preferred slices of shape dynamics give us a candidate for a global notion of time needed to provide an objective distinction between past, present, and future – and hence makes temporal naturalism a possible position to hold – consistent with current scientific knowledge.

Relational purism

A relational purist believes that once background structures are eliminated physics will be reduced to a description of nature purely in terms of relationships. An important example is the causal set program [26], which aims to develop a complete theory of quantum gravity – and hence nature – on the basis of an ontology of discrete events, the only attributes of which are bare causal relations. These are bare in the sense that *event A is a cause of event B* is a primitive. The causal set program denies there are any further properties, P of A and Q of B, such that *P of A causes Q of B.*

The aspiration of the causal set program is to construct the geometry of a Lorentzian spacetime approximately satisfying the Einstein equations as emergent only from a discrete set of events and their bare causal relations. To date this has not been realized except in trivial cases where the causal set is constructed by randomly sprinkling Minkowski spacetime with discrete events.

Impure relationalism: a role for intrinsic properties

Completion of the program of eliminating background structures does not imply that there can be no further properties of events except for their causal and other relations with other events. In an events ontology, you may eliminate all background structures – as the causal set program very nearly does – and still be left with an event having properties which are not specified when you know all the relations with other events. We can call such properties "intrinsic properties."

Intrinsic properties can be dynamical, in that they play a role in the laws of motion. For example, in an events ontology, energy and momentum can be intrinsic properties of events. They can play a role in dynamics and be transferred by causal links.

This view is realized in the energetic causal set framework [23, 24] according to which momentum and energy are fundamental and intrinsic and defined prior to spacetime. Indeed, in this approach dynamics is formulated strictly in terms of momentum and energy and causal relations. Position in spacetime is emergent and comes in at first just as Lagrange multipliers to enforce conservation of energy and momentum at events.

Dynamical pairings and relational versus intrinsic properties

I would like to argue that it is natural to suppose that energy and momentum are intrinsic in a world in which space and time are relational.[5] This is based on the fact that physics has a particular structure in which spacetime variables are paired with the dynamical variables, momentum and energy. This dual pairing is expressed by the Poisson brackets,

[5] Before going on it will be helpful to clear up two terminological confusions.

Intrinsic versus internal. If a property of an event is intrinsic it can be defined without regard to any relations to other events. That does not mean it plays no role in the dynamical equations of the theory. Let us reserve the term internal for a property of an event or a particle that plays no role in the laws of physics. Momentum can be intrinsic, but it is not internal. Qualia are intrinsic and appear to be internal.

Structural versus relational. By structural properties philosophers seem to mean the same thing that we physicists mean by relational properties. I prefer the term relational as structure seems to denote something static and hence timeless. A structural property seems to be one that transcends time or history, but temporal naturalism asserts there may be no such transcendental properties of nature. Structuralism seems to be a form of timeless naturalism which asserts that what is really real are structures which transcend particularity of time and place.

$$\{x^a, p_b\} = \delta^a_b \tag{1}$$

and has its most profound implication in Noether's theorem, which says that if there is a symmetry of a physical system under translation in a physical spatial coordinate x^a then the corresponding momentum p_a is conserved. Moreover, in the canonical formalism p_a is the generator of translations in x^a

$$\delta x^a = \alpha^b \{x^a, p_b\} = \alpha^a \tag{2}$$

This can be interpreted to say that if position is absolute and so has symmetries (i.e. nature is perfectly unchanged under translations in x^a), then the corresponding momentum p_a can be defined relationally, in terms of translations in x^a. But note that if space is defined relationally then there can be no perfect symmetry under translations in a space coordinate. The reason is that the identity of the indiscernible rules out symmetries because a symmetry is by definition a transformation from one physical state of a system to a distinct state which has identical physical properties. But Leibniz's principle asserts that no system can have two distinct identical states.[6] So if space is relational, we lose the relational definition of momentum as the generator of translations. So if space is relational, momentum can be intrinsic.

We can also turn this around and take the view that momentum is the primary quantity and is intrinsic, and define position relationally as the generator of translations in momentum space. This is the point of view taken by the framework of relative locality [30].

Adding intrinsic momentum and energy variables to the causal set description has an immediate advantage which is to resolve a longstanding problem with the purist causal set approach, which is to get a low-dimensional spacetime to emerge from a network of pure causal relations [23, 24].

[6] This reasoning does not rule out gauge symmetries which relate different mathematical descriptions of the same physical state.

So I would like to propose that generally we take momentum and energy as intrinsic quantities, defined at a level prior to the introduction or emergence of spacetime. Support for this comes from the Einstein equations:

$$R_{ab} - \frac{1}{2}g_{ab}R = 8GT_{ab} \tag{3}$$

The left-hand side is composed of geometric quantities that in general relativity are defined relationally. The right-hand side contains the energy–momentum tensor, which describes the distribution of energy and momentum on spacetime. Ever since Einstein began working on unified field theories, generations of theorists, down to late twentieth-century string theorists, have speculated that progress is to be achieved by reducing the right-hand side to geometry as well, so that physics can be expressed in a purely geometric structure. But perhaps that is mistaken – which would account for its not having worked definitively. Instead, we can posit that it is the left-hand side that is emergent from a more fundamental description in which energy and momentum are among the primary quantities, perhaps along with causal relations.

The Newtonian paradigm from the viewpoint of temporal naturalism

We can summarize the last few points by describing the proper role of the Newtonian paradigm for a temporal naturalist.

On cosmological scales the universe is unique and laws evolve; so the Newtonian paradigm breaks down. On fundamental scales events are also unique; so the Newtonian paradigm breaks down here also. Events are distinguished by their relational properties and thus must be fundamentally unique: there can be no simple and general laws on the fundamental scale.

Repeatable laws only arise on intermediate scales by coarse graining, which forgets information that makes events unique and allows them to be modeled as simple classes which come in vast numbers of instances. Hence the Newtonian paradigm works only on intermediate scales.

We can also see from this that intermediate-scale physics must be statistical, because similarity arises from neglect of information. It is interesting to wonder whether this might be the origin of quantum uncertainty. That is, the hidden variables needed to complete quantum theory, if we are to explain why individual events take place, must be relational. They must arise in adding the information needed to distinguish each event uniquely from all the others. Note that because the question of distinguishing individual events from others requires a comparison with others, such relational hidden variables must be non-local.

Finally, it may happen that uniqueness might sometimes not wash out on intermediate scales, leading to a breakdown of lawfulness, arising from novel states or events. This idea is developed below as the principle of precedence [13].

3 The setting: the puzzles of contemporary cosmology

The crisis of contemporary physics and cosmology begins with the triumphs of the standard model of particle physics and its counterpart in cosmology. The crisis arises out of our failure to go beyond the successes of these models to a deeper understanding of nature. As I will argue in detail, these failures have a common cause, which is the breakdown of the Newtonian paradigm when faced with cosmological questions. The questions left unanswered by these models then serve as the primary challenges to the science framed by the new, cosmological principles we have just outlined.

THE MESSAGE OF THE DATA FROM PARTICLE PHYSICS

What we know about the elementary particles and forces is neatly summarized in the standard model of particle physics, which has been tested by numerous experiments since first proposed in 1973. As of this date, experiments at Fermilab and CERN have so far failed to discover any phenomena not accounted for by the standard model.

The standard model describes the strong, weak, and electromagnetic interactions in terms of gauge fields, whose dynamics is determined by the gauge principle. More specifically, it is a quantum Yang–Mills theory coupled to chiral fermions and Higgs scalar fields. Two principles that may be derived from mathematical consistency, perturbative renormalizability and unitarity, restrict us to this class of theories in 3+1 spacetime dimensions. There are plausible arguments that this kind of theory would emerge from a much more general class of theories at energies low compared to the fundamental scale.

A key issue is that there is a great deal of arbitrariness in the standard model, even given it is a Yang–Mills theory. Three different kinds of choices must be made to get from the general class of gauge theories to a particular model.

The theory is first specified by fixing the gauge group. In the case of the standard model, the gauge group is $G^{SM} = SU(3) \times SU(2) \times U(1)$. The next choice is to pick the representations of that group that the matter fields come in. The standard model has three generations of fermions in a particular representation of G and an $SU(2)$ doublet of Higgs. The principle of mathematical consistency imposes some constraints on this choice, due to a phenomenon called gauge anomalies, but this is one of an infinite number of possible consistent choices.

Another somewhat selective principle is also satisfied by the standard model; this is that the theory is classically conformally invariant and chirally invariant, with an important exception which is the Higgs scale. As a result, all the coupling constants but one are dimensionless numbers and all the masses come from spontaneous symmetry breaking.

One then has to specify the values of all the coupling constants allowed to the theory by the principle of perturbative renormalizability. In the case of the standard model of particle physics, that is 29 dimensionless parameters – counting the masses and mixing angles of the neutrinos. There appears to be no principles of consistency which limit these choices within wide ranges.

We have no understanding of why the particular choices we observe were the ones made. For each of these three choices we require sufficient reason and we don't have it. There seems to be no reason so far put forward for the choice of the gauge group and representations.[1] When we come to the choice of the parameters of the standard model we face a very peculiar situation which is that the actual values appear to be highly unusual – or unnatural – in two senses.

Many of the dimensionless parameters are very small numbers. This strongly suggests a non-random distribution. These include the fine-structure constant, the cosmological constant, the quantum

[1] For attempts to explain the choices of gauge groups of the standard model see [31] and [32].

chromo dynamics (QCD) and weak scale in ratios to the Planck units, and the couplings between the Higgs and the fermions that determine the fermion masses. This is called the hierarchy problem. These unnatural choices are responsible for the very wide range of strengths and ranges of the different forces. These in turn make possible a universe with complex structures over a very wide range of scales. Many of the couplings appear to be finely tuned to give a universe with far more structure than would be present with random values of the parameters [8]. Our world has longlived stars, supernovas, a mix of helium and hydrogen, and on the order of a hundred stable nuclei with a wide variety of chemical properties, including carbon and oxygen, necessary for organic chemistry. None of this would be true were many of the parameters not fine-tuned to special values. This is called the special tuning problem.

The hierarchy problem and the special tuning problem pose unique challenges for proposing a sufficient and scientific reason for the laws of physics as we find them in our universe.

THE MESSAGE OF THE LARGE-SCALE ASTRONOMICAL DATA

Let us now turn from particle physics to cosmology, where we find similar issues of hierarchy and fine tuning.

The data tells a story of a universe with structure and complexity over a very wide range of scales [8]. To interpret it, we can rely on cosmological models, which are based on known physics. A model is not a theory, but it is very helpful as it tests how much of the data is explainable by known physics, and how much remains for a combination of the initial conditions and new physics closer to the initial singularity. The restriction to known physics can take us back to an era when the temperature was as high as that so far probed in high-energy experiments. This is a time before nucleosynthesis but after the era when inflation is usually conjectured to have occurred. We will discuss inflation later as a hypothesis which is posited to explain features of the initial conditions for the models.

On the largest observable scales (larger than 300 Mpc) the history of the universe appears to be well approximated by a Friedmann–Robertson–Walker (FRW) model, which is a homogeneous and isotropic solution to the Einstein equations, coupled to matter. The model has several parameters, including a positive cosmological constant, and the matter involves at least three components: electromagnetic radiation, baryonic matter, and dark matter. As in all homogeneous models there is a preferred cosmological time coordinate. Very remarkably, to within experimental accuracy, the spatial curvature of the constant-time surfaces appears to vanish. If there is a radius of spatial curvature, R, it is greater than the present Hubble scale.

Even at this very rough level, there are remarkable features of the data that suggest the initial conditions were very special. In addition to the vanishing of the spatial curvature, these include the fact that the cosmological constant is extremely small for a fundamental parameter of the laws of physics; in Planck units it is 10^{-120}. These also include two coincidences of time scales which concern the times of transitions between eras when different kinds of matter dominate the expansion rate. The first is that the transition from radiation to matter domination, t_m, took place at roughly the time, t_d, when the electromagnetic field decoupled from the matter because the universe had cooled sufficiently for atoms to form.

The second coincidence is that the present age of the universe, t, now appears roughly to be the crossover time, t_c, from matter to cosmological constant domination.

Another very unusual feature of the initial conditions is that the universe must be modeled as starting off very hot, where the matter is concerned, but very cold where the gravitational radiation is concerned. Out of the infinite number of solutions to the Einstein equations, which show a universe expanding from an initial singularity, the data suggests that the simplest suffice, which are featureless solutions that are homogeneous and isotropic. This means that there are no gravitational waves initially, so all gravitational waves that are detected can be assumed to have been radiated by matter sources.

Penrose has proposed a principle to characterize the absence of initial gravitational radiation, which he calls the Weyl curvature hypothesis [35]. This is that the Weyl curvature, which measures the strength of gravitational radiation, vanishes initially, not just in the models, but as a principle to be imposed on the true spacetime geometry. This is a time-asymmetric condition, as it is to be imposed only on initial, and not on final, singularities; otherwise it would forbid the formation of black holes.

One can comment that current knowledge is consistent with a similar condition being extended also to electromagnetic radiation. This is that there is no initial source-free electromagnetic radiation. Indeed, as pointed out by Weinstein [36], no electromagnetic radiation has ever been observed that does not plausibly point back to matter sources. If imposed, this condition would be time-irreversible as well, and would account for the electromagnetic arrow of time, independent of the thermodynamic arrow.

If one goes to smaller, but still very large, scales one sees a history of growth of large-scale structure, which appears to be driven by the dark matter. This is captured by more detailed models, some of them numerical simulations, which permit the study of the behavior of perturbations around the homogeneous solutions. These models are nonetheless still models, and have restricted domains of validity. These appear however to allow reliable conclusions to be drawn about the large-scale structure down to the scales of galaxies.

In these models, the baryons fall into potential wells formed by the growing structure in the dark matter distribution to form galaxies, whose distribution is highly clustered. The galaxies are in the present day dominated by dark matter, which is necessary to explain both their rotation curves and how they are bound into clusters. It appears that the seeds of this structure formation are imprinted in the cosmic microwave background (CMB) and that the whole story of large-scale structure is remarkably simple, because it evolves from an initial distribution of fluctuations in the density of dark matter which has the following characteristics.

- The fluctuations are small, with initially $\delta\rho/\rho \approx 10^{-5}$.
- The fluctuations are nearly scale-invariant.
- The fluctuations are Gaussian, which is to say they have no other structures.

A major challenge of cosmological theories is to explain the origin and features of these fluctuations.

I should caution that the dark matter has so far not been directly detected, so it is prudent to consider the possibility that it doesn't exist, but instead the Einstein equations are modified so as to produce the effects attributed to dark matter. This option has been investigated and, on the scale of individual galaxies, does remarkably well in explaining the observed rotation curves of large numbers of galaxies [37]. But so far there has not been a version of it that tells a compelling story about how the large-scale structure formed.

We next come down to the scale of individual galaxies. Here a lot more physics becomes relevant because of all the nonlinearities and chemistry associated with star formation, the dynamics of gas and dust, feedback due to star light, supernovas, etc. What is remarkable is the range of scales over which nonlinear and non-equilibrium phenomena, such as feedback governing star formation, is relevant. The result is to drive further nonlinear and non-equilibrium phenomena all the way down to molecular scales, up to the origin of life and its continuation. One of the things to be explained is why the whole universe from the largest scales down to the smallest produces a context that is friendly for life. This includes stable, longlived stars, needed to keep the surfaces of planets out of equilibrium for the billions of years life needs to develop, and plentiful production of carbon, oxygen, and the other chemical elements needed for life. I mention this because the question of why the universe is so biofriendly is properly a cosmological question, for this fact depends on many coincidences in the choices of laws and initial conditions.

WHAT QUESTIONS ARE IMPERATIVE, GIVEN THE DATA?

While there are many details of the standard model of cosmology which remain to be worked out, the key questions for us concern the choice of initial conditions. We have already mentioned some of them, but there are a few more to mention.

There are three main cosmological puzzles which motivate attempts to go beyond the picture I've just sketched.

- *The horizon problem*. The observations of the cosmic microwave background provide a snapshot of the state of the universe at a time called decoupling, roughly a million years after the big bang, when the universe first cooled sufficiently to allow hydrogen atoms to remain bound. Before this the electrons streamed free and the universe was a plasma. Observations show that, up to fluctuations of a few parts in a hundred thousand, the universe was then in thermal equilibrium, with a constant temperature coming from all directions in the sky. However, if we run solutions of general relativity with ordinary radiation back from that point, there was not enough time for the whole sky to have been in causal contact between the cosmological singularity and the time of decoupling. The largest regions which could have been in causal contact are now disks about 2 degrees on the sky. So by what processes has the universe as a whole come to thermal equilibrium at a single temperature?

  This problem is deepened by observations of the patterns in the fluctuations around equilibrium, some correlations of which extend up to 60 degrees on the sky. How could these patterns have been formed if causal processes could only connect regions closer than 2 degrees in the sky?
- *The flatness problem*. There is good evidence that on sufficiently large scales the universe is on average homogeneous. This means that the spatial geometry is characterized by a single number, which is its radius of curvature. If the surfaces are of constant time, which are homogeneous spheres, then these are measures of the size of those spheres. Observations suggest that this number is larger than the distances we can observe, corresponding to the sphere being so large that the apparent

geometry is flat. The question is why? This is an aspect of the specialness of the initial conditions. It is especially acute because it scales the wrong way: if you go back in time and ask how special the initial conditions would have had to have been, the answer is that the conditions get less probable rather than more probable as we go back in time. One way to parameterize this is with the parameter Ω, which is the ratio of the total energy density to that on the boundary between a closed and open universe.

- *The defects problem.* If the standard model is replaced by a further unification that governs the very early stages, then there may have been one or more phase transitions in the early universe. These will have created large numbers of defects like monopoles. But none are observed. Where are they?

We can add to this list those already discussed.

- The two coincidence problems.
- The initial spectrum of density fluctuations.
- The lack of incoming free gravitational and electromagnetic radiation.

We can note that the initial conditions of the universe would require explanation, whatever they were deduced to be. But it still could be the case that the observed initial conditions were generic in some well-defined sense, so nothing would require explanation except a lack of features requiring explanation. But our universe seems far from generic; on the contrary, the initial conditions of our universe seem extraordinarily special.

To this we can add the problem of giving a sufficient reason for the selection of the effective laws that govern our universe. To put these in context we now discuss the following.

WHAT FEATURES OF THE STANDARD COSMOLOGICAL MODEL ARE UNCONSTRAINED BY THE DATA?

The models we have been discussing fail to be theories in three senses. First and most obviously, they are based on truncation of the known laws of physics to a very coarse grained description with a handful of degrees of freedom. In the simplest case the infinite number of degrees of freedom of the dynamical geometry described by Einstein's equations are reduced to a single function, which is the expansion rate as a function of a single parameter, the scale size of the universe.

The second sense is that quantum mechanics is hardly included, as it is sufficient for most of the range of phenomena modeled to treat matter as a classical fluid.

The third sense is that our knowledge of the laws of physics is incomplete, so that the expected unification of gravity with quantum physics and with the other fundamental forces is not considered. This is not necessary either for the phenomena modeled, but it will be important to recall that at best our cosmological models are based on effective theories, which are understood to be approximations to an unknown fundamental theory, valid within some range of scales. That range of scales is bounded both below and above, in energy, time, and distance scales.

It is essential to understand this when attempting to extrapolate the cosmological models outside the domain we have described, where it is constrained by observation. There are three big questions about cosmology that are constrained neither by data nor by theory, so far, that are nonetheless crucial for any attempt to extend the cosmological models we have been discussing to true cosmological theories.

What happened at very early times, closer to the initial singularity? What will happen to our universe in the far future? What is there very far away from us, outside our cosmological horizon?

Let us discuss each of these, in turn.

What happened at very early times?

The laws used in the cosmological models are no longer justified by independent experimental check when the temperature exceeds 1 TeV = 10^3 GeV. If we extrapolate the models before that, the temperature continues to rise. It passes two scales of great theoretical interest, the unification scale, of perhaps 10^{15} GeV, where there may be unification of the gauge groups of the standard model of particle physics, and the Planck scale, $E_P = 10^{19}$ GeV, where quantum gravity becomes unavoidable. Then within a Planck time (10^{-43} s) before that the models all predict a singularity, where time stops.

The key question any unification of physics must address is whether the singularity is really there, or whether it is eliminated, so the universe had a past before the big bang. If the singularity is real, then we must confront the demand to give a sufficient reason for the arbitrary choice of initial conditions there. If the singularity is absent, then the sufficient reason for choices of initial conditions and laws may lie in the world before the big bang.

Brief review of the singularity theorems

Thus, the key results which frame the choice we have available when we seek to extend the cosmological model to a theory are the singularity theorems of classical general relativity; for a general review of singularity theorems, see [38]. The possibility of such theorems was discovered by Roger Penrose [39], who proved the first of a succession of theorems establishing that generic solutions to general relativity are singular, in a sense we will define below. Penrose's original theorem applied to black hole singularities, shortly after he and Hawking applied a similar method to prove a cosmological singularity theorem [40].

A solution to the field equations of general relativity is given by a metric on spacetime, together with matter fields that satisfy additional equations of motion. The metric specifies the causal structure, which specifies, for each two events, whether one is to the causal future of the other, or whether they are causally unrelated. Two events that are causally related can be connected either by a time-like or a null curve. From the metric one can also compute the curvature of spacetime, from which the tidal forces can be deduced.

There are several distinct senses in which a solution to the Einstein equations can be singular. The most intuitive is if components of the curvature tensor, normalized appropriately to refer to physically measurable quantities, become infinite. The energy density of matter can also become infinite; typically, but not necessarily, these will occur together. When the curvature is sufficiently singular the set on which it is singular is usually a boundary of the spacetime, because the field equations cannot be usefully extended past those points.

The singularities themselves have causal properties. A singularity may, for example, be spacelike. This means that the set of events a very small distance from the singularity is a spacelike surface, i.e. the points of that surface are causally unrelated.

A second notion of singularity follows from the expectation that singularities of curvature induce a boundary. This is that timelike or null geodesics cannot be extended beyond a singular boundary. It is sufficient to detect such a boundary if there are timelike geodesics which cannot be extended into the past or future more than a finite amount of proper time.

The Hawking–Penrose theorem applies to cosmological solutions of the Einstein equations. These are solutions where the spatial surfaces are compact without boundary. The theorem states that any cosmological solution which satisfies a list of conditions will have timelike geodesics that cannot be extended arbitrarily far into the past.

The conditions include (neglecting technical subtleties) the following:

1. The energy density of matter is everywhere positive.
2. There is a spacelike surface at which the universe is everywhere expanding.
3. The solution is sufficiently generic, i.e. without special symmetries.
4. The field equations of general relativity hold everywhere.

We note that these are all physically plausible, given present knowledge and observations. The consequence is quite strong, because the assumptions are so weak, plausible, and general.

The idea that a timelike geodesic cannot be extended arbitrarily to the past can be interpreted as saying that there are observers, carrying a clock, whose past history stops at a finite reading of that clock. We note that the theorem does not, strictly speaking, imply that the curvature or density of energy becomes singular. It is possible there are other mechanisms that could induce a boundary that prevents the observer from having a past that is arbitrarily long. But in all known generic cases, the cause of the incompleteness of the observers' histories is that the curvature and density of energy have become infinite in a finite time.

It is important to dispel some false impressions about the cosmological singularity theorems which are widely spread due to misleading accounts in some popularizations.

- The singularity does not occur at a point, from which the universe expands. Cosmological singularities are entire spacelike surfaces. The curvature and energy density become infinite all over space simultaneously, a finite time to the past of typical observers. A universe can even be infinite in spatial volume an arbitrarily short amount of time after the singularity.
- The singularity is not a moment of frozen time. The singular set is not in fact part of the spacetime geometry modeled by the metric. The singular set is a boundary which is a set of limit points of the spacetime geometry. There is no set in the spacetime geometry where time is not flowing.
- The singularity does not restrict the solution of the Einstein equations. The whole point is that generic solutions are singular, which is to say that there are an infinite number of solutions to the Einstein equations which look like large expanding universes at late times, have an initial cosmological singularity, but differ by details of the geometry just after the singularity. There may, for example, be lots of gravitational waves, and black holes, present just after the singularity. So the singularity does not eliminate the need to specify an infinite number of initial conditions to determine which solution of the Einstein equations describes our universe.
- There is no event, force or influence which starts the universe evolving. The cosmological singularities are simply boundaries to the extension of a spacetime history to the past. There is nothing there, before the singularity, which starts the universe going.

At the time they were published the singularity theorems came as a shock, as there was a widely held expectation, among experts, that the singularities which were present in the solutions then known were artifacts of the very large amount of symmetry of those solutions.

The meaning of the singularity theorems

There are two very different conclusions that might be drawn from the generality and power of the singularity theorems. The first option is that we have discovered a surprising fact about time, which is that it must have a beginning. The second option is that they mark a limitation to the validity of general relativity. Either the language of description of spacetime, the Einstein field equations, or both, are ceasing to become valid. This limitation of the domain of applicability of Einstein's theory is then not a sign that time has a beginning; it is a sign that the conditions are predicted to get so extreme that additional laws or principles must come into play.

We find the first option highly implausible for several interrelated reasons.

First, the history of science tells us that whenever we have had a choice between an extreme metaphysical conclusion and the modest realization that a theory is being extended past its domain of validity, the latter has turned out to be the case. We know of no reason why the resolution of the present issue should be different. We should only be driven to the first option were there no chance of the second working out.

We know for a fact that the laws that imply the singularity theorems are incomplete in at least one way: because quantum effects are neglected. It was then hypothesized long ago, by deWitt and Wheeler, that quantum effects eliminate the singularities, leading to time becoming extendible into the past – before where classical general relativity becomes singular. We will call this the deWitt–Wheeler hypothesis.

There are at least three precedents for singularities (in the sense of infinite or runaway divergences) in non-quantum theories being resolved by the introduction of quantum effects. The ultraviolet catastrophe in the thermodynamics of radiation found by Jeans in the 1890s was resolved by Planck's quantum hypothesis in 1900. In classical atoms, the electron would spiral in a finite time into the nucleus emitting an infinite amount of radiation; this was a crisis when Rutherford discovered the nucleus was much smaller than the atom, but it was quickly resolved by Bohr's application of the quantum hypothesis to the orbits of electrons in atoms. Third, in classical electromagnetism, point particles carry around electric fields that contain an infinite amount of energy. This is first reduced greatly by quantum effects to a logarithmic divergence and then eliminated by the procedure of renormalization.

The hypothesis that quantum effects eliminate spacetime singularities is highly plausible in light of these historical precedents.

There is a large literature investigating the deWitt–Wheeler hypothesis, using increasingly sophisticated mathematical tools to investigate the implications of a merger of quantum physics and general relativity. While the problem of quantum gravity is not yet completely solved, the answer cannot be considered definitive. But there are a large class of models in which the hypothesis may be investigated, and in all cases where the model is sufficiently rigorously or carefully studied to provide a useful answer – the answer is positive – the singularity is replaced by a history that extends into the past [41]. What these models show is that before the singularity the universe was contracting, then it passes through a region in which quantum effects are very determinate, after which the solution expands as in the early stages of a cosmological solution to the Einstein equations. These are called bounce solutions.

These require no special hypotheses beyond the application of quantum dynamics to models of cosmology.[2]

[2] But several approaches to quantum gravity, such as string theory, are based on additional hypotheses. In this context as well, there are plausible arguments that singularities are eliminated and replaced by bounces [42].

But the strongest reason for betting on the second option over the first is that a number of key cosmological mysteries are not solvable under the first option, but have clear possibilities for solution under the second, as we will describe below.

WHAT WILL HAPPEN TO THE FAR FUTURE OF OUR UNIVERSE?

According to the standard cosmological model we have been discussing, the universe is presently in a transition from a phase of domination by matter to a phase of domination by dark energy. The simplest models treat the dark energy as a cosmological constant, which means that its density is unchanged in space and time. All other forms of energy density dilute as the universe expands, so that in some billions of years the energy density of the universe will be dominated by the cosmological constant.

If that is the true physics, then the rest of the future of the universe is both bleak and paradoxical.

The future is bleak because the universe expands exponentially for all time, so that the dominant trend is an exponentially increasing dilution. After not too long the galaxies mostly go out of sight of each other, then all the stars die, then not much happens except cooling relics. Long after that black holes evaporate down to whatever state quantum gravity allows. Then an eternity of almost empty nothingness described to a very good approximation by empty de Sitter spacetime.

The future is paradoxical because of a phenomena akin to Hawking radiation in which the universe is filled with a thermal bath of photons and other quanta at a temperature, T, which is a function of the Hubble scale. That scale is constant and set by the cosmological constant. Thus if these predictions are right the late universe is not quite empty. One then has a cosmological scenario reminiscent of the cosmological speculations of Boltzmann: an eternity spent in a thermal bath.

The future is then much like Boltzmann discussed more than a century ago. The argument employs the fact that while the heat bath is in thermal equilibrium, and hence overall a state of maximal entropy or disorder, there are in any such thermal ensemble fluctuations by which a small region of space can have any lower entropy for a short period of time, till the entropy is again extremized. This, it is claimed, allows any arbitrarily improbably structured configuration to exist as a fluctuation in a small region of space, for a short span of time.

The claim is that since the thermal bath occupies an infinite spacetime volume, no configuration of a small subsystem is so improbably structured as to fail to exist somewhere in space and time. In fact, any improbable configuration will occur an infinite number of times in the eternal future of the hot universe.

In this scenario there are two ways to make a human brain, together with a full spectrum of memory and, briefly – it may be claimed – thought. One is the way we believe we arose, through biological evolution following chemical evolution, early in the history of the universe, long before it reached the dilute de Sitter phase. A certain finite number of such brains will have existed during this phase (making up an ensemble of all biologically evolved human beings who will ever live). The second way to make a brain with memories and thoughts is via a thermal fluctuation.

Now let us consider the ensemble of all brains that will ever exist. This is an infinite ensemble, almost every member of which is the second kind – the result of a brief improbable statistical fluctuation in a heat bath. This is because there are an infinite number of those, while the number of biological humans who will ever live will be finite. Therefore any argument of typicality of our situation as observers or human beings, or intelligent beings, etc., is short-circuited and leads to the prediction that you and I should most likely be experiencing a brief moment of existence as a random fluctuation in a universe filled mainly with radiation at temperature T.

But this manifestly disagrees with observation. We are not random fluctuations, we are, amazingly, creatures who evolved biologically in an evolving, active universe, far from thermal equilibrium.

Something must be incorrect or fallacious about the argument just presented. Of course there is no reason to believe that the cosmological constant is really constant. There are cosmological models in which a scalar field has a slowly varying vacuum expectation value, leading to a value of the dark energy which changes slowly on cosmological time scales. An ambitious program of observations is underway to try to capture the resulting evolution in time of the dark energy.

If the dark energy is dynamical, then the far future of our universe can be very different from that just discussed for a constant cosmological constant. Unfortunately, given present knowledge, it is not possible to predict reliably the far future of our universe.

There is another issue regarding the future of our universe which must be discussed, which is the fate of the future singularities in the black holes which form during the universe's history. This is a real and pressing question, as there is little doubt our universe forms many black holes, perhaps 10^{18}. Classical general relativity predicts that these each contain future singularities; indeed there is a singularity theorem, first proved by Penrose, that establishes this on very general grounds [39]. It also followed from a short list of assumptions, except that the second assumption of the cosmological singularity theorem is replaced by the assumption that there is a region containing *future trapped surfaces*. This is a two-dimensional surface, from which the light rays leaving it from both sides, going into the future, are converging. This is common to the interior of the event horizon of a black hole.

The fate of these singularities can only be decided by knowing the correct unification of general relativity with quantum theory. As in the cosmological singularities, there are models, computations and arguments that support the hypothesis that black hole singularities are removed by quantum effects [43]. If they are, then the future of the regions that were almost singular is part of the future of the universe.

One possibility is that the black hole singularities bounce to create new expanding universes [8, 6]. If this is the case then the future of our universe includes the histories of all the universes formed from black holes it contains. It is natural also to hypothesize that our universe is the product of a bouncing singularity in a black hole in a previous universe.

This is of course not the only possibility. It is also possible that the black hole singularity bounces to produce a quantum region of spacetime which eventually reconnects to the universe when the horizon of the black hole evaporates away [44]. It is even possible that some black holes lead to new universes, while others evaporate away. Which of these very different possible futures of our universe will be the case depends on the details of the dynamics of quantum geometry and hence on the quantum theory of gravity.

WHAT IS VERY FAR AWAY FROM US, OUTSIDE THE COSMOLOGICAL HORIZON?

Our view of the universe is constrained by the constancy of the speed of light to a region about 46 billion light years in radius. This defines the cosmological, or Hubble, horizon. So far as we can see, the hypothesis of approximate homogeneity holds up till there. But there is no way to tell from observations what happens further away.

There are two gross features of the cosmological models which are not determined by the observational data:

- Does the universe continue to be homogeneous on larger scales or do new features appear on scales larger than our current horizon size?
- Does the universe have a closed or open spatial topology? And if closed, what is the topology?

With respect to the first, we are on tricky ground. There is simply no way to observationally confirm any hypothesis about the structure of the universe outside of our Hubble horizon. However, the cosmological models all require that the universe be larger than our horizon. So long as there is an initial singularity a finite time in the past, it will never be possible to observe most of the spatial extent of the universe.

This means that there is a great temptation to make hypotheses about the universe outside of our horizon which are neither verifiable nor falsifiable. If we do not do this our cosmological models remain incomplete, in the sense that they tell us of the existence of a part of the world which is unknown. But if we do make hypotheses about the universe outside of our horizon, we risk making claims that are not checkable by observations.

The homogeneous cosmological models implicitly make an assumption about the universe outside of our horizon, because they assume all the spatial slices are homogeneous. This should be understood as an expedient rather than a principle, for the use of these models is to interpret data from within our cosmological horizon. The empirical content of these models is unchanged if they are truncated to the spacetime within our backwards light cone.

Many cosmologists have proposed the cosmological principle, which states that our position in the universe is not special. This implies that if our universe appears to be approximately homogeneous and isotropic from our location, it will be from any location. When applied within our horizon, this is an empirical hypothesis which can be compared with data. When it is applied outside of our horizon it becomes an unscientific metaphysical principle of the kind that science should avoid.

The more scientific response is simply to accept that there are regions of the universe that we will not be able to observe, at least so far as there is an initial singularity. This gives us reason to hope that the singularity is an artifact of classical general relativity and will be removed in a more correct theory, making it perhaps possible to observe the whole universe.

With regard to the topology of space, general relativity gives us two choices. The universe can be closed, without boundary. It may then have the topology of a sphere, a torus, or something more exotic. The universe may nonetheless be so big that we cannot see far enough (again assuming an initial singularity) to see observationally that it is closed, or it may be so small that we could find evidence that we are seeing all the way around it.

The other possibility allowed by general relativity is that the topology is open. But this brings with it several issues because the set of limit points infinitely far away from any observer constitute a boundary of an infinite, open universe. Careful analysis of the Einstein equations shows that the boundary at infinity cannot be ignored. To define the field equations and specify their solutions, boundary conditions must be imposed at the boundary at infinity. Otherwise the solutions of the theory do not come from a variational principle, which is the foundation of both classical and quantum dynamics. There must, it turns out, also be special boundary terms in the action that defines the variational principle, otherwise there are no solutions.

This is an unacceptable setting for a true cosmological theory. Boundary conditions and boundary terms involve the specification of information and conditions coming from outside the system studied. They are proper in the study of isolated systems, such as stars, galaxies and black holes, where the observations may be modeled as being made at infinity. But a cosmological theory in principle describes a system that is not the truncation of anything larger. Several of our principles rule out applying general relativity with boundary terms and conditions to the whole universe, including no ideal or external elements and explanatory closure.

This in no way implies that our past light cone cannot be modeled for simplicity by a homogeneous model with an open topology, for nothing at all would be changed, were the solution imbedded in a closed spatial topology, by making identifications outside our light cone. The use of an open topology in these models is just a gesture of convenience. Its successes in no way commit us to predict that the universe is spatially open or infinite.

THE OPTIONS: PLURALITY OR SUCCESSION

The knowledge we have of our universe does not suffice to answer the questions we have raised here. Given what has just been said, there are two directions to look for more facts about the world which will help us answer those questions:

1. In the past, particularly if the initial singularity is removed. This is the option of *succession*, in which our universe goes through several eras, separated by time, of which ours is one.
2. Outside of our light cone. This is the option of plurality, in which our observable universe exists simultaneously with a population of regions or universes outside of causal contact.

As we argue in detail below, only the first leads to hypotheses checkable by experiment. The second leads to multiverse cosmology which does not yield falsifiable predictions and so diminishes the hopes of science to continue to progress.

Our first principle, the uniqueness of the universe, forbids this disaster and restricts us to the first option.

4 Hypotheses for a new cosmology

In Chapter 1, we surveyed the experimental and observational situation in cosmology and proposed five principles which we believe must constrain the construction of any properly cosmological theory:

1. The principle of differential sufficient reason.
2. The principle of the identity of the indiscernible.
3. Explanatory closure.
4. No unreciprocated action.
5. Falsifiability and strong confirmability.

Having set the scene, we now propose three hypotheses which we suggest should guide the discovery of a cosmological theory which satisfies these principles. I will state the hypotheses and then discuss each in turn.

1. The uniqueness of the universe.
2. The reality of time.
3. Mathematics as the study of systems of evoked relationships, inspired by observations of nature.

Someone might call these also principles, but I want to stress that the first two are hypotheses about nature, which might be confirmed or disconfirmed as science progresses. They have force because they suggest different kinds of experiments and different results than approaches which deny them or embrace hypotheses which conflict with them.

Now I elaborate on each of these.

THE UNIQUENESS OF THE UNIVERSE

There is a single causally connected and causally closed universe. The universe is not a member of an ensemble of other simultaneously existing universes, nor does it have any copies.

This hypothesis has obvious aspects and implications that are more subtle. To elaborate on it we need to specify what we mean by the universe.

The universe, we will assume, has a history which is a single causally connected set of events. This implies that the set of events forms a causal set, by which we mean a partially ordered set. By causally connected we mean that any two events have at least one event which is in their common causal pasts. We also require that the universe contain all the causes of its events so that it satisfies the principle of causal closure.

When we insist that the universe have no copies we mean by this neither in a material nor in any other sense. In particular we deny the possibility that there exists a mathematical object which is isomorphic to the history of the universe, where by isomorphic we mean that every property of the universe or its history is mirrored in a property of the mathematical object.

To show this it suffices to exhibit one property of reality which is not a property of any mathematical object, which is that in the real world it is always some present moment.

We also deny the Mathematical Universe Hypothesis of Tegmark [5], according to which the universe is one mathematical object existing in an ensemble of all possible mathematical objects, all of which exist. This contradicts the hypothesis that the universe is unique and is not part of any ensemble.

To assert that there is no mathematical object isomorphic to the universe does not imply we must deny the evident usefulness of mathematics in physics. But it does challenge us to describe the actual role of mathematical physics, a challenge we accept in Chapter 5.

THE REALITY OF TIME

By the assertion of the reality of time we mean that all that is real is real in a present moment which is one of a series of moments. This implies that all that is true about reality is true about a property or aspect of a moment.

The world has then no properties which are not properties of a moment. To the extent that a general law is true, that is a property of a moment or moments. That is to say that a physical law is at best part of or an aspect of the state of the universe at a given time.

The future is not now real and there can be no definite facts of the matter about the future.

The past is also not real, but is different from the future because it has been real. Consequently, there can be facts of the matter about past moments. This is possible within temporal naturalism if we regard something having once been real sufficient for facts to exist about the properties of once real events. However, we can have evidence about the truth or falsity of propositions about the past only to the extent that records, fossils, memories or remnants of it are parts of a present moment. When we propose and investigate hypotheses about general laws we seek to confirm or falsify them by comparing them to records of past experiments and observations. The most secure knowledge we can have about a general law is that it has been confirmed by records of past experiments and observations – and this is a property of a present moment.

A law of motion is a property of a present moment because it is a summary of or explanation of records of past experiments, records which are themselves aspects of the state of the world at a present moment.

This does not preclude the possibility that the universe could have properties – such as a law – which hold in all moments. But there is no special category of timeless truths about nature. More to the point we know we need a new paradigm for a cosmological law to avoid the cosmological dilemma and cosmological fallacy.

Within this present perspective what requires explanation is why objects and other features of the universe persist. Furthermore, the fact that laws do succeed to explain features of the universe persistent over billions of years suggests that there are physical processes which facilitate, if not guarantee, the persistence of causes. To explain these, the novel paradigm we seek must rest on a hypothesis that there are causal processes which relate present events and properties to past events and properties.

Causality does not however imply determinism, as we discuss below. To avoid the cosmological dilemma and the cosmological fallacy, as well as to address the "why these laws" and "why these initial conditions" questions, we need a paradigm for a dynamical law, operating on a cosmological scale, that does not fit the Newtonian paradigm. However, it is difficult to avoid the idea that there is a state of the universe at a given moment that contains a description of its properties. The state must be different at different times, because the universe is – and also because to have two times with the universe in the same state would be to violate the principle of the identity of the indiscernible. If change is to be comprehensible there must be something like a law of motion. But its form must avoid the Newtonian paradigm. To achieve this, at least one of the timeless structures that define the Newtonian paradigm – the state space and the dynamical law – must be time-dependent. Thus, we are forced to adopt the notion that laws of nature evolve in time.

This conclusion is not new and was argued for by Peirce in 1892 [14] when he asserted that:

> *To suppose universal laws of nature capable of being apprehended by the mind and yet having no reason for their special forms, but standing inexplicable and irrational, is hardly a justifiable position. Uniformities are precisely the sort of facts that need to be accounted for... Law is par excellence the thing that wants a reason. Now the only possible way of accounting for the laws of nature and for uniformity in general is to suppose them results of evolution.*

How and when the laws change is then a scientific question which is a large part of the problem of inventing a scientific cosmological theory. It is possible that the laws change slowly over time, but there is only very weak and contested evidence to suggest this may be the case. It is more reasonable to suppose with Wheeler that the laws change at events such as our big bang, which is best interpreted not as the beginning of time but as a transition from an earlier stage of the universe – a transition during which the laws changed.

The notion of evolving laws brings with it a danger that must be circumvented, which is to re-create the Newtonian paradigm at the level of the laws. It is natural to represent a set of possible laws as points in a timeless landscape of laws, on which some meta-law acts to evolve the laws through time. The problem is that this reproduces the problems extending the Newtonian paradigm to cosmology, for one has to ask for a justification of the meta-law. Also the current laws will depend on initial conditions in the space of laws, so one also reproduces the initial conditions problem. Hence, the appeal to evolving laws will not solve the problems that led us to abandon the Newtonian paradigm if we have a formulation of evolving laws that reproduces that paradigm. Sufficient reason appears to be thwarted.

On the other hand if the evolution of laws is itself lawless we also do not further our understanding of the "why these laws" or "why these initial conditions" problem. This is the *meta-laws dilemma.*

Fortunately there are approaches to evolving laws that avoid this dilemma, which will be discussed in Chapter 6.

DOES A REAL TIME CONFLICT WITH THE RELATIVITY OF SIMULTANEITY?

The assertion that what is real is real in a moment conflicts with the relativity of simultaneity according to which the definition of simultaneous but distant events depends on the motion of an observer. Unless we want to retreat to a kind of event or observer solipsism in which what is real is relative to observers or events, we need a real and global notion of the present.

A well-known argument shows that this is impossible in special or general relativity [45]. To run the argument, assume that there is a transitive and symmetric notion of "as real as" or "equally real." If *A* and *B* are equally real then either *A* and *B* are both real or *A* and *B* are both unreal. To see why we must require this notion to be transitive, consider the situation in which *A* and *B* are equally real and *B* and *C* are equally real, but *A* is real while *C* is unreal. This implies that *B* is both real and unreal, i.e. this gives reality a context dependence that is inconsistent with a notion of objective reality.

For the notion of "equally real" to be objective we must also assume it is observer-independent, so that if one observer sees that two events are equally real, all observers will agree.

Next, we want to assert that two events *A* and *B* are equally real if they occur at the same moment of time. Then according to the definition of the reality of time given above, they will be both real during that moment, but unreal before and after. From an operational perspective this can be translated to the assertion that two events *A* and *B* are equally real if there is an observer, Bob, who sees them to be simultaneous.

This however runs afoul of the relativity of simultaneity. To see this, consider any two events *A* and *C* in Minkowski spacetime such that *C* is in the causal future of *A*. Then there will be an event *D*, spacelike to both *A* and *C*, such that there is an observer, Alice, who sees *A* and *D* to be simultaneous and another observer, Eve, who sees *D* to be simultaneous with *C*. But this implies that *A* and *D* are equally real, as are *C* and *D*. Since the relation "equally real" is transitive, it means that *A* and *C* are equally real. But this implies that the past and future are as real as the present. Hence we contradict our assertion that *A* and *B* are only real if they are in the present moment. We reach instead the conclusion that all events in spacetime are equally real, which implies the block-universe picture in which there is no ontological difference between past, present and future.

The principles of special and general relativity are confirmed to very high accuracy in many experiments. Predictions of special relativity are confirmed up to gamma factors of 10^{11} [46]; and recently the OPERA experiment pushes this up to 10^{12} [47]. Test of the breakdown of Lorentz invariance confirms special relativity up to the order of corrections proportional to energies in Planck units. These constrain the extent to which the relativity of inertial observers can fail for experiments local in spacetime.

To give up the relativity of simultaneity is a major step, but we have very good reason to consider it, if we are to escape the cosmological dilemma and fallacy and move toward a cosmological theory that could give sufficient reason for choices of laws and initial conditions.

A global preferred time would have to be relational, in that it would be determined by the dynamics and state of the universe as a whole. It would thus not be determinable in terms of information local to an observer. Such a relational local time could then be consistent with the relativity of simultaneity holding locally in regions of spacetime.

There is precedent for such a relational, dynamically determined global time in the Barbour–Bertotti model [48]. This raises the question of whether general relativity can be reformulated as a theory with a preferred dynamically determined global time. The answer is yes; this is shown by the existence of a formulation of general relativity as a theory defined on a fixed three-surface which evolves in a global time coordinate. This formulation, called shape dynamics [28], shares with general relativity diffeomorphism invariance on the three-dimensional spacelike surfaces but replaces the many-fingered time invariance with a new local gauge invariance which is invariance under local three-dimensional conformal transformations. These transformations however are restricted to preserve the volume of the universe. The spatial volume then becomes an observable and can be used as a time parameter.

There is then a theorem that shows that shape dynamics is equivalent, up to a gauge transformation, to general relativity in a particular choice of slicing of spacetime into an evolving three-dimensional space. That slicing is called constant-mean curvature slicing and it has been shown to exist in a large and generic set of solutions to the Einstein equations, which includes spatially compact solutions, external to horizons. This partial equivalence is sufficient for the description of nature. There then can be no objection in principle to taking the view that our universe is described by shape dynamics, which is a theory with a dynamically determined preferred global time.

The existence of a theory with a preferred global time which is equivalent to general relativity is sufficient to remove the objection to the reality of time coming from the relativity of simultaneity. But there are important advantages as well because a global time consistent with the successes of relativity theory opens up new avenues of research on several key foundational questions in physics.

In the area of quantum gravity the whole complex of problems associated with the absence of time in the canonical formulation of quantum gravity can now be resolved by quantizing the theory in the global physics time of shape dynamics. The quantum dynamics can be formulated conventionally, in terms of a Schrödinger equation for evolution of the quantum state in the preferred time. Note that this is not an example of the cosmological fallacy in which one formulated cosmological dynamics in terms of an absolute external time unmeasurable in the universe, because the preferred time is relational by virtue of the fact that it is dynamically determined. This also does not conflict with the success of the relativity of simultaneity locally or in small regions because local measurements do not suffice to pick out the observers whose clocks register the preferred time.

New directions are also opened up for quantum foundations because it is known that any theory that goes beyond quantum mechanics to give a precise description of individual processes must break the relativity of simultaneity [49]. As these theories agree with the statistical predictions of quantum mechanics, their predictions for probabilities for outcomes of experiments agree also with special relativity. But when they go beyond quantum mechanics to describe individual processes they require the specification of a preferred frame of reference.

But most importantly, a physical global time opens up the possibility that the laws of physics can evolve, because it suggests that there is a meaning to time that can transcend any particular theory.

5 Mathematics

If the view proposed in previous chapters is to have a chance of succeeding it must resolve several puzzles connected with the nature of mathematics and its role in physics. The problem is that our two principles – that there is one real world and that time is real and goes all the way down – make trouble for our received accounts of mathematics and of its role in the scientific study of nature. According to the view most commonly held among physicists and mathematicians, mathematics is the study of a timeless but real realm of mathematical objects. This contradicts our principles twice over, both because there is no real realm other than our one universe and because there is nothing real or true that is timeless.

A NEW CONCEPTION: MATHEMATICS AS EVOKED REALITY

The choice between whether mathematics is discovered or invented is a false choice. Discovered implies something already exists and it also implies we have no choice about what we find. Invented means that it did not exist before AND we have choice about what we invent.

So these are not opposites. These are two out of four possibilities on a square whose dimensions are choice or not and already existed or not.

Why could something come to exist, which did not exist before, and, nonetheless, once it comes to exist, there is no choice about how its properties come out?

Let us call this possibility *evoked*. Maybe mathematics is evoked. The four possibilities are indicated in the following diagram:

Has rigid properties? \ Existed prior?	Yes	No
Yes	Discovered	Evoked
No	Fictional	Invented

There are many things that did not exist before we bring them into existence but about which we have no choice, or our choices are highly constrained, once it does exist. So the notion of evocation applies to many things besides mathematics.

For example, there are an infinite number of games we might invent. We invent the rules but, once invented, there is a set of possible plays of the game which the rules allow. We can explore the space of possible games by playing them, and we can also in some cases deduce general theorems about the outcomes of games.

It feels like we are exploring a pre-existing territory as we often have little or no choice, because there are often surprises and incredibly beautiful insights into the structure of the game we created. But there is no reason to think that game existed before we invented the rules. What could that even mean?

There are many other classes of things that are evoked. There are forms of poetry and music that have rigid rules which define vast or countably infinite sets of possible realizations. They were invented. It is absurd to think that haiku or the Blues existed before particular people made the first one. Once defined there are many discoveries to be made exploring the landscape of possible realizations of the rules. A master may experience the senses of discovery, beauty and wonder, but these are not arguments for the prior or timeless existence of the art form independent of human creativity.

It just happens to be a true fact about the world that it is possible to invent novel games, or forms which, once brought into existence, have constraints or rules which define a vast or infinite space of realizations.

When a game like chess is invented a whole bundle of facts become demonstrable, some of which indeed are theorems that become provable through straightforward mathematical reasoning. As we do not believe in timeless Platonic realities, we do not want to say that chess always existed – in our view of the world, chess came into existence at the moment the rules were codified. This means we have to say that all the facts about it became not only demonstrable, but true, at that moment as well. Our time-bound world is just like that: there are things that spring into existence, along with a large and sometimes even infinite set of true properties. This is what the word evoked means to convey: the facts about chess are evoked into existence by the invention of the game.

The concept of evoked truths depends essentially on the reality of time because it has built into it the distinction between past, present, and future.

Once evoked, the facts about chess are objective, in that if any one person can demonstrate one, anyone can. And they are independent of time or particular context: they will be the same facts no matter who considers them or when they are considered. Furthermore, they are facts about our one world, just the same as facts about how many legs some insect has or which species can fly. The latter facts were evoked by evolution acting through natural selection; the facts about chess were evoked by the invention of the game as a step in the evolution of human culture.

One consequence of the Platonic view is to deny the possibility of novelty. No game, construction or theorem is ever new because anything that humans discover or invent existed already timelessly in the Platonic realm. The alternative to believing in the timeless reality of any potential game or species whose existence is logically possible is believing in the reality of novelty. Things come into existence and facts become true all the time. This is one meaning of the reality of time. Nature has within it the capacity to create kinds of events, or processes or forms, which have no prior precedent. We human beings can partake of this ability by the evocation of novel games and mathematical systems.

So it is not just human beings that have the power to evoke novel structures, which bring along with them novel facts of the matter that have definite truth values from then on. Nature has this capacity as well and uses it on a range of scales from the emergence of novel phenomena which are described by novel laws to the emergence of novel biological species which play novel games to dominate novel niches.

The notion of novel patterns or games evoked into existence gives a precise and strong meaning to the concept of emergence. In a timeless world in the context of the Newtonian paradigm, emergence is always at best an approximate and inessential description because one can always descend to the timeless fundamental level of description according to which all that happens is the rearrangement of particles with timeless properties under timeless laws. But once we admit the actuality of the emergence of novel games and structures with evoked properties, emergence has a fundamental, irreducible meaning.

In fact, biological evolution proceeds by a sequence of evokings of novel games and structures. Once cells with DNA and the standard biochemistry come into existence there is a vast landscape of possible species and ecologies. As the biosphere evolves it discovers many niches where species may thrive. New innovations appear from time to time like eukaryotic cells, multicellularity, oxygen breathing, plants, etc., which define further constraints, which in turn make possible new variations, niches, and innovations. All this is truly a wonder but it would add nothing and explain nothing to posit that there is a timeless Platonic world of possible DNA sequences, species, niches, ecologies that are being realized. Such a belief would explain nothing about how the real biosphere evolved and raise many additional questions whose answers, if they had answers, would add nothing to our understanding of the history of life or allow us to predict features of future life more than we already can.

What applies to biology also applies to mathematics itself. There is a potential infinity of formal axiomatic systems (FASs). Once one is evoked it can be explored and there are many discoveries to be made about it. But that statement does not imply that it, or all the infinite number of possible formal axiomatic systems, existed before they were evoked.

Indeed, it's hard to think what belief in the prior existence of an FAS would add. Once evoked, an FAS has many properties which can be proved about which there is no choice – that itself is a property that can be established. This implies there are many discoveries to be made about it. In fact, many FASs once evoked imply a countably infinite number of true properties, which can be proved. But the claim that the FAS existed before being evoked is not needed to deduce the true fact that, once evoked, there are an infinite number of true facts about it to be discovered and proved. Nor does the claim of prior or timeless existence explain the existence of those true properties because it involves belief in something that itself needs explanation. If the FAS existed prior or timelessly, what brought it into existence? How can something exist now but also exist timelessly. For if it only existed "outside time," would or could we, who are time-bound, and only come into contact with other things that live in time, ever know of it? How can something exist and not be made of matter? How can something that is not made of matter be known about, explored or influence us, who are made of matter?

So the postulation of prior or timeless existence explains nothing that is not explained by the notion of being evoked. It raises several questions including the ones just mentioned that are even more difficult to answer, and which centuries of attempts by very bright people have not answered.

Since the notion of evocation is sufficient to explain why an FAS once evoked has rigid properties to be discovered, the notion of prior or timeless existence is not needed, and it is not helpful. Also it requires us to believe in a whole class of existence, as well as belief in the existence of an infinite number of FASs, for which there is no evidence. By Occam's razor, this is not plausible.

As Roberto Mangabeira Unger has already remarked, Barry Mazur, in a very helpful essay ("Mathematical Platonism and its opposites [50]"), asserts that any answer to Platonism has to say something about the nature of proof. First of all, proof is a specialization of rational argument. It is a true fact about the world of possibilities brought into being when we humans evolved that in many situations we can rely on rational argument to lead to unambiguous conclusions. It happens to be the case that there are classes of questions that can be decided unambiguously by rational argument based on public evidence.

This fact of the reliability of rational deduction cannot be explained by pointing to an imagined world of timeless but existing logical forms, as that would raise more unanswerable questions of the above kind. So it has to be taken as a simple brute fact about the world that experience has long validated.

The process of rational deduction has itself been formalized, so rational argument from evidence is also a formal game whose rules have been defined in a way that in some classes of questions defines constraints sufficient to lead to unambiguous conclusions.

Among these classes are mathematical systems defined by rules, or FASs. Proofs are first of all just instances of rational argument applied to FASs to deduce true properties of them. Once evoked an FAS has many, often an infinite number of, true properties that can be so established.

Proofs can be formalized, and there may be different ways of doing that. Each formalization is itself a formal game which is evoked, after which it can also be studied and explored. One can then raise and answer questions about how different formal methods of proof are related to each other.

The bottom line is this: we have a choice between simple wonder and mystification. We can wonder at the vast complexity and beauty that is created by novel games, ideas, formal systems, etc. when they are evoked. That there is such possibility of novel systems to explore is a true fact about the world we find ourselves in, which is properly a source of wonder.

Or we can make mystical pronouncements that attempt to explain the infinite possibilities that might be evoked by imagining they all exist in a timeless reality apart from what we see physically exists around us. But these mystical beliefs add nothing and explain nothing and indeed, as indicated above, involve us in a pile of questions that, unlike questions about mathematics, cannot be answered by rational argument from public evidence. Moreover, to assert that one's avocation is the exploration of some timeless unphysical reality is presumptuous and seems like a claim to special knowledge or authority that, in fact, contradicts the fact that mathematical arguments are just finely disciplined cases of the usual rational thinking that all humans constantly engage in to understand their world.

Honest wonder about our world seems a better stance than mysticism, especially when what is involved is the highest form of rational creativity. For that reason it seems better to believe in the possibility of evocation to create novel realms of truth to be explored that did not exist before than to believe in a special ability to gain knowledge about a timeless realm disconnected from physical existence.

THE REASONABLE EFFECTIVENESS OF MATHEMATICS IN PHYSICS

So the answer to Wigner's question is that mathematics is reasonably effective in physics, which is to say that, wherever it is effective, there is reason for it. But mathematics does not of itself lead to discoveries about nature, nor is physics the search for a mathematical object isomorphic to the world or its history. There will never be discovered a mathematical object whose study can replace the experimental study of nature. There is no mathematical discovery in our future that will render moot from then on the experimental and observational basis of science. It will always be the case that the use of mathematics to model nature will be partial – because no mathematical object is a perfect match for nature. The use of mathematics in nature also involves a large degree of arbitrariness, because those mathematical objects that provide partial mirrors of parts of the world are a small, finite subset of the potentially infinite number of mathematical objects that might be evoked. So the effectiveness of mathematics in physics is limited to what is reasonable.

Moreover, any view about the role of mathematics in physics has to deal with the troubling issue of underdetermination of the choice of mathematical models of physical systems. Most mathematical laws used in physics do not uniquely model the phenomena they describe. In most cases the equation describing the law could be complicated by the addition of extra terms, consistent with the symmetries and principles expressed, whose effects are merely too small to measure given the state-of-the-art technology. These "correction terms" may be ignored because they don't measurably affect the predictions, but only complicate the analysis. That this is the right thing to do methodologically does not, however, change the fact that every one of the famous equations we use is merely the simplest of a bundle of possible forms of the laws which express the same ideas, symmetries, and principles, and have the same empirical content.

This fact of underdetermination is a real problem for those views which assert that nature is mathematical or that there is a mathematical object which is an exact mirror of nature, for only one out of the bundle of equations can be the true reality or mirror. Often we assert that the right one is the simplest, evoking a necessarily mystical faith in "the simplicity of nature." The problem is that it never turns out to be the case that the simplest version of a law is the right one. If we wait long enough we always discover that the simplest version is in fact wrong, because the theory is superseded by a new theory. The old equation turns out still to hold approximately, but with corrections which take a form that could not always have been guessed or anticipated prior to the invention of the new theory.

Thus, Newton's laws were found to be corrected by terms from special relativity, and then corrected again by terms from general relativity. Maxwell's equations received corrections that describe light scattering from light – a quantum effect that could have been modeled, but never anticipated by Maxwell. And so on.

The radical underdeterminacy of the mathematical representation of physics is however no problem for the view proposed here. It is rather exactly what you would expect, if mathematics is a powerful tool for modeling data and discovering approximate and ultimately temporary regularities which emerge from large amalgamations of elementary unique events. In this context we use the simplest equation that expresses a law, not because we believe nature is simple but because it is a convenience for us – it makes a better tool, much as a hammer with a handle moulded to the hand is a better tool. Moreover in this context every theory, is an effective theory which means that the limitations on the domain of applicability are always explicit and the correction terms are always there and ready to be exploited when a boundary of the domain of applicability is approached.

THE UNREASONABLE EFFECTIVENESS OF MATHEMATICS IN MATHEMATICS

Besides the unreasonable effectiveness of mathematics in physics, a satisfactory view of mathematics must also explain the unreasonable effectiveness of mathematics in mathematics itself. Why do developments in the elaboration of one core concept, say number, so often turn out to yield insight into another, say geometry? Why does algebra turn out to be so powerful a tool in the study of topology? Why do the different division algebras organize the classification of the possible symmetry groups of continuous geometries? If mathematics is just the free exploration of arbitrary ideas and axiom systems, why do these explorations so often intersect, and why are these intersections so productive of insight?

A short answer is that the contents of mathematics is far from arbitrary. While an infinite number of mathematical objects might potentially be evoked, the small and finite number that do prove interesting – even on purely mathematical grounds – develop a very small number of core concepts. These core concepts are not arbitrary – they are elaborations of structures which are discovered during the study of nature.

There are four of these core concepts: number, geometry, algebra and logic. They each capture a key aspect of the world and our interaction with it. Number captures the fact that the world contains distinguishable objects which can be counted. Geometry captures the fact that objects are found to take up space and form shapes. Algebra captures the fact that objects and number can be transformed, by processes carried out in time. And logic is the distillation of the fact that we can reason about the first three concepts, and so deduce predictions for future observations from properties of past observations.

The bulk of mathematics consists of elaborations of these four core concepts. In the course of these elaborations we often find that developments of one bear on another. These intersections tell us that these concepts go back to nature, which is a unity. For example, the elaboration of the concepts of space and number often intersect because space and number are both features of nature and hence are highly interrelated from the start. Hence, the discovery that a relation among numbers represents or is isomorphic to a relation among another strand of mathematics is often a discovery of a relation that is a true property of the one world.

There is no necessity to limit the study of mathematical systems to those that elaborate these four core concepts. But those that do display a vast richness of consequences and interconnections exactly because they are elaborations of core concepts that come from nature.

One may then even define mathematics as the study of systems of evoked relationships inspired by observations of nature.

This definition is consonant with and complementary to Roberto Mangabeira Unger's conception of mathematics as the study of nature stripped of particularity and time. It is stripped of particularity because a vast number of natural phenomena can be organized in terms of concepts of number, geometry (or space), transformation (or time) and logic. It is stripped of time, but only partly, because the subject is matters of fact about evoked structures and games, which come to be true at particular moments when the corresponding games or structures first arise, either through causal processes in nature or by human invention.

THE STAGES OF DEVELOPMENT OF MATHEMATICS

The concept of mathematics we have just sketched situates mathematics in time while stressing its objectivity and partial inevitability or lack of arbitrariness. A consequence is that insight into the nature of mathematical research can be obtained by stressing rather than ignoring the fact that mathematical research is a process carried out in time. This process has characteristic stages by which the core concepts are evoked and elaborated.

The history of mathematics shows, in each of the core concepts, a characteristic development consisting of a sequence of stages. At least the first few of these stages are reproduced also in the acquisition of mathematical knowledge by each of us. Each stage is characterized by a particular mode of thought. This development shows the extent to which mathematics develops from the study of nature with particularity and time removed. It also shows how much clarity one gets as to the nature of mathematics and mathematical research once one has a conception of mathematics that both situates mathematical truths in time while keeping them about objective matters of fact.

At the first stage, there is the study of the structure of our world, by examination of examples and relations between them, coming from the properties of physical objects or processes and their relations. In the case of space, one begins with establishing some of the elementary properties of different kinds of shapes: triangles, squares, circles, straight lines, etc. Geometry, as it was studied by the Greeks and as it is still encountered first by children, is very much a study of properties of things in our world, in which particularity is abstracted away. We learn that it is very rewarding not to study each example of a circle or a triangle, but circles and triangles in general. In the case of number we all begin by counting, as did our ancestors. And, as they did, we establish the validity of the basic operations of arithmetic by manipulation of physical objects and pictures of them.

We can refer to this as the *naturalistic phase of the study of mathematics*, or the phase of *exploration of the natural case.*

This first phase consisted in history, and in each of our own histories, largely of the exploration of the core concepts of number and geometry. But it is important to emphasize that these core concepts are also not themselves timeless, but were evoked at stages in the evolution of nature. For according to the current standard models of physics and cosmology, there was a stage of the world where there were no elementary particles, but only vacuum states of quantum fields. Number was evoked when the first particles were created in the decay of unstable vacuum states.

There is also increasing evidence that space is emergent and not fundamental, and emerged at a still earlier stage in the universe's evolution. If so, then the facts of geometry were evoked into existence by the emergence of space.

The second stage is the organization of the knowledge acquired in the naturalistic phase. One makes the discovery that all the knowledge gathered by examination of cases in nature can be reproduced by deduction from a small set of axioms. This is the phase of the *formalization of natural knowledge*. Note that the progression from the first to the second phase is a consequence of the invention of a new mode of reasoning, which is the axiomatic method, including methods for proofs of theorems from axioms.

In some accounts mathematics is defined by the use of the axiomatic method. This is incorrect, as in fact much of mathematics has been developed without the use of the axiomatic method. This is not only true historically; many contemporary mathematicians do not work axiomatically and many mathematical papers do not employ axiomatization. Furthermore, arithmetic was only axiomatized in the nineteenth century; we would not want to say that the enormous elaboration of the concept of number that preceded this was not mathematics. Finally, we have to recognize the fact that Gödel's theorem implies that the body of true facts about infinite systems of mathematical objects, once evoked, cannot be fully captured by what can be deduced by any single finite system of axioms.

What is significant about axiomatizion is that it is the first example – of several – of how the development of mathematics is catalyzed by the invention of novel methods of thought. This shows that the scope of mathematics cannot be restricted to the study of a fixed body of facts, because new modes of reasoning make possible the formulation and demonstrations of facts that could not have been previously conceived.

Before the axiomatic method was introduced, there were mathematical facts about circles and triangles, but there were no mathematical facts about the logical dependence or independence of systems of axioms. These are objective facts, but they require the invention of the axiomatic method to evoke them into being, for, apart from the invention of a particular method of reasoning, they have no meaning. So facts about the properties of particular systems of axioms are among the time-bound, yet objective, truths that are evoked during the development of mathematics.

Facts about the relations of axioms, or axioms and theorems, can be considered second-order; they are not facts about objects in nature themselves, but facts about logical relations among properties of objects. Nonetheless, they are facts about the one world we inhabit.

During each of these stages progress is driven by an internal dynamic, which stems from the fact that solutions to open questions and problems often bring with them new problems and questions to be addressed. We might describe this by saying that research in mathematics is progressive. In a progressive area of research, the solution of a problem, rather than being the end of research, is a door that opens onto new lines of research. When research is non-progressive, the solution of a problem is like an ascent of a mountain, after which you can only defend the summit from attack. In a progressive area, there is never a summit, just an unending series of ledges and ridges. With the conquest of each, the problem of how to go forward opens up.

This progressive nature of mathematics is mysterious in a Platonic or Pythagorean view according to which the fact that mathematical research is carried out in time is an inconvenient, even embarrassing circumstance that must be minimized. It can only be fully appreciated within a view of mathematics itself that allows mathematical truth to be objective, yet time-bound.

Natural science as a whole is progressive, but overspecialization can lead to narrow research programs losing their progressivity and degenerating into the defense of minor hill tops. Art is progressive; even if it is not well known to laypeople, the conception of their subject many artists have is that developments in art are driven by a process of research in which problems posed by an existing style of work are solved by the invention of new styles.

Of course, the solutions of some problems in mathematics are more generative of new research directions than others. Mathematical research is thus not an arbitrary search through the possible theorems that might follow from a system of definitions or axioms – this is the way one might program a computer to produce theorems, but this is not the way we do it. At each stage, mathematicians act on their judgment as to which questions and problems are more central to the progress of mathematics. For example, they will focus on problems that they judge will be generative of new ideas or methods and which will be applicable to a wide set of problems. Mathematicians strategize by assigning importance to problems, and mathematical progress tends to follow from the solution of these key problems. Other problems, whose solutions are so obvious they are unlikely to generate new modes of thought or areas of research, are deemed by mathematicians to be more or less trivial.

If the progress of research were arbitrary there would never be a perception of research advancing quickly or being frustratingly slow – for there are always trivial open problems to solve. The perception mathematicians have that the pace in a field is fast or slow is due to their having an understanding of which problems must be solved to lead to bursts of significant new insights.

At the next, or third, stage in the development of mathematics, several mechanisms of growth of mathematical knowledge come into effect which are internal to mathematics, as they no longer require the study of examples in nature to proceed. The first system of axioms to be formalized will always be one that describes the natural case, i.e. the examples most apparent in nature. But once there is any formal system there open up ways to quickly expand the scope of study by several common means: (a) The invention and study of new examples allowed by the axiom system, not (at least yet) noticed in nature. For example, one can define and study conic sections in general and not just circles and spheres. (b) The formulation of conjectures, which stimulates attempts to prove or disprove them. (c) The study of formal relations within the axiom system, for example, most famously, the issue of the independence of the fifth postulate of Euclid. (d) The posing and solutions of problems of all kinds. We can call this third stage the stage of *exploration of the formalized natural case*.

Apart from these developments of the natural case there is also available at this stage new modes of development, which come from varying the natural case. One does this by dropping, modifying or adding to the system of axioms. For example, one gets an infinite series of Euclidean geometries by generalizing the axioms to a general dimension, and not just two or three. That is, having removed a lot of particularity from the relations among objects in space, leading to the formalization of Euclidean geometry in three dimensions, one can go a step further and remove the particularity of three dimensions, leading to the study of Euclidean spaces of arbitrary dimension.

We learn from this that the actual dimension of our world is not logically entailed by the concepts of Euclidean geometry. This is interesting and gives rise to a real, so far unanswered question: Why does the world have three spatial dimensions? But the fact that this is an open question does not imply that there are other worlds with other dimensions. There is one world and it has three spatial dimensions, for a reason we do not yet understand.

So at this third stage there is a continual evocation of novel classes of mathematical objects, such as Euclidean geometries of arbitrary dimensions. The evocation of these cases is followed by exploration of their properties. This again illustrates the progressive nature of mathematical research, which is continually evoking into time-bound existence novel mathematical objects open to our exploration.

Note also that at each such stage there is a potential infinity of novel structures and mathematical objects that may be evoked and explored. Because there are a finite number of mathematicians working, each with a finite lifetime, there is always an infinite sea of possible mathematical objects that may be evoked which at any stage dwarf the finite number of objects that have been explicitly evoked and explored. Thus, there is always, at each stage, infinitely more mathematics in the adjacent possible of next objects that might be evoked than the finite set of results which constitute each present's mathematical knowledge.

This raises a puzzle, which is easily resolved within this time-bound viewpoint. At present the storehouse of evoked and studied mathematical objects includes spheres of arbitrary finite dimension, S^n, about which many theorems have been proved for arbitrary finite n. But there are many, perhaps an infinite number of, particular finite n, whose spheres have properties which only hold for that n, which have never been explicitly considered. All were evoked into existence when the concept of a sphere of arbitrary dimension was first considered. The puzzle is how properties of the general case can be understood when that general case includes many specific cases yet to be explored. This is puzzling only if you think that mathematics is the discovery of pre-existing objects in a timeless realm. But the infinite number of S^n whose properties will never be fully explored by any finite time are like the infinite number of chess games that by that time will not yet have been played. They indicate the inexhaustible nature of mathematics as a subject whose object of study continually grows by infinite steps as new concepts and structures are evoked. In this situation, it is just the case that more can be known about a general class of posited objects of thought than may be known about particular objects in the class.

The example of generalizing from three-dimensional Euclidean geometry to the study of geometry with arbitrary numbers of dimensions also shows how the internal development of mathematics sometimes brings forth new questions about nature. Without the generalization available of the notion of the dimension of space, there could not be a vivid formulation of the question why physical space has three dimensions. This is an early and easy example, but there are many examples of this phenomenon. A recent one is why the forces between elementary particles are governed by particular gauge groups.

More non-trivial examples of varying the natural case are found by altering one of the postulates. Famously, modification of the fifth postulate gave rise to the non-Euclidean geometries. This is the fourth stage, that of the *evocation and study of variations on the natural case.*

We can remark that now the process iterates, because each such variation generates a vast class of possible examples, conjectures, and problems which can be posed for study.

However, we should note that non-Euclidean geometries first arose not by variation of the axioms of Euclidean geometry, but from the study of examples within Euclidean geometry. The first appearances of non-Euclidean geometries are as surfaces imbedded within Euclidean space. Then a new question arises: Can the geometries of these surfaces be described without reference to those embeddings? So the combination of problem solving within established subjects and the invention of new questions together are part of the dynamics that drives the invention of new subjects and the evocation of novel mathematical objects.

In the case of arithmetic, the second and fourth stages were reversed. In the case of number the natural case is the natural numbers. The solutions of problems formulated in terms of the natural numbers had forced the Pythagorean mathematicians to expand the concept of a number. This occurred before Euclid and indeed there was not an axiomatization of the integers until the nineteenth century. So without an axiomatization, successively novel modes of reasoning gave rise in turn to the rationals, reals, and complex numbers. This proceeded further to the invention of the quaternions, octonions and other number systems and hence to developments in algebra.

But however far mathematics progresses along these phases, three things remain true. First, the core concern remains the study of a general feature of our world, as first captured by the natural case. However far geometry gets from three-dimensional Euclidean space, one is still studying possible ramifications and conceptions of space, and the interest and power of that study comes from the fact that space is a core attribute of nature. Second, one is still studying the world by stripping the properties of objects and relations in our world of aspects of their particularity – including time. The study of non-Euclidean geometries is the study of geometry with the particularity of the fifth postulate removed. Third, there is no reason at all to believe that the mathematical systems and objects obtained by variation of the natural case exist in any timeless sense, outside of the single, unique universe. They are inventions of mathematicians' minds, stimulated by the desire to understand as deeply and generally as possible concepts whose interest arises from their being derived from observations of nature. They are part of the one, time-bound world, evoked into existence by the inventions of novel sets of axioms or systems of relationships.

A fifth stage of development is the *invention and development of new modes of thought, new concepts and new methodologies in the study of an area*. These can greatly progress an area as new kinds of facts become definable and discussable. We have already mentioned the invention of the axiomatic method as an example of how the progress of mathematics is greatly stimulated by the invention of new methods.

There are many such examples. In the nineteenth century, the study of geometry was greatly enhanced by two inventions. The first was the study of symmetries of spaces, which in turn led to the invention of group theory, and the theory of group representations. These stimulated also developments in the subject of algebra. This shows that a new subject of mathematics can emerge from the invention of a new method for the study of an older area. And by doing so it can deepen our understanding of an old area by bringing into existence new facts about old subjects. For example, only once there is the invention of the idea of a symmetry group could we pose and answer the question: "What is the symmetry group of Euclidean three-space?"

Thus was evoked into existence the third of our four core concepts drawn from nature: the concept of transformation, realized by systems of algebras. Even if its invention followed the invention of number and geometry by thousands of years, the study of transformations deserves to be classed as an equal core concept. Its centrality is shown by the myriads of ways in which the concept of transformation illuminates the other core concepts in mathematics as well as deepening our representation of time (at least within the Newtonian paradigm.)

How can we then know that the list I gave of four core concepts is complete? Can we be sure that the development of mathematical research might not lead to the evoking of a novel core concept, as central to developments from then on as number, geometry, transformation, and logic have been up till now? We cannot. The future of mathematics is genuinely open to the future.

A second innovation in mathematical thinking in nineteenth-century geometry was the invention of topology. This brought new kinds of facts into existence, such as that the Euclidean plane is open, the sphere is closed, and there is no isomorphism between the plane and the sphere. This shows that the invention of new modes of thinking can create new facts about the natural cases in the core of the subject, as well as cases developed since. That this can happen at any time suggests that mathematics does not deal with a fixed set of facts, which is closed in the older subjects.

New methods can also increase the significance of old examples. Tori were known before Poincaré's invention of topology, but their significance as an example of a space with non-trivial topology only became apparent then.

Another nice example of how new modes of reasoning create new mathematical facts is Cantor's work on transfinite arithmetic. It is common to say that these are shown to exist by application of the diagonal argument. We would like to stress that without that argument – or its logical equivalent – there is no way to reason about the transfinite numbers. Hence I would suggest that it is not meaningful to talk about the "existence" of the transfinite cardinals before the invention of the diagonal argument. Rather, I would propose to describe the transfinite numbers as having been evoked into existence by Cantor's invention of the diagonal argument.

Once there are a variety of cases developed by variation of the natural case, a sixth stage of development can play a role, which is *to define new kinds of objects by unification of diverse cases.* For example, the different Euclidean and non-Euclidean geometries are all unified within Riemannian geometry. This often coincides with innovations in method, as indeed was the case here. New tools became available to study geometry locally, such as connections and curvatures. This leads to new insight into old examples – we can compute the curvature of Euclidean space and find it vanishes! And it also leads to the study of still further examples.

Once again, the process can iterate, as a significant generalization can, once invented, play a role analogous to that which the natural case played originally in further research. Indeed, according to general relativity, Riemannian geometry is the best description we have of the geometry of space and spacetime, so it replaces Euclidean geometry as the natural case in contemporary thinking. Variations of its axioms now are giving rise to non-commutative geometry, complex manifold theory, and other novel constructions. All these are recent additions to the storehouse of mathematical objects, evoked into time-bound existence by recent research.

Finally, mathematics develops through two more kinds of discoveries, one external and one internal. The first is that a construction, example or case developed in the path flowing out of one of the core concerns can turn out to illuminate or apply to knowledge in another stream of development. Developments in geometry can illuminate problems in number theory and vice versa. For example, once the complex numbers are in hand, they are used to represent rotations in the Euclidean plane. Seeking to generalize this, Hamilton invented the quaternions and found they could be used to represent rotations in Euclidean three-space. And far more non-trivially are the intricate connections between the octonions and the exceptional continuous transformation groups. We can call these the *discovery of relationships between constructions generated autonomously within mathematics.*

Lastly, examples, cases or modes of reasoning invented due to the internal development of mathematics can surprisingly turn out to be applicable to the study of nature. We can call this the *discovery of the applicability to nature of knowledge resulting from internal development within mathematics*. Examples include the application of complex numbers in quantum mechanics or the applications of the quaternions in Dirac's description of the relativistic quantum electron. Tantalizingly open is the possibility that octonions may be keys to the unification of the elementary particles and forces [32].

To summarize, we have discussed eight stages by which mathematics develops from the study of relations among natural objects:

1. Exploration of the natural case.
2. Formalization of natural knowledge.
3. Exploration of the formalized natural case.
4. Evocation and study of variations on the natural case.
5. Invention and application of new modes of reasoning.
6. Unifications of cases within more general frameworks.
7. Discovery of relationships between constructions generated autonomously within mathematics.
8. Discovery of the applicability to nature of knowledge developed internally.

Driving progress within each of these stages is the progressive dynamic of mathematical research, which is that the solution of each problem, the resolution of each question and the proof of each theorem lead normally to new problems, questions or conjectures. Furthermore, the first and last of these have to do with nature; the others are modes of internal development.

Each of the core areas of mathematics – space and number – has passed through these stages or modes of research. These eight stages or modes are not logically necessary, and they need not always occur, nor always in this order. But nonetheless, these stages have characterized the developments of the core areas.

Nor is this meant to be an exclusive list, as it is always possible that new stages in the development of mathematics and new ways of progressing our mathematical knowledge will occur. So this is a view of the development of mathematics, which is always open to future developments.

There is a confusion about the notion of evoking structures into existence which must be clarified. One can speak both of nature's power to evoke novel kinds of structures into existence by the emergence of novel phenomena and the mathematician's ability to evoke new classes of facts into existence by the invention of novel mathematical objects. Both are valid uses of the concept of evocation. At each moment there is a time-bound set of natural structures that exist and at each stage in the development of mathematics there is a time-bound storehouse of mathematical objects evoked into a kind of conceptual existence by the invention of a mathematician. Both these results of acts of nature and acts of imagination are part of the single, time-bound universe. But there is no necessary relationship between them. Thus, it can be the case that the exploration of a mathematical concept – for example, Euclidean geometry – can predict some properties of physical space.

But it is never going to be the case that mathematical structures can serve as oracles to properties of nature, such that purely mathematical research can obviate the need for empirical research and checking. For our knowledge of nature is always provisional. In this example, it was discovered in the last century that Euclidean geometry is not a perfect mirror of physical space. So the use of Euclidean geometry to explore properties of physical space was always just a tool, useful in, but subservient to, the experimental method. So, even if the evocation of a property of nature precedes the evocation of a property of a mathematical object posited to mirror it, it is never the case that the exploration of the mathematical object involves the discovery or recollection of properties of the physical object. The correspondence between physical objects and mathematical objects is always provisional and approximate.

For this reason, it is perhaps better to speak of novel systems in nature as having emergent structures or properties to contrast this with evoked concepts and mathematical objects. But I want to suggest that our ability to evoke novel concepts and games is a consequence of the more general fact that nature evokes into existence novel properties by the emergence of novel kinds of systems and structures.

We can contrast this view of the development of mathematics with an older view, which is that there will sometimes be discovered a single formal axiomatic system from which all of mathematics can be derived and which will from that point on serve as the foundations of mathematics. This was the goal of Russell and Whitehead as they attempted to found all of mathematics on a single axiomatization of logic, and it was also part of the program of Hilbert to formalize mathematical knowledge once and for all. These programs failed, for reasons which are internal to them, among them being the paradoxes of naive set theory and the incompleteness results of Gödel. Indeed, these failures were most productive, giving way to new developments in mathematics. Since then there have continued to be attempts to propose foundations for mathematics with the intention of basing the entire subject on a single fixed axiomatization. Our view maintains that there can be no final foundations for mathematics, because its subject is no fixed or pre-existing set of facts. There is rather the aim and the tradition of seeking to construct increasingly general and useful understandings of the nature of our universe, when time and particularity are stripped away. New modes of thinking or investigation and new inventions can expand indefinitely the set of facts that are relevant for our understanding of space and number, so that all that is possible is a unification, at any one time, of the mathematics known up till that time.

Finally, this description of the stages of mathematical understanding emphasizes the point made by Roberto Mangabeira Unger that neither invention nor discovery conveys the right meaning for how we are to understand the arising of new mathematical examples, systems, and facts. Since we do not believe that mathematics is the exploration of an existing, static timeless realm of knowledge, "discovery" is not the right word. But neither is it the case that mathematics developments take place arbitrarily, with the freedom of the artist or the poet, combining at whim ideas and materials. So "invention" is not the right word either. As we have discussed, the development of mathematics is constrained at the different stages by both external and internal factors. Externally, it is constrained as its core concepts come from the distillation or abduction of properties or relations of things in the natural world. It is constrained as well in the internal stages of its development, by the unfolding logic of concepts as the solution of one problem almost always gives rise to further problems and challenges.

Thus, it seems that, contrary to the Platonic view, in which all mathematics is discovery of pre-existing timeless facts, the truly novel construction, mode of thought and example plays a necessary role in the development of mathematics. But it is a novelty very much constrained by the progressive character of mathematical research.

The novel examples, methods, and results were not known before and, if there is no separate Platonic world, they did not exist before. But they are neither free creations nor subjective; nor are they, in any sense, social constructs. They are required by the unfolding and progressive logic of the project of mathematics, which is the exploration of notions of number and space necessary to the framework of the universe we live in.

We can summarize this view of mathematics as follows: Mathematics is a system of objective facts, that is nonetheless time-bound and open to unpredictable developments in the future.

WHY IS MATHEMATICS EFFECTIVE IN PHYSICS?

If we give up the idea that there is a mathematical object existing in a timeless Platonic realm which is isomorphic to the history of the universe, we still have to explain why mathematics is so effective in physics. It will be sufficient to point to an interpretation of the use of mathematics in physics that is consistent with the view of mathematics just presented. Here is one such interpretation: Mathematics is useful as providing models that summarize the content of records of past observations. When we test a theory in the Newtonian paradigm we make and record observations of motions, which consist of values of observables that we represent as the coordinates of the configuration space of a system. These records are static, in that once taken they do not change in time. Or, more precisely, they may change, by being degraded or erased, but once they do they cease to function as records of past experiments. They can be compared to a trajectory in the configuration space which, being a mathematical object, is also static.

We can propose that the main effectiveness of mathematics in physics consists of these kinds of correspondences between records of past observations or, more precisely, patterns inherent in such records, and properties of mathematical objects that are constructed as representations of models of the evolution of such systems. This view does not require the postulation either that physical reality is timeless or that mathematical objects exist in a separate timeless realm. It is sufficient that records of past observations are static and that the properties of a mathematical object are, once evoked into existence by their invention, static. Both the records and the mathematical objects are human constructions which are brought into existence by exercises of human will; neither has any transcendental existence. Both are static, not in the sense of existing outside of time, but in the weak sense that, once they come to exist, they don't change.

Another use of mathematics in physics is as an aid to imagination; we invent a model of a physical system and imagine that it models aspects of that system. By seeing the mathematical object in our minds we aid and stimulate a process by which we imagine to ourselves what is going on in nature, at least within that system. Apart from its practical benefits, this kind of imagining has an aesthetic element: it can stimulate our admiration and appreciation of the beauty of patterns which we represent in our model and project onto nature.

We should only be wary of mistaking the way our imaginative vision of a physical system can be aided by a mathematical model with a transcendental insight into some timeless essence of that system.

6 Approaches to solving the meta-law dilemma

The principles and hypotheses we present in this book become a research program when we see that they can be implemented in particular theories and models. As we have argued, such a theory must be based on the idea that the laws of nature evolve in a real, global, cosmological time. These must avoid the cosmological dilemma and fallacy and so cannot be expressed within the Newtonian paradigm, yet they have the task of providing sufficient reason for the laws and initial conditions that govern subsystems of the universe. We can call the problem of framing this new paradigm of explanation the *meta-law problem*, because the issue is to discover how and why laws evolve. This must be done in a way that avoids the meta-law dilemma.

It is natural to describe the evolution of laws by means of an imagined space of possible laws. The evolution of laws can then be visualized and studied as evolution of either an individual universe or a population of universes on this space of possible laws. In the first case, for example, we have a sequence of points representing the laws that hold in different eras of a universe. The space of possible laws has come to be called the landscape; this terminology was first introduced in the context of cosmological natural selection and was chosen to evoke thoughts of the fitness landscapes that are studied in models of population biology [8]. In some, but not all, work on the landscape, it is assumed that the possible laws represented by points of the landscape are perturbative string theories, each an expansion around a vacuum state of string theory, which is in turn a solution to a meta-theory such as *M* theory. In these cases we refer to the landscape of string theory.

As useful as the metaphor of the landscape has been, there is a great danger its use will lead us into the meta-laws dilemma. This will be the fate of our theorizing if we assume that the landscape is itself timeless and that the evolution on it is governed by a timeless meta-law. In this case we reproduce the Newtonian paradigm and its issues. The *why these laws problem* just becomes the *why this meta-law problem*. But these assumptions can be avoided in several ways.

The landscape does not have to be specifiable timelessly in advance – it can grow and evolve as the universe evolves. In biology, the possible laws that govern biological phenomena emerge with those phenomena. We can then follow Stuart Kauffman and speak of evolution into the adjacent possible – because it may be that only the next possible steps are specifiable in the past [51]. The evolution on the landscape can be stochastic, even random. But as we shall see in cosmological natural selection (and in a very different way with the principle of precedence), sufficient reason for the choice of laws as well as falsifiable predictions can both result. The law can merge with the state, and both be evolved by a universal dynamics.

These options for avoiding the meta-laws dilemma frame several approaches to the meta-law problem, which I will now describe. The correct solution may or may not be among them, but this suffices to show that there is a fertile research program with many leads to be pursued.

One way to approach the meta-law problem is to assert that our observable universe does not contain enough information to answer the two big why questions: Why these laws and not others? and Why those initial conditions? But if we insist on the principle of explanatory closure then the universe must contain enough information to answer any query that can be made about its properties – including these two questions. The answer must lie in regions of the universe that we have not so far observed directly. For it is likely that the universe is bigger and older than the region we can observe.

In our analysis of the "why these laws" question we concluded that laws must have evolved dynamically to be explained. This implies that there were dynamical processes in our past by which the laws evolved. As we do not see any evidence that the fundamental laws or their parameters evolved in the observable past, these processes must have gone on in regions yet unresolved observationally. This accords with the intuitive picture that the effective laws may have evolved in events that involved energies or energy densities much in excess of those in our observable universe.

We now face several choices:

- Was there a bounce or singularity to our past?
- Did the evolution of laws happen all at once, or incrementally over many stages? That is, do we live in a first-generation universe or does our universe have a long chain of ancestry?
- Was the chain of ancestry linear, so that each universe gives rise to a single progeny, or does it branch, with each universe giving rise to many progeny. That is, is the solution to the meta-laws problem to be found in expanding the universe by succession, so it extends into the past, or by plurality, so that there is a population of simultaneously existing universes?

Let us investigate the different options. These give rise to three classes of global cosmological scenarios.

THREE OPTIONS FOR GLOBAL STRUCTURE OF THE LARGER UNIVERSE

If we posit that the initial singularity was really the first moment of time, then there is a brief time available for the evolution of the laws to have taken place. In this case there is unlikely to have been time for incremental evolution through many epochs. Our universe then probably arose from some primordial state in one or a few steps.

One early suggestion for such a cosmological setting for variation of the laws was Vilenkin [52] and Linde's [53] eternal inflation scenario, within which an infinite number of universes are born as bubbles in phase transitions from a primordial eternally inflating medium. In the simplest version of this framework our observable universe is one of an infinite number of universes each produced in a single step from a primordial state of eternal inflation. (It is also possible that there are bubbles within bubbles but these chains of descent are not taken as central to the explanatory power of the scenario as they are in cosmological natural selection.[1] There can even be tunneling back to the initial false vacuum leading to a recycling of the universe [55].)

The resulting multiverse scenario posits that the infinite numbers of universes are mostly causally disjoint from each other. A bubble universe may have collided with other bubbles, but almost certainly any pair of bubbles in the population of universes are causally disjoint.

An observer in a bubble will see a finite number of bubbles colliding with theirs in the past. Eventually, given infinite time, a bubble may collide with an infinite number of other bubbles but this will still be an infinitesimal fraction of the infinite collection of bubbles. We can call this a *pluralistic cosmological scenario.*

Although eternal inflation was proposed before the realization of the string landscape, it has become the setting in which much research on dynamical evolution of laws on the landscape has been carried out.

On the other hand, by positing that the singularity was replaced by a bounce we endow our universe with a deep past during which there may have been many epochs of classical universes. These would have allowed the effective laws to evolve incrementally over many generations, all in our causal past. These may be called *cosmological scenarios with succession.*

There are again two choices, depending on what bounced. The big bang may have followed a complete collapse of a prior universe. So we arrive at the scenario of a cyclic universe [56, 57].

The big crunch of a cyclic universe may have given rise to a single progeny – or it may have given rise to many. The latter may be the case if there is a selection effect whereby regions of the crunch must be sufficiently homogeneous to bounce. Hence we have to distinguish between linear cyclic cosmologies, in which a universe has a single progeny, and branching cyclic cosmologies, in which there will be many [58].

[1] I thank Mathew Johnson for a conversation on this and other fine points on eternal inflation.

The other possibility was that the big bang was the result of the bounce of a black hole singularity. If black hole singularities bounce then a universe may have many progeny, each the result of a collapse to a black hole. Indeed, our universe can be estimated to have at least 10^{18} black holes and hence at least as many progeny. Hence scenarios in which black hole singularities bounce are branching cosmological scenarios.

The scenario of bouncing black hole singularities is the setting for the framework of cosmological natural selection which will be discussed below [8, 6].

The only kinds of singularities which are generic in solutions to the Einstein equations are cosmological and black hole singularities. So these are the only options for cosmological scenarios in which singularities are replaced by bounces.

So we have the following options for a global cosmological model:

1. Pluralistic scenarios such as eternal inflation in which there is a population of universes, all derived from a primordial state by a one-stage process, largely if not completely causally distinct from each other.
2. Linear cyclic scenarios in which there is a succession of universes, each with a single parent and a single ancestor.
3. Branching scenarios in which each universe has a single parent but many progeny.

We now investigate the options for explaining the selection of laws in our universe in each of these three kinds of scenarios.

PROSPECTS FOR A SOLUTION OF THE LANDSCAPE PROBLEM IN THE THREE SCENARIOS

Before we analyze the possible solutions to the landscape problem offered by the three kinds of scenarios we should be mindful of a few key issues:

- In any landscape scenario – whether in biology or physics – there are two landscapes: the landscape of fundamental parameters and the landscape of parameters of effective low-energy theories. There can be a rather complicated relationship between them. In biology these are the spaces of genotypes – the actual DNA sequences – and the space of phenotypes – the space of actual features of creatures that natural selection acts on. In physics these may be the landscape of string theories and the landscape of parameters of the standard model. In biology, as well as in physics, the explanatory power of a scenario depends partly on how well understood are the relationships between the two kinds of landscapes.

- The bounces are very high energy processes, but there is evidence for a lot of fine tuning at the level of the low-energy parameters. How can the bounces then play a role in selecting for fine tuning of the low-energy parameters?
- We can observe only what is in our past light cone. If a cosmological scenario posits an ensemble of universes outside of causal contact with our own, then we risk a situation where the characterization of the other members of the ensemble is free from check by observation. There is a great danger then of just making stuff up to get answers we want. The only way to constrain an ensemble of causally disconnected universes by observation is if there is a dynamical principle that makes it possible to deduce that every or almost every universe in the ensemble shares some property P. Then an observation that P is not seen would falsify the theory.

Mindful of these cautions, we can now examine what opportunities our three kinds of cosmological scenarios offer for a solution of the landscape problem.

Linear cyclic models

The linear cyclic models have a great advantage over the other two scenarios in that all the epochs or universes they posit are in the causal past of ours. There is then abundant opportunities for making predictions that are subject to observational check. So far two kinds of cyclic models have been studied, and both offer falsifiable predictions. The ekpyrotic models of Steinhardt, Turok, and collaborators predict that there will be no observable tensor modes in the CMB [56]. The conformal cyclic cosmology of Penrose predicts the existence of concentric circles in the CMB due to gravitational waves formed by colliding black holes in the previous era [57]. Claims by Penrose and Gurzadyan that these have been observed [59] are presently controversial [60].

What prospects then do the linear cyclic models have to explain the selection of laws? One can easily hypothesize that at each bounce there are changes in the effective laws, perhaps brought about by phase transitions among vacua of string theory or whatever the fundamental theory is. This will give us a series of points in the landscape, representing the effective laws in each epoch. However, to explain the choice of laws there must be an attractor in the landscape. Otherwise the progression of laws through the epochs will just be random, and nothing about the present choice of laws will be explicable.

For the evolution on the landscape to converge to an attractor, the changes in each generation must be small. Also, to explain the choice of parameters of the low-energy theory by a series of transitions in the fundamental theory, it must be that small changes in the fundamental landscape give rise to small changes in the landscape of the low-energy effective theory.

Furthermore, that attractor must somehow be determined by properties of low-energy physics, otherwise the fine tunings of the standard model will not be explicable.

Branching models

Branching models share one good property with linear cyclic models, which is that there are long chains of descent. This can make possible incremental accumulation of good properties through slow, stable, evolution to attractors. However, they deviate from linear cyclic models in giving rise to a growing population of causally disconnected universes. These can lead to predictions about our universe only to the extent that it can be predicted that there will be properties, *P*, shared by all or almost all members of the ensemble.

This is illustrated by the two examples we have of branching models.

Branching cyclic cosmologies

In the branching cyclic cosmologies it can be hypothesized that only regions of the collapsing universe that are sufficiently spatially homogeneous will bounce to make new expanding universes [58]. Because the region must be very homogeneous to bounce, each new universe will be very homogeneous. Homogeneity is then a property *P* that is shared by all members of the ensemble – hence it is predicted for our universe. One can hope that more detailed modeling of the bounces may lead to new predictions for our universe of this kind which may be falsifiable.

We can note that a great advantage of cyclic cosmologies in general is that they eliminate the need for inflation to explain the specialness of the cosmological initial conditions.

Can the branching cyclic cosmologies explain the selection of the low-energy physics? In this regard the answer is the same as with regard to the linear branching cosmologies: the changes in both the fundamental and effective laws must be small from generation to generation and there must be an attractor in the landscape of the low-energy theory for the evolution to converge to.

COSMOLOGICAL NATURAL SELECTION

Cosmological natural selection was invented to give an answer to the landscape problem that explained the reasons for the fine tunings of the standard model without making use of the anthropic principle [8, 6]. The idea was to invent a cosmological scenario that naturally explained why the universe is fine tuned for complex structures such as longlived stars, spiral galaxies and organic molecules – using the same mechanism that biology uses to generate improbable complex structure.

This suggested that there would be in cosmology an analogue of biological fitness – the number of progeny of a universe as a function of its low-energy parameters. This analogy inspired the suggestion that there would be an evolution of effective field theories on a landscape of parameters analogous to the fitness landscapes studied by population biologists.

This was inspired by an analogy between selection of effective laws in a cosmological setting and natural selection in a biological setting. The theory is based on two hypotheses:

- **(H1)** Universes reproduce when black hole singularities bounce to become new regions of spacetime.
- **(H2)** During the bounce, the excursions through a violent interlude at the Planck scale induce small random changes in the parameters of the effective field theories that govern physics before and after the transition.

The analogue of biological fitness is then the average number of black holes produced in a universe, seen as a function of the parameters of the standard models of physics and cosmology. We can call this function on the landscape the cosmological fitness. Combinations of parameters that are local maxima of this fitness function are attractors on the landscape. After many generations the population of universes becomes clumped in the regions near these local maxima.

The great advantage of cosmological natural selection over linear cyclic cosmologies is then that it creates attractors on the landscape.

This comes about because the effective laws which are most common in the ensemble of universes are those that reproduce the most, which means they have the most black holes. Thus, a property *P* shared by almost all members of the ensemble will be, after many generations, the following: small changes in the parameters of the effective landscape will almost always lead to universes which produce fewer black holes.

Another way to say this is the following. If we define the fitness of a point in the landscape by the average number of black holes produced by a universe with those parameters, then after many generations almost every member of the ensemble will be near a local maximum of the fitness.

This explains the specialness of the tunings of the parameters of the standard model, because it turns out that several aspects of those tunings enhance the production of black holes. These include the following [8, 6, 61]:

1. The large ratios required for the existence of longlived stable stars, including $\frac{m_{\mathrm{proton}}}{m_{\mathrm{Planck}}}$, $\frac{m_{\mathrm{electron}}}{m_{\mathrm{proton}}}$ and $\frac{m_{\mathrm{neutrina}}}{m_{\mathrm{proton}}}$.
2. The coincidences among the proton–neutron mass difference, electron and pion masses, making nuclear fusion possible, as well as the sign of the proton–neutron mass difference.
3. The strength of the weak interaction which appears fine tuned both for nucleosynthesis and for supernovas to inject energy into the interstellar medium, catalyzing the production of massive stars whose remnants include black holes.
4. The fine tunings which result in the stability and plentiful production of carbon and oxygen. These appear to be necessary to cool the giant molecular clouds, from which form the massive stars which are the progenitors of black holes, as well as to provide insulation to keep the clouds cold.

It should be emphasized that cosmological natural selection is the only one of our scenarios that explains the fine tunings of the parameters of the standard model. It does so because the cosmological scenario makes low-energy physics causative of structure on a vast scale – that of the population of universes. It does so by strongly influencing the distribution of parameters in that population.

This feature could be mimicked by the branching cyclic models, but only if there were some reason why having something like our present low-energy physics could lead a universe to have more regions which were sufficiently homogeneous to bounce. This is unlikely because the conditions in the final crunch are not going to be sensitive to details of the choices of parameters of low-energy physics. What cosmological natural selection accomplishes, apparently uniquely, is to make the population of universes delicately sensitive to the parameters of low-energy physics. It does this naturally and necessarily, because it takes delicate tunings of parameters to produce a large number of black holes.

Because of this coupling between cosmology and low-energy physics, cosmological natural selection makes a few predictions that are vulnerable to falsification by present observations. It is instructive to review three of them.

To maximize the number of black holes produced, the upper mass limit (UML) for stable neutron stars should be as low as possible. As pointed out by Brown and collaborators [62], the UML would be lower if neutron stars contain kaon condensates in their cores. That is,

$$UML_{\text{kaon}} < UML_{\text{conventional}} \tag{4}$$

This requires that the kaon mass, and hence the strange quark mass, be sufficiently low. Since none of the other physics leading to black hole production is sensitive to the strange quark mass (within the relevant range), cosmological natural selection then implies that the strange quark mass has been tuned so that neutron stars have kaon condensate cores.

Both the theoretical understanding of the nuclear physics of kaon condensate stars and the observational situation has evolved since this prediction was published in 1992.

Bethe and Brown [62] argued that a kaon condensate neutron star would have a $UML_{\text{kaon}} \approx 1.6M_{\text{solar}}$, so that is the figure I used initially. However, as emphasized recently by Lattimer and Prakash [63], there is actually a range of predictions for UML_{kaon}. These depend on assumptions about the equation of state and range upward to two solar masses. So in the light of current knowledge the correct prediction is

$$UML_{\text{kaon}} < 2M_{\text{solar}} \tag{5}$$

The present experimental situation is summarized in [63]. There is an observation of a neutron star with a mass of 1.97 solar masses, to good accuracy. This is just inside the range consistent with the prediction that neutron stars have lowered upper mass limits due to having kaon condensate cores. However, there are observations of neutron stars with wider error bars of around 2.4 solar masses. This, if confirmed, would be inconsistent with the prediction of cosmological natural selection.

So while it is disappointing that the observation of a 1.97 solar mass neutron star cannot be taken as a falsification of cosmological natural selection, that theory remains highly vulnerable to falsification in the near future.

One question often raised is why cosmological natural selection is not ruled out by the possibility of changing a cosmological parameter to greatly increase the production of primordial black holes. This could be done by turning up the scale of the density fluctuations, $\delta\rho/\rho$, which has been measured to be around 10^{-5}.

An answer can be given in the context of single field – single parameter inflation [8, 6, 61]. In that theory $\delta\rho/\rho$ is determined by λ, the strength of the self-coupling of the inflaton field. This controls the slope of the inflaton potential and hence the number of foldings grows with decreasing λ as

$$N \approx \lambda^{-1/2} \tag{6}$$

This means that the volume of the universe, and hence the number of ordinary black holes produced, scales as

$$V \approx \mathrm{e}^{3N} \approx \mathrm{e}^{3\lambda^{-1/2}} \tag{7}$$

Hence, there is a competition between raising the number of primordial black holes while exponentially shrinking the universe and so decreasing the number of black holes produced by stellar evolution. The exponential dominates and the result is that cosmological natural selection predicts the smallest possible $\delta\rho/\rho$ consistent with galaxy formation. One makes more black holes overall by having an exponentially bigger universe and making them later from stars than one does by having a lot of primordial black holes in a tiny universe.

However this argument only works in the simplest model of inflation. In more complex models with more fields and parameters, $\delta\rho/\rho$ is uncoupled from N, and one can have a large universe whose black hole production is dominated by primordial black holes. Hence, cosmological natural selection predicts that inflation, if true, must be single field–single inflation whose potential is governed by a single parameter. This is so far consistent with all observations, but it could be falsified by future observations, for example if high levels of non-Gaussianity are confirmed.

Once $\delta\rho/\rho$ is fixed in this way, cosmological natural selection makes a prediction for the value of the cosmological constant. This is because, if $\delta\rho/\rho$ is small, as is observed in our universe, there is a critical value Λ_0, of the cosmological constant, Λ, such that, for $\Lambda > \Lambda_0$, the universe would expand too fast for galaxies to form. But without galaxies there would not be many massive stars, which are the pathway to most black holes in our universe. Hence cosmological natural selection predicts $\delta\rho/\rho$ small and $\Lambda < \Lambda_0$.

The next question to be addressed is whether the number of black holes will strongly depend on Λ in the region below Λ_0. There are two competing effects to consider. As Λ is decreased from its present value, the time at which dark energy begins to accelerate the expansion is pushed later. This gives more time for structures to form, and this may lead to the birth of additional galaxies from the collapse into dark matter halos of baryons that at our present time are in the intergalactic medium. This could lead to more overall star formation and hence more black holes.

The competing effect is that with a delayed acceleration of the expansion there may be expected to be more collisions of spiral galaxies. The result of a collision involving a spiral galaxy is to heat the gas in the disk of the galaxy, turning off star formation, and converting the galaxy into an elliptical galaxy. Indeed, elliptical galaxies, which lack disks and have no active star formation, are believed to be the results of mergers and collisions among spiral galaxies. Hence, one effect of decreasing Λ below its present value may be to decrease overall star and black hole formation by leading to increased mergers of spiral galaxies into ellipticals.

Which of these two competing effects dominate cannot be determined without detailed modeling. It is tempting to conjecture that modeling would show that the present value of Λ is at the crossover point where the total black hole production is maximized.

Note finally that the choice of initial conditions is not so far explained by the scenario of cosmological natural selection. This is challenging as new universes arise from black hole singularities which are generically very inhomogeneous. Thus, cosmological natural selection probably requires inflation to make sense of the specialness of the initial conditions.

PLURALISTIC COSMOLOGICAL SCENARIOS

Let us finally turn to the pluralistic scenarios, of which eternal inflation is the main example. In this scenario an infinite population of universes is produced in one step from the formation of bubbles in an eternally inflating primordial phase. At least in its simplest form, this lacks the strengths of either the cyclic or the branching scenarios.

While a few other bubbles may have collided with our universe – giving a chance to confirm but not falsify predictions of the scenario [64] – almost all the universes in the population are causally disconnected from our own. It is usually assumed that the universes that are created randomly sample the points in the landscape of the fundamental theory so there are almost no properties common to all universes. The only property P put forward as satisfied in all universes is that the curvature should be slightly negative (slightly negative curvature from eternal inflation). However, this will be difficult to confirm or falsify with near-future observations because it will require a great deal of precision to distinguish this from vanishing curvature.

Moreover there are no long chains of descent so even if there were attractors in the landscape there is no mechanism to reach them by the single step by which universes are created from the primordial inflating state. This can be put more strongly. There may be cascades of decays by which bubbles form within bubbles, leading to some chains of descent. But for this to function to lead the ensemble to dominance by an attractor it must be that almost every member of the infinite ensemble is the result of such a long chain of descent. For dominance by an attractor to work this would have to be shown to be a consequence of dynamics on the landscape.

In addition, the formation of bubbles takes place at very high energies, typically grand unified scales where the details of the parameters of low-energy physics are not going to matter. So there is no mechanism for a coupling between the fine tunings of low-energy physics and the dynamics that produces the ensemble. In its absence, it has to be concluded that universes like ours with fine tunings of low-energy parameters are very rare.

There are other candidates for a property P, common to most members of the string landscape, that might be contemplated. These are motivated by the results of surveys so far carried out of the properties of theories in the string landscape [34].[2] For example it has been suggested that some kinds of particles which might appear in an extension of the standard model, and which would be consistent with the principles of quantum field theory, cannot arise from a string vacuum. Examples of this include gauge groups too large to fit into $E8 \times E8$ or too high dimensional representations of grand unified gauge groups, such as $SO(10)$. If we were more confident than present knowledge allows that the absence of theories with these features in the parts of the string landscape so far explored extends to the whole landscape, we could make their absence a falsifiable prediction of string theory.

But notice that, while this would be a genuine prediction of string theory, it is in no way a prediction of a pluralistic cosmological

[2] I am grateful to Paul Langacker for a conversation about this.

scenario such as eternal inflation. This is because the hypothesis that there is more than one universe plays no role in the argument leading to the putative prediction. The prediction would follow only from the hypothesis that our universe is described by a string theory, and is independent of whether our universe is unique or not.

By comparing this case to that of cosmological natural selection we see that the occurrence of a property P common to all members of a theory landscape becomes a prediction of a multiverse scenario only when the mechanism that constructs the ensemble of universes is necessary to P holding in almost all its members. This is the case in cosmological natural selection, but it is not the case in the example just considered.

For a different kind of case, consider now a different prediction of string theory we might be able to make if we knew the results of surveys of the landscape to date are representative of the whole landscape. It has been observed [34] that nearly every string vacuum that contains the minimally supersymmetric standard model (MSSM) also contains exotic particles in their low-energy spectra such as leptoquarks or additional generations. While no exotic states are common to all models, the absence of all kinds of exotica is exceedingly uncommon; therefore, for all practical purposes, a falsifiable prediction of string theory plus the anthropic landscape is that some kind of exotica must be seen at the TeV scale.

This, however, is not a prediction of string theory alone because there are known examples of string vacua that contain the MSSM but no exotics. Even if these are rare they imply that the lack of discovery of exotics would not falsify string theory. But this would falsify the combination of string theory and the hypothesis that the string vacua we observe are picked randomly from the set of anthropically allowed vacua, because that requires our universe to be governed by a law that is typical of anthropically allowed extensions of the standard model. But, at least if the landscape surveys so far carried out are representative, typical anthropically allowed vacua which contain the standard model also contain exotics.

In the absence of any large set of properties P common to the ensemble, proponents of eternal inflation have to fall back on the anthropic principle [4, 33]. This has so far not led to any genuine predictions, and it is pretty clear why this is unlikely. The properties of a universe can be divided into two classes. The first class consists of properties that play a role in making a universe friendly to life. Examples include the values of the fine-structure constant and the proton–neutron mass difference. Class two consists of properties that do not strongly influence the biofriendliness of a universe. These include the masses of the second- and third-generation fermions (so long as they stay sufficiently heavier than the first generation).

The first class of properties must hold and their verification does not provide evidence for any cosmological scenario – because we already know the universe is biofriendly. That is to be explained, by an argument that is not circular, i.e. does not assume our existence. The second class are assumed to be randomly distributed in the ensemble – hence, since they are uncoupled from biofriendliness they will be randomly distributed in the ensemble of biofriendly universes. Hence no prediction can be made for them.

These kinds of arguments, developed in more detail elsewhere [8, 7, 10, 9], make it very unlikely that the anthropic principle can ever be the basis for a prediction by which a cosmological scenario could be falsified or strongly verified.

What are we to make then of the claims that there have been successful predictions made based on the anthropic principle? In fact, such claims must be fallacious, and they have been shown to be. This is discussed in detail in previous books and papers [8, 7, 10, 9], but I can mention briefly here that there are basically two kinds of fallacies in these claims.

First, a statement that X is essential for life is added to an already correct argument involving X. For example, Hoyle argued successfully that if carbon is produced in stars there must be a certain energy level in a nucleus [66]. He based this successful argument on the observation that carbon is abundant in the universe. The fact that carbon plays a role in life plays absolutely no role in the argument.

Later Weinberg argued that if there were to be an ensemble of universes with random values of the cosmological constant, that Λ would be seen to have a value within an order or two of magnitude below a critical value, Λ_0, above which no galaxies form [67]. This had nothing to do with life, as galaxies are observed to be plentiful. It is true that the observed value came to within a factor of one over twenty of the Λ_0 Weinberg used.

However, Weinberg's argument was also fallacious, because his estimate for Λ_0 depended on an unverifiable assumption, for which there is no justification, about the ensemble of universes, which is that Λ is the only constant that varies in the ensemble of universes. If other parameters are allowed to vary, the estimate of Λ_0 greatly increases, making the prediction far less successful [68]. For example if Λ and $\delta\rho/\rho$ are both allowed to vary the chances are quite small that their values are both as small as observed.

The point is not that an ensemble with only Λ varying is more likely than an ensemble where both Λ and $\delta\rho/\rho$ vary. The point is that one has to be very cautious about reasoning from the properties of a posited but unobservable ensemble because one can just make things up to fit the data. Without any independent check on the properties of the ensemble the fact that one can manipulate the assumptions you make about the ensemble to make an outcome seem probable does not in any way constitute evidence for the existence of that ensemble. For example, Garriga and Vilenkin observed that the argument comes out looking the best if one considers varying a different parameter [69]. But this doesn't add any strength to the claim, because, when a false argument has many possible versions, there will always be one that fits the data best. The flexibility of tuning a false argument to fit the data better does not provide evidence that its underlying assumptions are true.

We can contrast this with the argument made above in the context of cosmological natural selection, where there is an independent argument for $\delta\rho/\rho$ to be small.

Weinberg's prediction was made a decade before the discovery of dark energy, and in science this is not nothing; sometimes a strong intuition can produce a correct prediction even if the logic can be objected to. But this cannot be used as evidence for the assumption that there really is an ensemble of universes, as the argument from that assumption to the prediction was fallacious, for the reason just explained. One should also credit Sorkin for correctly predicting the value of the cosmological constant, but for most theorists this doesn't strongly increase their confidence in the causal set theory on which Sorkin's prediction was based [70].

These concerns are deepened by the measure problem in eternal inflation [71]. This arises because there are an infinite number of bubble universes created. When one has infinite ensembles then assertions of predictions based on relative frequencies become highly problematic. Any claim that outcome A is more probable than outcome B is problematic when the numbers of A and B are infinite. The ratios of relative frequencies, $N(A) / N(B)$, are then undefined.

There is a literature whose authors experiment with different measures on these infinite sets which give definitions of the ratios and hence relative frequencies. The challenge is to avoid various paradoxes, some of which bedevil any application of probability theory to infinite sets, others of which are special to cosmology. However, even if a measure that succeeds in avoiding all the paradoxes were found, that would in no way serve to increase the likelihood that the eternal inflation scenario is correct – it would just be another instance of making up the specification of an unobservable ensemble in order to get what one wants from it. The fact that there may be a best version of a false claim does not increase the credibility of that claim, in the absence of any independent verification of it.

PRINCIPLE OF PRECEDENCE

Another kind of evolving law is possible in the context of quantum theory.[3]

[3] The results described in this subsection are described in more detail in [13].

We are used to thinking that the laws of physics are deterministic and that this precludes the occurrence of genuine novelty in the universe. All that happens is rearrangements of elementary particles with unchanging properties by unchanging laws. This is also usually taken to imply that the notion that human beings have freedom or free will must be an illusion.

But must this really be the case? We need determinism only in a limited set of circumstances, which is where an experiment has been repeated many times. In these cases we have learned that it is reliable to predict the outcome of future instances of the same experiment as we have seen it to give in the past.

Usually we take this to be explained by the existence of fundamental timeless laws which control all change. But this is an overinterpretation of the evidence. What we need is only that there be a principle that measurements which repeat processes which have taken place many times in the past yield the same outcomes as were seen in the past.[4] Such a *principle of precedence* would explain all the instances where determinism by laws works without restricting novel processes to yield predictable outcomes [13]. There could be at least a small element of freedom in the evolution of novel states without contradicting the application of laws to states which have been produced plentifully in the past.

But are there any truly novel states in nature?

It is fair to say that classical mechanics precludes the existence of genuine novelty, because for certain all that happens is the motion of particles under fixed laws. But quantum mechanics is different, in two ways. First, quantum mechanics does not give unique predictions for how the future will resemble the past. It gives from past instances only a statistical distribution of possible outcomes of future measurements.

Second, in quantum physics there is the phenomenon of entanglement which involves novel properties shared between subsystems which are not just properties of the individual subsystems. The free

[4] This idea was to some extent anticipated by Peirce in [72].

will theorem of Conway and Kochen [12] tells us that in these cases systems respond to measurements in a way that can be considered free, in the sense that the result of an individual measurement on one element of an entangled system could not be predicted by any knowledge of the past.

An entangled state can be novel in that it can be formed from a composition of particles into a state never before occurring in the prior history of the universe. This is common for example in biology where natural selection can give rise to novel proteins and sequences of nucleic acids which almost certainly, due to the combinatorial vastness of the number of possibilities, have not existed before.

There is then the possibility that novel states can behave unpredictably because they are without precedent. Only after they have been created enough times to create ample precedent would the behavior of these novel states become lawful.

Hence we can have a conception of a law which is sufficient to account for the repeatability of experiments without restricting novel states from being free from constraints from deterministic laws. In essence the laws evolve with the states. The first several iterations of a novel state are not determined by any law. Only after sufficient precedent has been established does a law take hold, and only for statistical predictions. Individual outcomes can be largely unconstrained.

Quantum physics allows this possibility because the generic single measurement is not determined by quantum dynamics. Only if the system is prepared in an eigenstate of the measurement being made is the result determined. But these require fine tuning and are hence non-generic. Otherwise it is stochastic so that no outcome of a single generic observation can disagree with predictions of quantum mechanics.

There are aspects of measurements that are not predicted by quantum mechanics which offer scope for genuine novelty and freedom from deterministic evolution. Imagine a double-slit experiment with a very slow source of photons. The measurement gives a sequence of positions to which the photons fall on the screen, $x_1, x_2, \ldots, x_N$. Each individual photon can end up anywhere on the screen. Quantum mechanics predicts the overall statistical ensemble that accumulates after many photons, $\rho(x)$. But it does not, for example, restrict the order by which they fall. Quantum mechanics is equally consistent with a record in which the x_i are permuted, from one random sequence to another.

Macroscopic outcomes could depend on the order of positions; for example, if someone chooses to make a career in science or politics based on whether the 13th photon falls to the left or right side of the screen.

The basic idea of the formulation of quantum theory proposed is that (1) systems with no precedents have outcomes not determined by prior law, (2) when there is sufficient precedence the outcome of an experiment is determined by making a random selection from the ensemble of prior cases, and (3) the outcome of measurements on systems with no or few precedents is as free as possible, in a sense that needs to be defined precisely. Stated more carefully these become the principles of this approach to quantum theory, to be enunciated below.

Thus, this is a twist on the real ensemble interpretation proposed earlier [73]. The principle proposed there was that whenever probabilities appear in quantum physics they must be relative frequencies within ensembles every element of which really exists. In the original version of this idea the ensemble associated with a quantum state existed simultaneously with it. In the current version the ensembles exist in the past of the process they influence.

How much precedence is necessary to turn freedom into deterministic dynamics? There must for each system be an answer to this question.

If the first instance of a measurement made on a novel state is undetermined, but a measurement with a great deal of precedence is tightly determined, there is, for any system, a number of prior preparations which is needed to determine as well as can be done any future outcomes of measurements made on future iterations of that system.

This is the number of degrees of freedom of the system, to be denoted K. There is also the dimension or capacity of the system, which is the number, N, of outcomes that can be distinguished by measurements on the system. These numbers and their relation must play a crucial role because they determine when there is sufficient precedent for future cases to be determined as possible.

We show in [13] that there is a precise sense in which quantum kinematics is specified by requiring that K be as large as possible, given N, consistent with a small set of reasonable general axioms. This means that there is the maximal amount of information needed per distinguishable outcome to predict the statistical distribution of outcomes for any experiment. As a result, we can say that the responses of quantum systems to individual measurements are maximally free from the constraints of determinism from prior cases.

To illuminate this idea we can make use of an axiomatic formulation of quantum theory, given by Masanes and Müller [74]. (The idea of formulating quantum mechanics in terms of simple operational axioms was introduced by Hardy [75].) They give four axioms for how probabilities for outcomes behave when systems are combined into composite systems, or subsystems are projected out of larger systems, and proves that they imply quantum mechanics or classical probability theory. To these we add a new, fifth, axiom which picks out the quantum case. These five postulates define the kinematics of quantum systems.

The hard work needed to show this has already been done by Masanes and Müller [74]; my observation that these five postulates determine quantum theory is a trivial consequence of their work. Informally stated these five postulates are as follows:

1. The state of a composite system is characterized by the statistics of measurements on the individual components.
2. All systems that effectively carry the same amount of information have equivalent state spaces.
3. Every pure state of a system can be transformed into every other by a reversible transformation.

4. In systems that carry one bit of information, all measurements which give non-negative probabilities are allowed by the theory.
5. Quantum systems are maximally free, in that a specification of their statistical state, sufficient for predicting the outcome of all future measurements, requires the maximal amount of information to specify, relative to the number of outcomes of an individual measurement.

To these we add a postulate about quantum dynamics. To motivate it let us ask how we measure the statistical state of a system. The answer is that in the past we have prepared an ensemble of systems with the same preparation and subjected each to a measurement. There are a number of distinct measurements required to completely determine the outcome of any measurement we might make on a similarly prepared system in the future. The probabilities for the outcomes of these measurements make up the statistical state. Informally stated, the principle of precedence says the following:

> *Principle of precedence*. In cases where a measurement of a quantum system has many precedents, in which an identically prepared system was subject to the same measurement in the past, the outcome of the present measurement is determined by picking randomly from the ensemble of precedents of that measurement.

UNIVERSALITY OF META-LAW: REDUCING THE CHOICE OF LAWS TO CHOICES OF INITIAL CONDITIONS

I now turn to a different approach to the meta-law problem.[5] Suppose that a large class of theories, which included the standard model as well as a large set of plausible alternatives, were actually equivalent to each other, in the sense that there were transformations that mapped the degrees of freedom of any two of these theories into each other.

[5] The approach discussed in this section was explored in a paper [76].

Of course not every theory in this class would have to resemble general relativity coupled to gauge theories and chiral fermions. It would be sufficient if only a subclass did.

If this were the case then there would be no sense in which any of these theories could be considered more fundamental than another, nor would there be any meaning that could be given to the claim that one, rather than another, was the true theory. The puzzle of the ambiguity of choice of initial conditions versus choice of theories would be resolved, because the only meaningful choices within the class would be choices of initial conditions. Furthermore the evolution from one theory to another could be understood in terms of quantum transitions between different semiclassical solutions of a single theory.

Before dismissing this possibility as crazy, let us take into account the various arguments that lead to the conclusion that quantum theory plus diffeomorphism invariance forces theories to be finite, so that there are finite numbers of degrees of freedom in every quantum theory containing gravity. In this case, each theory of gravity plus $SU(n)$ gauge fields in d space dimensions has, at least naively, roughly

$$N = \left(\frac{L}{l_{\text{Planck}}}\right)^d (n^2 + 1) \tag{8}$$

total degrees of freedom, where L is the infrared cutoff given by the cosmological constant and l_{Planck} is the ultraviolet cutoff given by the Planck scale. (We neglect fermions in the following to simplify the argument.) It seems plausible that theories with different N values cannot be equivalent. But could two gauge theories coupled to gravity be equivalent, with different dimensions and gauge groups, so long as they had the same N? The demonstration of such equivalences would involve mappings between their degrees of freedom that mix up spacetime with internal symmetries. That is, the transformations between theories would not respect locality. These transformations would not be apparent from the naive continuum expressions of the theories, but they would become apparent when they were expressed in cutoff forms with finite numbers of degrees of freedom. Were this true, the holographic principle might be a special case of a wider class of equivalences among theories.

I would like to suggest that there are indeed such large universality classes of cutoff theories of gauge fields and gravity. This is done by exhibiting a simple matrix model that has solutions and truncations which lead to a diverse set of cutoff gauge and gravitational theories, in different dimensions, with different gauge groups.

Before presenting this theory, let me mention three considerations which suggest the plausibility of this resolution of the search for a fundamental theory.

First, there are already examples of large equivalence classes among gauge theories of different types. Some of the best studied of these arise in supersymmetric gauge theories and string theory. These include conjectures of dualities between theories with different gauge groups and, as in the case of the anti-de Sitter/conformal field theory (AdS/CFT) conjectures, different numbers of dimensions. Others do not require supersymmetry but involve dualities among non-commutative and matrix formulations of gauge theories. It is then natural to ask if all of these conjectured dualities, supersymmetric and not, may be several tips of a single iceberg involving a much wider class of dualities. If so, the question is what principle underlies all these dualities.

Second, consider the consequences of two widely held beliefs, that spacetime is emergent and that the theory it is emergent from is finite. It follows that locality is also an emergent property. If different spacetimes emerge by constructing different effective field theories around different solutions of the fundamental theory, then it follows that whether two degrees of freedom are related by a translation in space or by an internal symmetry transformation will not be absolute, but will depend on the solution the effective description is based on. This makes it possible that theories with different spacetime dimensions and internal symmetries will emerge from the same fundamental dynamics.

Third, there have been a number of suggestions that physical processes are computations. However, the central result in computer science is the universality of computation, that all computers are equivalent to a universal computer, a Turing machine. Any computer can be simulated on any other computer, by writing an appropriate program. Might it be that there is also a universality class of dynamical theories, any solution of one may be represented by a solution of another by a precise choice of initial conditions?

The metaphor of "programming the universe," even if it is not precisely true, may give us guidance for how to proceed here. For, even if there is a large equivalence class of theories as described above, it may be easier to see this from one representative than another. What is needed is something like the Turing model, a very simple representative of the class, which is very helpful when proving the universality of computation. For one does not have to directly demonstrate the equivalence of any two computers, one just needs to show the equivalence of each to a Turing machine.

We then seek the equivalent of a Turing machine for gauge and gravitational theories, a simple theory from which a variety of different theories of gauge and gravitational interactions can be reproduced.

Let us then note that there are at least three ways that a dynamical theory, U, may give rise to another theory T. One can plug in an ansatz to the action for U, leading to an action for T. In this case we say that U truncates to T. Or the solutions to U can include solutions to T, in which case we say that U reduces to T. They are not the same because equations of motion will be missing in a truncation that are implied by the variation of the original action – and so have to be satisfied in a reduction. Thus, reduction is stronger than truncation. Another possible relation is for T to arise as a low-energy effective approximation to the expansion in terms of small deviations from either a truncation or a reduction of U.

We are then looking for a theory U that has the following characteristics:

> It has a very large but finite number of degrees of freedom. It should truncate or reduce to cutoff versions of a large variety of different

> theories, including general relativity in 3+1 dimensions, coupled to Yang–Mills theory for a variety of gauge groups, G. For reasons discussed above, the truncations or reductions will introduce notions of locality that are inconsistent with each other, as they lead to theories of different dimensionality. This suggests the theory should be truly background-independent, so that the spacetime manifold on which the metric, connection, and gauge fields are defined is not present when the dynamics is formulated, but emerges only from the study of special classes of solutions. The action and equations of motion of the theory should be extremely simple, so that their physical content is minimal and the specification of kinematics and dynamics arises only by the truncation of degrees of freedom or selection of a class of solutions of the meta-theory.

In [76], I proposed a candidate for such a universal meta-theory and provide evidence it has truncations with the required properties. The degrees of freedom are as simple as possible; they are an $N \times N$ Hermitian matrix, for a very large N, which will be called M.

The dynamics cannot be linear because we want its solutions to reproduce those of nonlinear field equations. The simplest nonlinear dynamics are quadratic equations, which arise from a cubic action. The simplest possible nonlinear action for matrices is

$$S = \mathrm{Tr}M^3 \tag{9}$$

The theory has a gauge symmetry under $U(N)$, the group of $N \times N$ unitary matrices. Let U be an element of $U(N)$; then the action is invariant under M goes to UMU^{-1}.

I was able to find evidence that this simple theory has reductions with the required properties. This simple model has truncations that yield many of the theories of connections that physicists study. These include topological field theories such as Chern–Simons theory for any $U(N)$ in three dimensions, as well as a class of topological field theories known as BF theories in four dimensions. Other truncations yield theories with local degrees of freedom including general relativity in four spacetime dimensions, and Yang–Mills theory in d equal to or greater than 4.

We also argued that when the one-loop effective action is taken into account, there are truncations, and perhaps reductions, which yield general relativity coupled to Yang–Mills fields for any $U(N)$ in four dimensions.

This form also has truncations that give the bosonic sectors of some of the cubic matrix models [77], studied before, which were proposed as background-independent forms of string theory. Thus, it appears possible that at least bosonic string theories are also contained in the class of theories that arise from truncations of this action.

One may ask how all these theories can arise from a simple cubic action. The answer is that when expressed in certain first-order forms, where auxiliary fields are used to write the actions so that only a single derivative appears, these theories all have cubic actions. For general relativity this requires writing the theory in connection variables, such as those given by Ashtekar [80] and Plebański [79]. This is a remarkable fact, whose significance for the project of unification has perhaps been insufficiently appreciated. There could not be a simpler form as the equations of motion are then all quadratic equations; any simpler theory would be linear.

At a non-perturbative level the simple first-order form of the actions make possible clean paths to quantization in which the Hamiltonian formulations are all polynomial and the path integral measures are determined by group theory [81]. Thus, the unification of all the above theories in a single, simple matrix model is another piece of evidence that these connection formulations of general relativity are more fundamental than the original metric formulation and are a necessary route to their quantization.

Once it is realized that these different theories all have cubic actions the idea of unifying them by writing them as truncations of a cubic matrix model naturally suggests itself.

One can then ask how such different theories may arise from a single matrix model. The answer is that the different truncations involve different tensor product decompositions of the space of matrices. This is analogous to the way in which emergent degrees of freedom and associated conservation laws may arise from symmetries associated with tensor product decompositions, giving rise to noiseless subsystems in quantum mechanics [82]. This has been proposed as the origin of the physical degrees of freedom of background-independent theories [83].

Finally, we may remark that there are four independent lines of argument that matrix models may underlie quantum mechanics as their ordinary statistical mechanics appears to naturally describe a non-local hidden-variables theory which approximates quantum mechanics [78].

THE UNIFICATION OF LAW AND STATE

In this section, I describe another approach to the meta-laws dilemma, which is that the distinction between states and laws breaks down [89]. This new proposal is also realized in a simple matrix model. Instead of timeless law determining evolution on a timeless space of states, we have a single evolution which cannot be precisely broken down into law and state. Formally, what this means is to imbed the configuration space of states and the landscape parameterizing laws into a single meta-configuration space. The distinction between law and state must then be both approximate and dependent on initial conditions.

There is, it must be granted, an evolution rule on the meta-configuration space, but we can choose an evolution rule that is almost entirely fixed by some natural assumptions. The remaining freedom is, I conjecture, accounted for by the principle of universality, which I just described. Because the complexity of the effective law is now coded into the state, the meta-law can be very simple, because all it has to do is to generate a sequence of matrices, in which the differences from one to the next are small. The meta-law dilemma is addressed by showing that the form of this rule is almost completely fixed by some natural assumptions, with the remaining freedom plausibly accounted for by universality.

In this model of a meta-theory, the meta-state is captured in a large matrix, X, which we take to be antisymmetric and valued in the integers. It might describe a labeled, directed graph. The meta-law is a simple algorithm that yields a sequence of matrices, X_n. The rule is that X_n is obtained by adding to a linear combination of X_{n-1} and X_{n-2} their commutator $[X_{n-1}, X_{n-2}]$:

$$X_n = aX_{n-1} + bX_{n-2} + [X_{n-1}, X_{n-2}] \tag{10}$$

Given the first two matrices, X_0 and X_1, the sequence is determined. This is more like a simple instruction in computer science than a law of physics, and we are able to argue it is almost unique, given a few simple conditions.

That almost unique evolution rule acts on a configuration space of matrices, whose interpretation depends on a separation of time scales. For certain initial configurations, there will be a long time scale, T_{Newton}, such that, for times shorter than T_{Newton}, the dynamics can be approximately described by a fixed law acting on a fixed space of states. Both that law and that state are coded into the X_n. But for longer times everything evolves, laws and states together, and it is impossible to cleanly separate what part of the evolution is changes in law and what part is changes in state. Furthermore, which information evolves slowly, and goes into the specification of the approximate time-independent law, and which evolves fast, and goes into the description of the time-dependent state, is determined by the initial conditions.

So the question of "why these laws" becomes subsumed into the question of "why these initial conditions" in a meta-theory. This does not yet solve the problem of explaining the particular features of the standard model and its parameters, but it gives a new methodology and strategy with which to search for the answer.

Starting from the standard model, one might move in the direction of a meta-theory by elevating all parameters to degrees of freedom. This is something like what happens in the string landscape. Here we make a simple model in which the meta-state is a large sparse matrix, perhaps representing the connections on a graph.

In the present era, observations constrain us to T_{Newton} greater than the present age of the universe. This may be due to the fact that in the present era the universe is, in Planck units, very close to its ground state (in the conventional state space). Another way to say this is that the meta-state is dominated by information that goes into the laws, while the information associated with the state constitutes a small perturbation. This accords with the sense that the initial conditions of the universe were very special, in a way that is characterized by a poverty of information.

The choice of this evolution rule is fixed by the following ideas.

The evolution rule should mimic second-order differential equations, as these are basic to the dynamics of physical systems. So two initial conditions should be required to generate the evolution. We should then need to specify X_0 and X_1 to generate the sequence. We are then interested in rules of the form $X_n = F(X_{n-1}, X_{n-2})$. The changes should be small from matrix to matrix, at least given suitable initial conditions. This is needed so that there can be a long time scale on which some of the information in the matrices is slowly varying. This makes it possible to extract a notion of slowly varying law, acting on a faster varying state. So we will ask that

$$X = F(X, Y) \tag{11}$$

We require that the evolution rule be nonlinear, because nonlinear laws are needed to code interactions in physics. But we can always use the basic trick of matrix models of introducing auxiliary variables, by expanding the matrix, in order to lower the degree of nonlinearity. This, as I described in the previous section ("Universality of meta-law: reducing the choice of laws to choices of initial conditions"), accords with the fact that the field equations of general relativity and Yang–Mills theory can, by the use of auxiliary variables, be expressed as quadratic equations, as for example in the Plebański action. The simplest nonlinear evolution rule will then suffice, so we require a quadratic evolution rule.

The last condition we impose is reversal invariance, but only at the linear level when laws and initial conditions separate, and not generally. A simple evolution rule that realizes these is

$$X_n = 2X_{n-1} - X_{n-2} + [X_{n-1}, X_{n-2}] \tag{12}$$

It is nearly unique, as shown in [89], to which the reader interested in more details is referred. There it is shown that the information carried by the sequence of matrices X_n can be divided for many time steps into those that characterize a law, which is slowly changing, and which evolves the second piece, which can be characterized as an evolving state. This model then captures the idea that law and state can be unified within a meta-state that evolves according to a universal meta-law, from which approximate notions of law and state emerge for large but finite times.

7 Implications of temporal naturalism for the philosophy of mind

In Chapter 1, I introduced the metaphysical folly as the tendency for unreflective naturalists to believe in an imagined nature constructed in their imaginations as being more real than the world we perceive with our senses.

A symptom of the metaphysical folly is the move from *Sense impressions give unreliable knowledge of nature, nature is instead truly X* to *Sense impressions are incompatible with the concept that the world is X,* so qualia must not exist. But the one thing we can be sure of is that qualia exist. Therefore, as Galen Strawson [85] and other philosophers of mind [86] emphasize, if we are naturalists and believe everything that exists is part of the natural world then qualia must be also part of the natural world. The right statement – if we are naturalists – must be:

> *X may provide a good description of some class of observations of the world, but the world cannot be X exactly because qualia are undeniably part of the world and X are not qualia.*

Here I would like to argue that it is much easier to conceive of qualia as part of the natural world in temporal naturalism than in timeless naturalism.

I can begin with two basic observations. First, every instance of a qualia occurs at a unique moment of time. Being conscious means being conscious of a moment. Being ordered and "drenched" in time is a fundamental attribute of conscious experience.

Second, facts about qualia being experienced now are not contingent. There are no facts of the form, "If there is a chicken in the road then I am now experiencing a brilliant red."

It follows that qualia cannot be real properties of a timelessly natural world, because all references to now in such a world are contingent and relational. Nor can qualia be real properties of a pluralistic simultaneity of moments because what distinguishes those moments from each other are relational and contingent facts.

Qualia can only be real properties of a world where "now" has an intrinsic meaning so that statements about now are true non-relationally and without contingency. These are the case only in a temporal natural world.

It has been objected that eternalists can see the history of the universe having "temporal parts" with intrinsic qualities. This misses the key point, which is that any reference to one of those timeless parts in a block-universe framework must be contingent and relational, whereas our knowledge of qualia are unqualified by either contingency or relation to any other fact.

That was the short version of the argument. Here is a longer version:

We have direct experience of the world in the present moment. Just as the fact that we experience is an undeniable feature of the natural world, it is also an undeniable feature of the natural world that qualia are experienced in moments which are experienced one at a time. This gives a privileged status to each moment of time, associated with each experience: this is the moment that is being experienced now. This means that we have direct access to a feature of the presently present moment that does not require relational and contingent addressing to define it. We can define and give truth values to statements about now which are not contingent on any further knowledge of the world.

How can these facts about nature – that each qualia is an aspect of a presently privileged present moment, that does not require contingent relational addressing to define or evaluate – be incorporated into our conception of the natural world? This fact fits comfortably in a temporal naturalist viewpoint, because in that viewpoint all facts about nature are situated in, or in the past of, presently privileged present moments and no relational and contingent addressing is required to define those that refer to the present.

This fact cannot fit into a timeless version of naturalism according to which there are no facts situated in presently privileged present moments, except when that can be defined timelessly through relational addressing. The same is the case for Barbour's moment pluralism.

We can draw a stronger conclusion from this. There is no physical observable in a block-universe interpretation of general relativity that corresponds to my ability to evaluate truth values of statements about now, without any need for further contingent and relational facts. The block universe cannot represent now because now is an intrinsic property and the block universe can only speak of relational properties. Hence the block universe is an incomplete description of the natural world.

That is, because qualia are undeniably real aspects of the natural world, and because an essential feature of them is their existing only in the present moment, qualia allow the presently present moment to be distinguished intrinsically without regard to relational addressing. Any description of nature that does not allow *now* to be intrinsically defined is an incomplete description of nature because it leaves out some undeniable facts about nature. Hence the block universe and timeless naturalism are incomplete, and hence they are wrong.

TWO SPECULATIVE PROPOSALS REGARDING QUALIA

I would like to offer two speculative proposals regarding the physical correlates of qualia.

Panpsychism asserts that some physical events have qualia as intrinsic properties, some of which are neural correlates of human consciousness. But it does not need to assert that all physical events have qualia. Might there be a physical characteristic which distinguishes those physical events that have qualia?

According to the principle of precedence which I discussed above, there are then two kinds of events or states in nature: those for which there is precedence, which hence follow laws, and those without precedence, which evoke genuinely novel events. My speculative proposal is that the correlate of qualia are those events without precedence.

It is commonplace to observe that habitual actions are unconscious in people. Maybe the same thing is true in nature. Maybe brains are systems where a lot of novel events take place.

Here is a second question raised by panpsychism: If brains have states which are neural correlates of consciousness, but consciousness is a general intrinsic property of matter, then what physical properties correlate to qualia? Or, to put it differently, in what way do the physical attributes of correlates of consciousness vary when the qualities of qualia vary?

Panpsychists argue that the elements of the physical world have structural properties and intrinsic and internal properties. By arguing that matter may have internal properties not describable in terms needed to express the laws of physics, panpsychists reserve a place for qualia as intrinsic, non-dynamical properties of matter. I would propose to cut the pie up differently. I would hold that events have relational and intrinsic properties, but relational properties include only causal relations and spacetime intervals which are derivative from them. Under intrinsic properties I would include the dynamical quantities: energy and momenta, together with qualia. I would go further and relate energy and qualia. I would point out that the experienced qualities of qualia correlate with changes of energy. Colors are a measure of energy, as are tones.

8 An agenda for science

The main test of the ideas we have argued for, as of any new scientific ideas, is whether they generate a new agenda for research in science that succeeds in generating new knowledge even as it instigates a new paradigm for the organization of our ideas about nature. In this last chapter of this book I show that it is already doing so. The main fields affected are cosmology, quantum gravity, and the foundations of quantum theory.

THE AGENDA FOR OBSERVATIONAL COSMOLOGY

The first field affected by our program is cosmology, where indeed there already is a split between those investigating pluralistic models of cosmology and those developing models based on a succession of universes. We have said enough about the failure of many-universe cosmologies to generate falsifiable predictions (but for those readers needing more convincing of this, see [7, 10, 9]) and need only contrast this with the genuine predictions generated by cosmological scenarios which assume the big bang is not the first moment of time, but a passage before which the universe existed, if possibly under different laws. Three examples, discussed above, suffice to demonstrate the claim that such successional hypotheses can and do generate falsifiable predictions for doable experiments.

- The cyclic cosmologies of Steinhardt, Turok, and collaborators make two predictions for the structure of the fluctuations in the CMB which strongly distinguish them from predictions of generic inflation models [56]. (The qualifier "generic" is necessary because inflation models can be fine-tuned to generate diverse predictions.) These are an absence of tensor modes and a significant non-Gaussianity. Both predictions are being tested in data from the Planck satellite which is being analyzed as of this writing.

- The alternative cyclic cosmology proposed by Penrose, called conformally cyclic cosmology (CCC), predicts there should be many concentric circles of elevated temperature visible in the CMB [57]. These are traces of bursts of gravitational waves emitted in the era prior to ours by collisions of very massive black holes in the centers of galaxies. Penrose and a collaborator claim to have observed such concentric sets of circles in the CMB data [59]. These claims are disputed by several cosmologists who argue that the signals claimed can be generated by chance in random data [60]. We do not have to take a point of view on the resulting controversy to note that it proves our point that successional cosmological scenarios make testable and even falsifiable predictions for doable experiments.
- Cosmological natural selection makes falsifiable predictions, as I illustrated with two examples, the upper mass limit for neutron stars and that inflation, if true, be generated by a single field and governed by a single parameter [8, 6, 61, 9]. Both are highly vulnerable to falsification. In addition, it should be noted that cosmological natural selection is the only scenario which proposes a genuine explanation for the fine tunings and present values of the parameters of the standard model of particle physics.

These examples show an agenda already in operation: propose scenarios for the universe to evolve through a succession of eras, whether linear or branching, and deduce testable consequences by which they can be tested.

CAN THE LAWS OF NATURE BE OBSERVED TO CHANGE?

Given the hypothesis that laws evolve with the universe we must distinguish several ways they may have been different in the past. These are distinguished by when changes in the laws might have occurred which would have observable consequences. We can distinguish four possibilities.

1. The laws might change during bounces which are transitions between different eras. In this case there can be indirect evidence of the process by which laws evolve of the kind just summarized in our discussions of cosmological natural selection.
2. The laws may have been different in the early universe. If inflation is true then the energetics of the early universe is dominated by a field, the inflaton – or possibly several coupled fields – of which there is no evidence today. This absence of evidence for the inflaton currently is natural in a quantum field theoretic setting if it is too massive to generate in accelerators. (Although we can ask if inflatons should be generated in collisions of ultra-high-energy cosmic rays, with energies of about a billionth of the Planck mass, with the atmosphere and whether those collisions might be observable in cosmic ray detectors such as AUGER.) If the standard model is part of a timeless law, the inflaton must be unified with the familiar particles. But if we allow the parameters of the standard model to evolve in time it is possible that the inflaton can be identified with the standard model Higgs, but with different mass and couplings.
3. The laws of nature may have changed over the history of the observable universe. Evidence of variation of a few constants of the standard model on cosmological time scales has been looked for, so far without complete success. Two examples are Newton's gravitational constant and the fine-structure constant. In the latter case, there have been claims by Webb *et al.* to observe changes in the fine-structure constant by measuring spectral lines in distant galaxies through observations of quasar spectra [87]. These claims are presently controversial.
4. The laws of nature might be observed to change in present experiments as in tests of the principle of precedence discussed above [13]. These arise in the field of quantum foundations, to which we now turn.

THE AGENDA FOR QUANTUM FOUNDATIONS

The field of quantum foundations is in the midst of a renaissance due to the resurgence of tests of quantum theory, which is the result of the emergence of the new field of quantum information science. Nonetheless, it remains too isolated from the other fields engaged in the search for new laws of physics such as elementary particle physics, cosmology, and quantum gravity. The field is split by an inner conflict between two missions. The first is to clarify conceptual and logical issues in the existing theory of quantum mechanics by resolving, within the fixed framework of quantum mechanics, the measurement problem and related issues. There are diverse proposals for how to do this which are called interpretations of quantum mechanics. The second mission is to propose a better (i.e. truer) theory of quantum phenomena that lacks the persistent issues with quantum mechanics such as the measurement problem. The aim here is to provide a complete description of individual quantum processes and experiments to replace the statistical description given by quantum mechanics. Such proposals are called hidden-variables theories.

The two missions are in conflict, as the success of either would moot the other, and the resulting creative tension in some cases animates and in other cases suppresses either effort, indeed within individual researchers, many of whom pursue both agendas, as well as in the field as a whole.

The ideas we have discussed in this book generate a new direction for research in quantum foundations framed by the following hypotheses:

1. Quantum mechanics is a theory of subsystems of the universe. As it is framed within the Newtonian paradigm it cannot be extended to a theory of the universe as a whole. Hence, there can be no quantum theory of cosmology within conventional quantum mechanics.
2. Quantum mechanics must then be an approximation to a truly cosmological theory, formulated outside of the Newtonian paradigm, derivable as an approximation to that theory by truncating it to a description of subsystems.
3. The hidden variables then do not refer to a more detailed description of an individual quantum system. They must instead be a description of

relationships between that subsystem and the rest of the universe that are lost in the truncation of the cosmological theory that yields quantum mechanics. This is consonant with the results of theorems by Bell and Kochen–Specter that any hidden variables be non-local and contextual.

4. The governing cosmological theory must be a relational theory, hence the hidden variables must concern relations between the subsystem and the rest of the universe.
5. The cosmological theory must have a distinguished global time. This is consistent with the result of Valentini that any hidden-variables theory have a preferred global time, which can be observed in experimental tests that distinguish the hidden-variables theory from quantum mechanics [49]. The global time can and generally is invisible to experiments in which the predictions of the hidden-variables theory and quantum mechanics coincide.
6. No reference to anything outside the universe should be required to explain anything within the universe. Applied to quantum mechanics this means that no imaginary ensemble can be utilized to explain any real experiment in nature.

We see that the research agenda in quantum foundations mandated by the view proposed in this book falls into the second stream of research – as the goal must be to discover the cosmological theory that quantum mechanics approximates, but it channels that search in a certain direction. It is notable that the hypothesis of a preferred global time coming from our program is consistent with the need for a global time to express a non-local hidden-variables theory. This has been seen explicitly in relativistic versions of Bohmian mechanics [88] and spontaneous collapse models [90], which reproduce the Poincaré invariance of the predictions of quantum field theory while breaking that invariance for predictions that diverge from those of standard quantum theory.

While there are clearly several directions in which one could set out to develop this agenda, in my own work I have taken as the starting point the question raised by point 6 above: to the extent that the quantum state refers to an ensemble of systems, it must be an ensemble of physically real systems. It cannot be a virtual or imaginary ensemble, because that would involve the action of the unreal on the real. So where exactly are the other real systems that constitute the ensemble that the wave function associated with a particular atom in a particular laboratory is to represent?

In the real ensemble interpretation proposed in [73] I give one possible answer: it is the ensemble of atoms existing at that time (in the preferred global time) in the universe with the same constituents and preparation, so that they would be described by the same quantum state. I propose a dynamics wherein the different copies of a system making up such an ensemble interact with each other by copying values of their "beables" and show that a particular form of that dynamics reproduces the predictions of quantum evolution via the Schrödinger equation.

An immediate consequence is that systems that have no such ensemble of copies will not evolve according to quantum dynamics; in particular, the superposition principle will fail for them. This immediately solves the measurement problem and explains why macroscopic bodies such as ourselves and our cats are not described by quantum mechanics, while the atoms we are made of, which have many copies, are. This novel view of the problems raised by quantum mechanics is a direct outcome of pursuing the research program set out here, and we can imagine that there are more to come.

The real ensemble formulation of quantum mechanics implies that quantum dynamics will fail for systems with no copies. This could be seen in experiments aimed at realizing quantum computation and communication which construct systems in pure quantum states that due to their complexity may not exist elsewhere in the universe. Such novel, artificial quantum systems could serve as the point where quantum mechanics fails an experimental test.

A different answer to the question of where are the systems making up the ensembles that quantum states characterize is given by another novel approach to quantum foundations that was also an early fruit of this research program – the principle of precedence which I described earlier [13]. In that case the answer is the ensemble of past systems with the same preparation. (In this case the past could be the causal past or the past in the preferred global time.)

Equally important, the principle of precedence serves to illustrate that the notion of timeless laws is not necessary to explain the regularities we observe in nature. The basic idea developed in [13] is that systems choose the outcomes of quantum measurements by picking randomly from the ensemble of results of similarly prepared systems being confronted by the same measurement. When there are no such precedents, a system will respond randomly.

The theory is in an early stage and so far incomplete. The principle of precedence can account for the seemingly lawful behavior of systems with many precedents, and a system without precedence must give a result which is not predictable based on knowledge of the past, no matter how complete. What the theory lacks is a hypothesis to describe what happens in between, as precedence builds up over the period when nature is confronted with the first several instances of a measurement.

Nonetheless it may be possible to test the basic idea with quantum technologies which are capable of constructing novel entangled systems of several atoms. Quantum mechanics makes definite predictions for such systems, whether novel or not, and no matter how complex, because we believe we know the Hamiltonian that describes quantum evolution. By the principle of reductionism, we assume that there are no forces apart from those that follow from the fundamental interactions of the elementary particles so that no new forces can arise applying to complex or novel systems.

The principle of precedence predicts that in situations where novel states encounter novel measurements, the Hamiltonian we would deduce by adding up the forces between the elementary constituents will not predict the right probability distribution for the measurement outcomes – rather the outcome would be random, i.e. not predictable from any knowledge of either the constituents or the past of the system.

It seems quite likely that this prediction can be tested with current quantum technologies, but specific proposals have not so far been made – this remains a task for future research.

THE EXISTENCE OF A PREFERRED GLOBAL TIME

The conclusion of our arguments that most challenges the well-tested and well-established physical theories is that there must exist a preferred and global conception of time. Some may query whether the arguments for the reality of time and change necessarily imply a preferred global time. It is difficult, however, to avoid that conclusion if one wants there to be an actual distinction between a real present and a yet-to-be-real future which holds over the whole universe and is observer-independent, and hence objectively real.

This assertion appears to challenge both the relativity of simultaneity of special relativity and the many-fingered time gauge invariance of general relativity. Both forbid the existence of a physically preferred global time, but in different ways, because the former is a global symmetry while the latter is a gauge symmetry.

In addition, as already mentioned, recent tests of Lorentz invariance bound violations of the relativity of inertial frames to less than the order of energy over the Planck energy [46]. These involve studying gamma rays or high-energy cosmic ray protons which have traveled for astronomical and cosmological distances, over which small effects on travel time or thresholds can build up to observable magnitudes. One test, which involves an effect that simultaneously breaks parity and Lorentz invariance, is bounded to several orders of magnitude past Planck scales [91].

An important agenda for science is then to reconcile the need for a preferred global time to realize the goals of a truly explanatory cosmological theory with the strong evidence for the principle of relativity on smaller scales.

It is then prudent to hypothesize that all theories of subsystems of the universe should be relativistically invariant so that the effects of the preferred global time, if it exists, are not detectible in experiments at less than the scale of the whole universe. As discussed above, shape dynamics provides a prototype of a theory of spacetime and gravitation with a global time which is relational and only detectible by making measurements spanning the whole universe. The predictions of the classical theory are in fact equivalent to general relativity.

But shape dynamics is not the only proposal for a global time in the context of general relativity; others are proposed by Ellis *et al.* [92] and Soo *et al.* [93].

Having said that the effects of the preferred global time may not be observable locally, it is also the case that experimentalists should continue to vigorously challenge Lorentz invariance by improving the limits on its possible violations.

THE AGENDA FOR EXPLAINING THE ARROWS OF TIME

Before we discuss this topic there is an important subtlety that is sometimes overlooked. It has long been clear that gravity plays an important role in keeping the universe out of thermal equilibrium. Gravitationally bound systems have negative specific heat, which means their internal velocities increase when energy is removed. Consider a system with many objects bound by gravity such as a globular cluster. Such systems do not evolve to homogeneous equilibrium states as they age. Instead they become increasingly heterogeneous as they fragment into subsystems. These subsystems, consisting of two or more bodies, become more tightly bound over time, releasing energy which goes into evaporating the cluster as individual stars, double stars, and larger bound systems leave it. A gravitationally bound system that starts off random and homogeneous ends up ordered and heterogeneous.

The laws that gravitationally bound systems obey, whether expressed in the language of Newton or Einstein, are invariant under time reversal. It is the same for shape dynamics. But shape dynamics has given us an important insight into how it happens that the most probable way for a gravitationally bound system to evolve is to become more structured and heterogeneous [94]. Roughly speaking, rather than evolve to homogeneous states of equilibria, which look the same with the clock run forward or backward, when gravity dominates the time reversal invariance is spontaneously broken so that most solutions have a strong arrow of time.

In [94] the problem of *N* point particles interacting gravitationally is studied in the context of Newtonian dynamics. The authors impose "Machian" boundary conditions which are that the total energy, momentum, and angular momentum are constrained to vanish. These are analogous to the condition imposed in general relativity that cosmological solutions be spatially compact. They find that a typical solution begins as widely separated individuals and bound pairs, which collapse to a dense and chaotic state. However, rather than staying in that maximally compressed state they separate again into bound systems of two or a small number of points, each dispersing again.

The behavior is roughly symmetric under reversal of time around a single moment, which is that of maximal compression. If you were to take the state at that time as the initial time, then the system becomes dispersed and heterogeneous as it evolves to the future. But if you reversed time, the behavior could be similarly characterized as you move into the past.

This helps to illuminate the question of the origin of the arrows of time, by explaining why our universe has not evolved to a structureless, homogeneous equilibrium. If we take the big bang to be that point of maximal compression, then there is a natural explanation for why the universe expands as it becomes more structured and more heterogeneous in time. But this only partially solves the problem. An explanation of why the universe starts off so drastically homogeneous and featureless is still missing. Why, indeed, is the strikingly simple past of the universe so unlike its complex and messy future?

If time is emergent from timeless law, and if those emergent laws are symmetric under time reversal or a natural extension, CPT[1] – as is

[1] CPT means the simultaneous transformation of the state under charge conjugation, parity, and time reversal.

the case with general relativity and relativistic quantum field theories which underlie the standard model – then there can be no fundamental difference between the future and the past. In this case the dominance of the universe in all eras up till now by irreversible processes – leading to the several arrows of time – can only be explained by the imposition of extremely improbable initial conditions. This is unsatisfactory as an explanation. The fact to be explained is why the universe, even 13.8 billion years after the big bang, has not reached equilibrium, which is by definition the most probable state, and it hardly suffices to explain this by asserting that the universe started in an even less probable state than the present one.

As I remarked in the second section in Chapter 3 ("The message of the large-scale astronomical data"), the initial state can be characterized as extremely homogeneous and absent of incoming gravitational and electromagnetic radiation [35, 36]. Nor do there appear to be any initial black or white holes. This means that a highly time-asymmetric universe – one in which, for example, information only propagates from the past to the future, and we recall only the past and, by our actions, affect only the future – is to be explained by imposing time-asymmetric initial conditions on time-symmetric laws. These pick a measure zero out of the possible solutions to the time-symmetric laws – those that have only retarded and no advanced propagation of radiation. This has become so natural to us in practice that we have to step back and reflect on what a drastic truncation of the solution space of Maxwell's theory is involved by setting to zero all the advanced fields.

On the other hand, if we take the view that time is real, in the sense that has been urged here, then we already are committed to the views that the future is fundamentally different from the past and that the future is constructed from the past and present, moment by moment. It is then perfectly consonant with this view to contemplate that there is a deeper level of fundamental law which is time-asymmetric under time reversal. This deeper law might, for example, reduce to Maxwell's theory but with only retarded solutions emerging from solutions to the deeper, time-asymmetric law.

An aspect of the empirical agenda for temporal naturalism is then to investigate the hypothesis that an effective, time-symmetric law could emerge from a deeper time-asymmetric law. Some first steps toward this have been taken in [23, 24]. But this is not a new idea, for Penrose proposed in [35] that the fundamental laws describing quantum gravity would be time-asymmetric in a way that would be realized at the effective level by time-asymmetric initial conditions.

Ever since Boltzmann we are familiar with attempts to explain the arrows of time by time asymmetric effective laws emerging from time symmetric laws, once time asymmetric initial conditions have been imposed. But there is a large class of systems in which the opposite happens: a reversible effective dynamics emerges from irreversible fundamental laws.[2] Consider a deterministic, discrete, dynamical system with a finite number of possible states. By deterministic we mean that each state has a unique successor state. But a given state can have several, or even no predecessors, so the dynamics is generally not reversible.

Now, let's start at an arbitrary state and follow the evolution, which generates a sequence of states. Since the total number of possible states is finite, sooner or later that sequence must repeat a previous state. But once that happens the system is constrained by its deterministic evolution rule to repeat a cycle endlessly. So after a finite time the system shrinks to a finite number of cycles. Restricted to the cycles, each state has a unique predecessor, so someone encountering the system once it has converged to cycles could easily make the mistake of representing the dynamics as reversible.

Suppose that, having made that mistake, a physicist encounters an ensemble of identical systems of this kind. They would be shocked to discover that all the members of the ensemble go around their cycles in the same direction. Thinking, mistakenly, that the underlying dynamics is reversible, they would have no choice but to explain this by a very improbable choice of initial conditions – whereas the right explanation is that they are observing the late time behavior of a

[2] L. Smolin, preprint in preparation, August 2014.

fundamentally irreversible dynamics. So my hypothesis is that this is the mistake we are making in fundamental physics.

THE AGENDA FOR QUANTUM GRAVITY

The last three decades have seen an explosion of results on quantum theories of gravity, along several directions. These can be divided into two classes: those that are background-dependent, in the sense that they study the quantization of small fluctuations around classical spacetimes, and those that are background-independent, because they do not utilize any such classical background. These roughly express the two sides of the old debate between absolute and relational approaches to spacetime. As the ideas of this book grow out of, and strongly support, the relational character of space and time, they mainly proscribe an agenda for background-independent theories.

Present research in background-independent approaches to quantum gravity incorporate several different models of quantum spacetime, whose dynamics are studied by a diverse range of methods. Both canonical and path integral approaches to quantum theory are developed, by methods that range from numerical studies to rigorous mathematical theorems. These include loop quantum gravity, studied by both canonical methods and path integral methods (also called spin foam models), group field theory, causal dynamical triangulations, causal sets, and quantum graphity. Within the last few years there have been major advances in all these directions, concerned with the emergence of classical spacetime as the semiclassical or low-energy limit [95]. There are also very encouraging results which recover the precise value of black hole entropy for generic black holes [96].

On the background-dependent side, there are also new developments in computing scattering amplitudes, which point to the possible discovery of new dynamical principles [25]. These arise from the marriage of two older research programs, the bootstrap and twistor theory.

There are two general contexts within which all this work is framed. The first are models of isolated systems, i.e. bounded regions of quantum geometry, surrounded by classical boundaries. These include many of the results on the emergence of classical general relativity in spin foam models, as well as the results on the conjectured AdS/CFT correspondence in string theory. In these contexts a classical notion of time, and a corresponding physical Hamiltonian, is present as expressed by the boundary conditions. These calculations then take place properly within the Newtonian paradigm, and the present ideas do not imply any strong modification of them.

This is definitely not the case with the second context, which is the quantization of gravity with cosmological boundary conditions, where the spatial manifold is compact without boundary. From the point of view advocated here, the framework for the study of most such models, which is the quantization of a fully constrained Hamiltonian by means of the Wheeler–DeWitt equation, must be abandoned. That equation expresses in the vanishing of the Hamiltonian the strongest argument that a consequence of the extension of the Newtonian paradigm to the universe as a whole is the elimination of time in the fundamental laws of nature. If the arguments of this book are correct, this approach to quantum cosmology must be replaced by a new approach based on a cosmological theory which

- has genuine evolution within a fundamental and global preferred time,
- is not quantum mechanics, but is a theory from which quantum theory emerges for small subsystems,
- resolves the meta-laws dilemma.

There is also good reason to hypothesize a fourth characteristic which is that

- space is not fundamental but is emergent from a more fundamental description.

This is suggested by the results in several background-independent approaches to quantum gravity including causal dynamical triangulations [97] and quantum graphity [98]. In these approaches space emerges from a more fundamental level of description which is combinatorial and algebraic, while time is taken to be fundamental.

Notice that there is a dovetailing of the agendas in quantum gravity and quantum foundations. Both require that quantum mechanics emerge from a cosmological theory which is not quantum mechanics. Both require a global time. Note that if space is emergent from a more fundamental level of description, locality must be emergent as well [99]. This can neatly resolve the mystery of the more fundamental theory being non-local, as required by the experimental tests of the Bell inequality, because the distinction between local and non-local interactions will be emergent and contingent.

THE MAIN CHALLENGE: RESOLVING THE META-LAWS AND COSMOLOGICAL DILEMMAS

The agendas for observational cosmology, quantum foundations, and quantum gravity come together in a single agenda which is to invent a new cosmological theory that can resolve the twin dilemmas we explored above: the cosmological dilemma and the meta-laws dilemma.

By resolving the meta-laws dilemma this new theory will be able to propose explanations for the choices of laws and initial conditions. By resolving the cosmological dilemma these explanations will have testable consequences in the form of falsifiable predictions for doable experiments.

Here is what we know about the theory we are searching for:

- It will not be based on the Newtonian paradigm. It will neither have a fixed, timeless configuration space nor describe evolution in terms of fixed, timeless laws.
- It will embrace the reality of time, in the form of the hypothesis that all that is real is real in a moment, which is one of a succession of moments. The objectivity of the distinction between the present and the future requires that this time manifest itself as a global, but relational, time coordinate.
- The elementary excitations and their interactions will be described by laws that evolve in this real time.
- The distinction between law and state will be relative and approximate, as both state and law must be properties of the present moment.

- The fundamental theory will not be quantum mechanical, but quantum mechanics will emerge in the case of small subsystems.
- The fundamental theory will not exist in space, but space will be emergent in some eras of the universe.
- The new theory will be framed by the principle of sufficient reason and its consequences, including the principle of the identity of the indiscernible, the principle of explanatory closure, the principle of no unreciprocated actions, and the absence of ideal or absolute elements.
- Mathematics will be a tool to formulate and develop aspects of it but in a way that makes it impossible to identify a mathematical object that is a complete mirror of the history of the universe.

The main agenda of theoretical cosmology in the present century is to invent candidates for this truly cosmological theory and develop and test its predictions for experiment and observation. While we do not presently have a full candidate of such a theory to propose we are confident the search for it is a fruitful direction of research. Part of the reason for our optimism is the fact that we have been able to invent models and hypotheses for theories that fall outside the Newtonian paradigm and address in different ways the meta-laws dilemma. These include cosmological natural selection, the principle of precedence, the unification of law and state, and the postulate of meta-law universality.

These ideas are a starting point for an investigation of cosmology guided by our three postulates: the uniqueness of the universe; the reality of time; and the reasonable, but limited role of mathematics in physics. As we seek to develop this new science we will discover that our success can be measured by the extent to which the future of cosmology becomes the cosmology of the future.

9 Concluding remarks

We close as we opened, with the crisis in cosmology. The growth of untestable scenarios about unobservable multiple universes or extra dimensions are not a cause of the crisis, they are a symptom of the need to change paradigms to avoid stumbling over unanswerable questions, or the proliferation of untestable hypotheses. The great universities and research institutes are full of theoretical physicists who would like to be Albert Einstein, but have no idea how to do it.[1] If so many people of undeniable talent and dedication are unable to make progress, the reason must be in a common mistake, a common shared assumption that is incorrect.

In this book we have aimed to identify the wrong assumptions and conceptual mistakes that are leading cosmology away from the disciplines of science into untestable speculations. They begin with the cosmological fallacy: the mistake of taking a scientific methodology, the Newtonian paradigm, outside of the domain where it can make contact with experiment and observations. The first step in our argument is the understanding that the Newtonian paradigm can only be used in the description of small subsystems of the universe.

The dismantling of the currently popular view continues by exposing several fallacies which have been used to argue that the anthropic principle has empirical content. These false arguments have been used fallaciously to argue that Hoyle and Weinberg made actual predictions [7, 10].

Still, some theorists come back with what would seem like an unanswerable assertion: "What if the world is just like that? What if it simply is the case that our universe is one of a vast or infinite ensemble

[1] "The world is currently filled with leaders who would like to be Franklin Roosevelt, but have no idea how to do it." Roberto Mangabeira Unger, talk at Centre for International Governance Innovation (CIGI) Conference on the Economic Crisis, 2008.

of universes, all unobservable by us? Isn't it arrogant of us to insist that the universe be organized in such a way that the answers to the questions we would most like to know are achievable by the methods of empirical science?"

The answer is that science is not about what might be the case. There are an infinite number of things that might be true of the universe, but which could never be observed. Multitudes of giant angels and unicorns might be hovering just outside our cosmological horizon. The dark matter might be tiny elves left over from the big bang.[2]

Science is only about what can be conclusively established on the basis of rational argument from public evidence [7]. This is why, if cosmology is to have a future as a science, it must begin with the principle that there is a single causally connected universe that contains all its causes. By all causes, I also mean that the laws themselves are explained in a way that has testable consequences. As we have argued at length in this book, this requires the laws to evolve in a real time. Only if that is the case will science be able to converge on a new paradigm about the nature of the universe. Only on that basis can cosmology continue to exist as a science.

[2] "I like the stories about angels, unicorns and elves
Now I like those stories as much as anybody else but
When I'm seeking knowledge either simple or abstract
The facts are with science
The facts are with science

A scientific theory isn't just a hunch or guess
It's more like a question that's been put through a lot of tests and
When a theory emerges consistent with the facts
The proof is with science
The truth is with science"

Lyrics: John Flansburgh and John Linnell (They Might Be Giants), from the song *Science is Real* on "Here Comes Science."

Acknowledgments

Besides my enormous debt to my co-author, which these twinned texts make obvious, I am grateful to several friends and collaborators for conversations, collaborations, and challenges over the many years it has taken to formulate the view presented here. These include David Albert, Abhay Ashtekar, Julian Barbour, Harvey Brown, Marina Cortes, Louis Crane, Laurent Freidel, Chris Isham, Jenann Ismael, Ted Jacobson, Stuart Kauffman, Jaron Lanier, Fotini Markopoulou, Carlo Rovelli, Simon Saunders, Paul Steinhardt, Max Tegmark, Neil Turok, and Steve Weinstein.

This research was supported in part by Perimeter Institute for Theoretical Physics. Research at Perimeter Institute is supported by the Government of Canada through Industry Canada and by the Province of Ontario through the Ministry of Research and Innovation. This research was also partly supported by grants from NSERC, FQXi, and the John Templeton Foundation.

References

[1] Smolin, L. "The culture of science divided against itself." In *Brick Magazine*, issue no. 88, 2012.

[2] Lanier, J. *You Are Not A Gadget*. Random House, 2010.

[3] Smolin, L. "Temporal naturalism." arXiv:1310.8539, 2013.

[4] Susskind, L. "The anthropic landscape of string theory." In *Universe or Multiverse*, ed. B. Carr, pp. 247–266. Cambridge University Press, 2007 (arXiv: hep-th/0302219, 2003).
Susskind, L. *The Cosmic Landscape: String Theory and the Illusion of Intelligent Design*. Little, Brown and Company, 2006.

[5] Tegmark, M. *Our Mathematical Universe: My Quest for the Ultimate Nature of Reality*. Random House, 2014.
Tegmark, M. "The mathematical universe." *Foundations of Physics* **38**(2) (2008) 101–150.

[6] Smolin, L. *The Life of the Cosmos*. Oxford University Press, 1997.

[7] Smolin, L. *The Trouble With Physics*. Houghton Mifflin, 2006.

[8] Smolin, L. *Time Reborn: From the Crisis of Physics to the Future of the Universe*. Houghton Mifflin Harcourt, 2013.

[9] Feyerabend, P. *Against Method: Outline of an Anarchistic Theory of Knowledge*. Humanities Press, 1975.

[10] Conway, J. and Kochen, S. "The strong free will theorem." arXiv:0807.3286, 2008.
Conway, J. and Kochen, S. "The free will theorem." arXiv:quant-ph/0604079, 2006.

[11] Smolin, L. "Precedence and freedom in quantum physics." arXiv:1205.3707, 2012.

[12] Leibniz, G. W. *The Monadology*, 1698. Translated by R. Latta, and available at http://oregonstate.edu/instruct/phl302/texts/leibniz/monadology.html.
Leibniz, G. W. *G. W. Leibniz, Philosophical Texts* (Oxford Philosophical Texts), trans. and eds R. S. Woolhouse and R. Francks. Oxford University Press, 1999.
Alexander, H. G. (ed.). *The Leibniz–Clarke Correspondence*. Manchester University Press, 1956. For an annotated selection, see http://www.bun.kyoto-u.ac.jp?suchii/leibniz-clarke.html.

[13] Mach, E. *The Science of Mechanics: A Critical and Historical Account of its Development*. Open Court, 1907.

[14] Einstein, A., in Einstein, A., Lorentz, H. A., Weyl, H. and Minkowski, H. *The Principle of Relativity: A Collection of Original Papers on the Special and General Theory of Relativity*. Dover, 1952.

[15] Barbour, J. B. *Absolute or Relative Motion?*, vol. 1, *The Discovery of Dynamics*. Cambridge University Press, 1989.

[16] Rovelli, C. "Relational quantum mechanics." *International Journal of Theoretical Physics* **35**(8) (1996) 1637–1678.

[17] Peirce, C. S. "The architecture of theories." In *The Monist*, 1891. Reprinted in *Philosophical Writings of Peirce*, ed. J. Buchler. Dover, 1955.

[18] Smolin, L. "The case for background independence." arXiv:hep-th/0507235, 2005.

[19] Kuchar, K. "General relativity: dynamics without symmetry." *Journal of Mathematical Physics* **22** (1981) 2640.

[20] Cortês, M. and Smolin, L. "The universe as a process of unique events." arXiv:1307.6167 [gr-qc], 2013.

[21] Cortês, M. and Smolin, L. "Energetic causal sets." arXiv:1308.2206 [gr-qc], 2013.

[22] Cachazo, F., Mason, L. and Skinner, D. "Gravity in twistor space and its Grassmannian formulation." arXiv:1207.4712 [hep-th], 2012.

[23] Bombelli, L., Lee, J., Meyer, D. and Sorkin, R. "Spacetime as a causal set." *Physical Review Letters* **59** (1987) 521–524.
Henson, J. "The causal set approach to quantum gravity." In *Approaches to Quantum Gravity – Towards a New Understanding of Space and Time*, ed. D. Oriti. Cambridge University Press, 2006 (arXiv:gr-qc/0601121, 2006).

[24] Stachel, J. "Einstein's search for general covariance, 1912–1915." In *Einstein and the History of General Relativity*, eds D. Howard and J. Stachel, vol. 1, pp. 63–100. Birkhauser, 1989.

[25] Gomes, H., Gryb, S. and Koslowski, T. "Einstein gravity as a 3D conformally invariant theory." *Classical and Quantum Gravity* **28** (2011) 045005 (arXiv:1010.2481, 2010).
Barbour, J. "Shape dynamics. An introduction." arXiv:1105.0183, 2011.

[26] Gomes, H. "A Birkhoff theorem for shape dynamics." arXiv:1305.0310 [gr-qc], 2013.
Gomes, H. and Herczeg, G. "A rotating black hole solution for shape dynamics." arXiv:1310.6095 [gr-qc], 2013.

[27] Amelino-Camelia, G., Freidel, L., Kowalski-Glikman, J. and Smolin, L. "The principle of relative locality." *Physical Review D* **84** (2011) 084010 (arXiv:1101.0931 [hep-th], 2011).

Amelino-Camelia, G., Freidel, L., Kowalski-Glikman, J. and Smolin, L. "Relative locality: a deepening of the relativity principle." *International Journal of Modern Physics D* **20** (2011) 14 (arXiv:1106.0313 [hep-th], 2011).

[28] Nielsen, H. B. "Random dynamics and relations between the number of fermion generations and the fine structure constants." *Acta Physica Polonica B* **20**(5) (1989) 427.

Foerster, D., Nielsen, H. B. and Ninomiya, M. "Dynamical stability of local gauge symmetry: creation of light from chaos." *Physics Letters B* **94**(2) (1980) 135–140.

Bennett, D. L., Brene, N. and Nielsen, H. B. "Random dynamics." *Physica Scripta* **15** (1987) 158.

[29] Furey, C. "Unified theory of ideals." arXiv:1002.1497, 2010.

[30] Penrose, R. "Singularities and time asymmetry." In *General Relativity: An Einstein Centenary Survey*, eds S. W. Hawking and W. Israel. Cambridge University Press, 1979.

[31] Weinstein, S. "Electromagnetism and time-asymmetry." *Modern Physics Letters A* **26** (2011) 815–818 (arXiv:1004.1346, 2010).

[32] Milgrom, M. "A modification of the Newtonian dynamics as a possible alternative to the hidden mass hypothesis." *Astrophysical Journal* **270** (1983) 365–370.

[33] Hawking, S. W. and Ellis, G. F. R. *The Large Scale Structure of Space-Time*, vol. 1. Cambridge University Press, 1973.

[34] Penrose, R. "Gravitational collapse and space-time singularities." *Physical Review Letters* **14**(3) (1965) 57–59.

[35] Hawking, S. W. and Penrose, R. "The singularities of gravitational collapse and cosmology." *Proceedings of the Royal Society of London A* **314**(1519) (1970) 529–548.

[36] Ashtekar, A. and Singh, P. "Loop quantum cosmology: a status report." *Classical and Quantum Gravity* **28** (2011) 213001 (arXiv:1108.0893, 2011).

Bojowald, M. "Isotropic loop quantum cosmology." *Classical and Quantum Gravity* **19** (2002) 2717–2742 (arXiv:gr-qc/0202077, 2002).

Bojowald, M. "Inflation from quantum geometry." arXiv:gr-qc/0206054, 2002.

Bojowald, M. "The semiclassical limit of loop quantum cosmology." *Classical and Quantum Gravity* **18** (2001) L109–L116 (arXiv:gr-qc/0105113, 2001).

Bojowald, M. "Dynamical initial conditions in quantum cosmology." *Physical Review Letters* **87** (2001) 121301 (arXiv:gr-qc/0104072, 2001).

Tsujikawa, S., Singh, P. and Maartens, R. "Loop quantum gravity effects on inflation and the CMB." arXiv:astro-ph/0311015, 2003.

Gambini, R. and Pullin, J. "Discrete quantum gravity: a mechanism for selecting the value of fundamental constants." *International Journal of Modern Physics D* **12** (2003) 1775–1782 (arXiv:gr-qc/0306095, 2003).

[37] Gasperini, M. and Veneziano, G. "The pre-big bang scenario in string cosmology." *Physics Reports* **373**(1) (2003) 1–212.

Frolov, V. P., Markov, M. A. and Mukhanov, M. A. "Through a black hole into a new universe?" *Physics Letters B* **216** (1989) 272–276.

Lawrence, A. and Martinec, E. "String field theory in curved space-time and the resolution of spacelike singularities." *Classical and Quantum Gravity* **13** (1996) 63 (arXiv:hep-th/9509149, 1995).

Martinec, E. "Spacelike singularities in string theory." *Classical and Quantum Gravity* **12** (1995) 941–950 (arXiv:hep-th/9412074, 1994).

[38] Modesto, L. "Disappearance of the black hole singularity in loop quantum gravity." *Physical Review D* **70**(12) (2004) 124009.

Modesto, L. "Loop quantum black hole." *Classical and Quantum Gravity* **23** (18) (2006) 5587.

[39] Smolin, L. "Did the universe evolve?" *Classical and Quantum Gravity* **9** (1992) 173–191.

[40] Ashtekar, A., Taveras, V. and Varadarajan, M. "Information is not lost in the evaporation of 2-dimensional black holes." *Physical Review Letters* **100** (2008) 211302 (arXiv:0801.1811, 2008).

[41] Putnam, H. "Time and physical geometry." *Journal of Philosophy* **64** (1967) 240–247.

[42] Amelino-Camelia, G. and Smolin, L. "Prospects for constraining quantum gravity dispersion with near term observations." *Physical Review D* **80** (2009) 084017 (arXiv:0906.3731, 2009).

[43] Amelino-Camelia, G., Freidel, L., Kowalski-Glikman, J. and Smolin, L. "OPERA neutrinos and deformed special relativity." *Modern Physics Letters A* **27**(10) (2012) 1250063.

[44] Barbour, J. B. and Bertotti, B. "Mach's principle and the structure of dynamical theories." *Proceedings of the Royal Society of London A* **382**(1783) (1982) 295–306.

[45] Valentini, A. "Signal-locality in hidden-variables theories." *Physics Letters A* **297**(5) (2002) 273–278.

[46] Mazur, B. "Mathematical Platonism and its opposites." In *Newsletter of the European Mathematical Society* no. 68 (2008) 19–21.

[47] Kauffman, S. A. *Investigations*. Oxford University Press, 2002.

[48] Vilenkin, A. "Birth of inflationary universes." *Physical Review D* **27** (1983) 2848.

[49] Linde, A. "The inflationary universe." *Reports on Progress in Physics* **47** (1984) 925.

Linde, A., Linde, D. and Mezhlumian, A. "From the big bang theory to the theory of a stationary universe." *Physical Review D* **49** (1994) 1783 (arXiv:gr-qc/9306035, 1993).

Garcia-Bellido, J. and Linde, A. "Stationarity of inflation and predictions of quantum cosmology." *Physical Review D* **51** (1995) 429 (arXiv:hep-th/9408023, 1994).

Garriga, J. and Vilenkin, A. "A prescription for probabilities in eternal inflation." *Physical Review D* **64** (2001) 023507 (arXiv:gr-qc/0102090, 2001).

[50] Garriga, J. and Vilenkin, A. "Recycling universe." *Physical Review D* **57** (1998) 2230 (arXiv:astro-ph/9707292, 1997).

[51] Steinhardt, P. J. and Turok, N. "A cyclic model of the universe." *Science* **296** (5572) (2002) 1436–1439.

[52] Penrose, R. *Cycles of Time: An Extraordinary New View of the Universe.* Random House, 2010.

[53] Lehners, J.-L. "Diversity in the phoenix universe." arXiv:1107.4551, 2011.

Lehners, J.-L., Steinhardt, P. J. and Turok, N. "The return of the phoenix universe." *International Journal of Modern Physics D* **18** (2009) 2231–2235 (arXiv:0910.0834, 2009).

[54] Gurzadyan, V. G. and Penrose, R. "CCC-predicted low-variance circles in CMB sky and LCDM." arXiv:1104.5675, 2011.

Gurzadyan, V. G. and Penrose, R. "More on the low variance circles in CMB sky." arXiv:1012.1486, 2010.

Gurzadyan, V. G. and Penrose, R. "Concentric circles in WMAP data may provide evidence of violent pre-Big-Bang activity." arXiv:1011.3706, 2010.

[55] Wehus, I. K. and Eriksen, H. K. "A search for concentric circles in the 7 year Wilkinson Microwave Anisotropy Probe temperature sky maps." *Astrophysical Journal Letters* **733** (2011) L29 (arXiv:1012.1268, 2010).

Moss, A., Scott, D. and Zibin, J. P. "No evidence for anomalously low variance circles on the sky." arXiv:1012.1305, 2010.

Hajian, A. "Are there echoes from the pre-big bang universe? A search for low variance circles in the CMB sky." arXiv:1012.1656, 2010.

[56] Smolin, L. "The status of cosmological natural selection." arXiv:hep-th/0612185, 2006.

[57] Brown, G. E. and Bethe, H. A. "A scenario for a large number of low-mass black holes in the galaxy." *Astrophysical Journal* **423** (1994) 659.

Brown, G. E. and Weingartner, J. C. "Accretion onto and radiation from the compact object formed in SN 1987A." *Astrophysical Journal* **436** (1994) 843.

Brown, G. E. "The equation of state of dense matter: supernovae, neutron stars and black holes." *Nuclear Physics A* **574** (1994) 217–230.

Brown, G. E. "Kaon condensation in dense matter." In *Bose–Einstein Condensation*, eds A. Grin, D. W. Snoke and S. Stringari, p. 438. Cambridge University Press, 1995.

Bethe, H. A. and Brown, G. E. "Observational constraints on the maximum neutron star mass." *Astrophysical Journal Letters* **445** (1995) L129.

[58] Lattimer, J. M. and Prakash, M. "What a two solar mass neutron star really means." In *From Nuclei to Stars. Festschrift in Honor of Gerald E Brown*, ed. S. Lee, p. 275. World Scientific, 2011 (arXiv:1012.3208, 2010).

[59] Feeney, S. M., Johnson, M. C., Mortlock, D. J. and Peiris, H. V. "First observational tests of eternal inflation: analysis methods and WMAP 7-year results." *Physical Review D* **84** (2011) 043507 (arXiv:1012.3667, 2010).

Aguirre, A. and Johnson, M. C. "A status report on the observability of cosmic bubble collisions." *Reports on Progress in Physics* **74** (2011) 074901 (arXiv:0908.4105, 2009).

[60] Barger, V., Langacker, P. and Shaughnessy, G. "TeV physics and the Planck scale." *New Journal of Physics* **9** (2007) 333 (arXiv:hep-ph/0702001, 2007).

[61] Carr, B. J. and Rees, M. J. "The anthropic principle and the structure of the physical world." *Nature* **278**(605) (1979) 23.

Barrow, J. D. and Tipler, F. J. *The Anthropic Cosmological Principle*. Oxford University Press, 1986.

[62] Smolin, L. "A perspective on the landscape problem." *Foundations of Physics* **43** (2013) 21–45.

[63] Kragh, H. "An anthropic myth: Fred Hoyle's carbon-12 resonance level." *Archive for History of Exact Sciences* **64**(6) (2010) 721–751.

[64] Weinberg, S. "Anthropic bound on the cosmological constant." *Physical Review Letters* **59** (1987) 2607.

Weinberg, S. "A priori probability distribution of the cosmological constant." *Physical Review D* **61** (2000) 103505 (arXiv:astro-ph/0002387, 2000).

Weinberg, S. "The cosmological constant problems." arXiv:astro-ph/0005265, 2000.

[65] Rees, M. J. "Anthropic reasoning." *Complexity* **3** (1997) 17–21.

Rees, M. J. "Numerical coincidences and 'tuning' in cosmology." In *Fred Hoyle's Universe*, eds N. C. Wickramasinghe, G. R. Burbidge and J. V. Narlikar, pp. 95–108. Kluwer, 2003 (arXiv:astro-ph/0401424, 2004).

Tegmark, M. and Rees, M. J. "Why is the CMB fluctuation level 10^{-5}?" *Astrophysical Journal* **499** (1998) 526–532 (arXiv:astro-ph/9709058, 1997).

Graesser, M. L., Hsu, S. D. H., Jenkins, A. and Wise, M. B. "Anthropic distribution for cosmological constant and primordial density perturbations." *Physics Letters B* **600** (2004) 15–21 (arXiv:hep-th/0407174, 2004).

[66] Garriga, J. and Vilenkin, A. "Anthropic prediction for Λ and the Q catastrophe." *Progress of Theoretical Physics Supplements* **163** (2006) 245–257 (arXiv:hep-th/0508005, 2005).

[67] Ahmed, M., Dodelson, S., Greene, P. B. and Sorkin, R. "Everpresent Λ." *Physical Review D* **69** (2004) 103523 (arXiv:astro-ph/0209274, 2002).

[68] Vilenkin, A. "A measure of the multiverse." *Journal of Physics A: Mathematical and Theoretical* **40**(25) (2007) 6777.

[69] Peirce, C. S. "A guess at the riddle." Available in, for example, *Charles S. Peirce: The Essential Writings*, ed. E. C. Moore. Prometheus, 1998.

[70] Smolin, L. "A real ensemble interpretation of quantum mechanics." arXiv:1104.2822, 2011.

[71] Müller, M. P. and Masanes, L. "Information-theoretic postulates for quantum theory." arXiv:1203.4516, 2012.

Masanes, L. and Müller, M. P. "A derivation of quantum theory from physical requirements." *New Journal of Physics* **13** (2011) 063001 (arXiv:1004.1483, 2010).

Dakić, B. and Brukner, C. "Quantum theory and beyond: is entanglement special?" In *Deep Beauty: Understanding the Quantum World Through Mathematical Innovation*, ed. H. Halvorson. Cambridge University Press, 2010 (arXiv:0911.0695v1, 2009).

[72] Hardy, L. "Quantum theory from five reasonable axioms." arXiv:quant-ph/0101012v4, 2001.

[73] Smolin, L. "Matrix universality of gauge and gravitational dynamics." arXiv:0803.2926, 2008.

[74] Smolin, L. "M theory as a matrix extension of Chern–Simons theory." *Nuclear Physics B* **591** (2000) 227–242 (arXiv:hep-th/0002009, 2000).

Smolin, L. "The cubic matrix model and a duality between strings and loops." arXiv:hep-th/0006137, 2000.

Smolin, L. "The exceptional Jordan algebra and the matrix string." arXiv:hep-th/0104050, 2001.

Livine, E. R. and Smolin, L. "BRST quantization of matrix Chern–Simons theory." arXiv:hep-th/0212043, 2002.

[75] Ashtekar, A. "New variables for classical and quantum gravity." *Physical Review Letters* **57** (1986) 2244–2247.

[76] Plebański, J. E. "On the separation of Einsteinian substructures." *Journal of Mathematical Physics* **18**(12) (1977) 2511.

[77] Rovelli, C. *Quantum Gravity*. Cambridge University Press, 2004.
Ashtekar, A. (in collaboration with R. S. Tate) *Lectures on Non-Perturbative Canonical Gravity*. World Scientific, 1991.
Gambini, R. and Pullin, J. *A First Course in Loop Quantum Gravity*. Oxford University Press, 2011.

[78] Zanardi, P. and Rasetti, M. "Noiseless quantum codes." *Physical Review Letters* **79** (1997) 3306.
Lidar, D. A., Chuang, I. L. and Whaley, K. B. "Decoherence-free subspaces for quantum computation." *Physical Review Letters* **81** (1998) 2594 (arXiv:quant-ph/9807004, 1998).
Knill, E., Laflamme, R. and Viola, L. "Theory of quantum error correction for general noise." *Physical Review Letters* **84** (2000) 2525.
Kempe, J., Bacon, D., Lidar, D. A. and Whaley, K. B. "Theory of decoherence-free fault-tolerant universal quantum computation." *Physical Review A* **63** (2001) 042307.

[79] Konopka, T. and Markopoulou, F. "Constrained mechanics and noiseless subsystems." arXiv:gr-qc/0601028, 2006.
Kribs, D. W. and Markopoulou, F. "Geometry from quantum particles." arXiv: gr-qc/0510052, 2005.
Markopoulou, F. "Towards gravity from the quantum." arXiv:hep-th/0604120, 2006.

[80] Alder, S. *Quantum Theory as an Emergent Phenomenon*. Cambridge University Press, 2004.
Alder, S. "Statistical dynamics of global unitary invariant matrix models as pre-quantum mechanics." arXiv:hep-th/0206120, 2002.
Starodubtsev, A. "A note on quantization of matrix models." *Nuclear Physics B* **674** (2003) 533–552 (arXiv:hep-th/0206097, 2002).
Starodubtsev, A. "Matrix models as non-local hidden variables theories." arXiv: hep-th/0201031, 2002.
Markopoulou, F. and Smolin, L. "Quantum theory from quantum gravity." *Physical Review D* **70** (2004) 124029 (arXiv:gr-qc/0311059, 2003).

[81] Smolin, L. "Unification of the state with the dynamical law." arXiv:1201.2632, 2012.

[82] Strawson, G. "Real naturalism." *London Review of Books* **35**(18) (2013) 28–30.

[83] Nagel, T. *Mind and Cosmos: Why the Materialist Neo-Darwinian Conception of Nature is Almost Certainly False*. Oxford University Press, 2012.

[84] Webb, J. K., Murphy, M. T., Flambaum, V. V., Dzuba, V. A., Barrow, J. D., Churchill, C. W., Prochaska, J. X. and Wolfe, A. M. "Further evidence for

cosmological evolution of the fine structure constant." *Physical Review Letters* **87**(9) (2001) 091301.

[85] Cushing, J. T., Fine, A. and Goldstein, S. (eds) *Bohmian Mechanics and Quantum Theory: An Appraisal,* Boston Studies in the Philosophy and History of Science, vol. 132. Springer, 1996.

[86] Tumulka, R. "A relativistic version of the Ghirardi–Rimini–Weber model." *Journal of Statistical Physics* **125**(4) (2006) 821–840.

[87] Gleiser, R. J. and Kozameh, C. N. "Astrophysical limits on quantum gravity motivated birefringence." arXiv:gr-qc/0102093, 2001.

[88] Ellis, G. F. R. and Goswami, R. "Space time and the passage of time." arXiv:1208.2611, 2012.
Ellis, G. F. R. "The arrow of time and the nature of spacetime." arXiv:1302.7291, 2013.

[89] Ó Murchadha, N., Soo, C. and Yu, H.-L. "Intrinsic time gravity and the Lichnerowicz–York equation." *Classical and Quantum Gravity* **30** (2013) 095016 (arXiv:1208.2525, 2012).

[90] Barbour, J., Koslowski, T. and Mercati, F. "A gravitational origin of the arrows of time." arXiv:1310.5167, 2013.

[91] Oriti, D. (ed.) *Approaches to Quantum Gravity: Toward a New Understanding of Space, Time and Matter.* Cambridge University Press, 2009.

[92] Bianchi, E. "Entropy of non-extremal black holes from loop gravity." arXiv:1204.5122, 2012.
Ghosh, A. and Perez, A. "Black hole entropy and isolated horizons thermodynamics." *Physical Review Letters* **107**(24) (2011) 241301.

[93] Ambjørn, J., Jurkiewicz, J. and Loll, R. "Emergence of a 4D world from causal quantum gravity." *Physical Review Letters* **93** (2004) 131301 (arXiv:hep-th/0404156, 2004).
Ambjørn, J., Jurkiewicz, J. and Loll, R. "Quantum gravity as sum over spacetimes." In *New Paths Towards Quantum Gravity*, pp. 59–124. Springer, 2010.

[94] Konopka, T., Markopoulou, F. and Severini, S. "Quantum graphity: a model of emergent locality." arXiv:0801.0861, 2008.

[95] Markopoulou, F. and Smolin, L. "Disordered locality in loop quantum gravity states." *Classical and Quantum Gravity* **24**(15) (2007) 3813.

A note concerning disagreements between our views

Roberto Mangabeira Unger and Lee Smolin

There follows a list, and a brief discussion, of points of difference between our views.

When these disagreements are presented together, as they now are here, they may seem so far-reaching in implication and so expansive in scope that they could not be compatible with adherence to the same direction of thought. They nevertheless pale in significance when compared with what unites us in the development of an approach in natural philosophy and cosmology, which is everything else. As these differences of substance or of emphasis show, there is more than one way to develop this approach.

1. *Temporal naturalism and the prerogatives of science.* In his account of temporal naturalism, LS describes science as our most reliable guide to the understanding of nature. For RMU, the claim that science is either the sole or even the most reliable source of insight into nature should exercise no influence on the temporal naturalism that we here espouse. There is no hierarchy of forms of inquiry that could entitle one to say that one such form trumps the others. If there were such a hierarchy, it could not be established within science or even within natural philosophy, which has nature rather than science as its proximate subject matter. It would have to be an extra-scientific view of science.

Scientism stands to science as militarism stands to the army. To assert, on behalf of science, a privilege over other sources of insight and experience does science no favor.

These objections do not apply to science defined as LS does (in "The Trouble with Physics" and "Time Reborn") as open, ethical communities, membership within which is defined by adherence to an ethic of seeking truth and minimizing error. The communities of science carry and teach methods of inquiry that have been found by long practice to be helpful to minimize error and mitigate against the human tendency to fool ourselves and each other. Communities of science are by definition open to the discovery of unexpected and novel methodologies and hypotheses, and everything else, apart from adherence to the search for knowledge, in good faith, is on the table.

2. *The principle of sufficient reason and the factitious character of the universe.* RMU rejects the principle of sufficient reason, at least if it is understood in anywhere close to the way in which its original proponent, Leibniz, defined it.

Causation is a primitive feature of nature. The laws, symmetries, and constants of nature are the characteristic form of causal connection in the cooled-down universe. They are not, RMU argues, the only form of causation: in extreme states of nature causal connection may fail to present itself in recurrent and therefore law-like form.

The principle of sufficient reason has implications outreaching the idea that causation is fundamental. It signals an ambition of completeness and closure of scientific insight, a seamless movement from science to metaphysics. It expresses a refusal to accept the factitiousness of the universe: that it just happens to be one way rather than another. Once we have exhausted our powers of inquiry, capable of indefinite strengthening and yet not unlimited, we must accept the sheer just-so-ness of nature. The most important feature of the universe is that it is what it is rather than something else. We cannot hope to show that it must be what it is. If we insist on doing so, we allow metaphysical rationalism to corrupt our understanding of nature.

Leibniz clearly described the stakes in his reflections "On the Ultimate Origins of the Universe." He remarks on the inadequacy of even a conception of the eternal existence of the world to satisfy the demands of the principle of sufficient reason. Suppose, he says, that a book on the elements of geometry has always existed, with each edition copied from an earlier one. Why does the book exist at all and why did it start out with the content that it did? There must be a reason. The reason must result from an understanding of why the world has to exist and to be what it is.

There is no such reason to be found, RMU believes, either in science or in our natural understanding beyond the boundaries of science. We cannot show, as the principle of sufficient reason would require, that the world has to be what it is or that it has to be at all. What we offer in this book, instead of Leibniz's series of copies, preceded by rational necessity, is the idea of an evolution, of a cumulative transformation, of both the structure and the regularities of nature, back into an indefinite past and forward into an indefinite future. History extends the field of causal inquiry. However, it fails to pass the test of the principle of sufficient reason because it does not show why there had to be a universe and why it had to have, from the outset, the history that it has.

There are many justifications, argued in this book, to look for a historical answer to the question why the laws and initial conditions of the universe are what they are and for completing the transformation of cosmology into a historical science. These justifications enlarge the domain of causal inquiry. They postpone our confrontation with the factitiousness of the universe. They fail, however, to avoid that confrontation. Natural philosophy should not have as its program to build a bridge between science and rationalist metaphysics. Consequently, it should reject the principle of sufficient reason, which would serve as such a bridge.

LS understands the principle of sufficient reason (PSR) as a heuristic guide to suggest questions to ask and strategies to follow, in the formulation of hypotheses for cosmological and physical theories. Chief among these strategies is the mandate to consider hypotheses that reduce references in theories to background structures. These are fixed non-dynamical elements that are necessary to define the observables measuring the dynamical degrees of freedom of a system. The paradigmatic example of such background structures is the absolute space and time of Newton. As Mach understood, these fixed elements refer implicitly to degrees of freedom outside the system being described, such as the distant stars and galaxies. The imperative of the PSR is to replace these implicit references, coded and idealized as fixed non-dynamical background structures, with hypotheses as to dynamical degrees of freedom outside the system.

The PSR has consequences such as the principle of the identity of indiscernibles (PII) and Einstein's principle of no unreciprocated action, which have also played key heuristic roles in interpreting general relativity and suggesting new hypotheses.

While there may never be complete sufficient reason, as a heuristic principle the PSR has been a very successful guide. LS would suggest it be refined to a differential principle of sufficient reason, i.e. to the imperative to always seek to increase sufficient reason for cosmological questions by eliminating background structures.

3. *The cosmological fallacies.* RMU and LS agree in rejecting what LS calls the cosmological fallacy and RMU labels the first cosmological fallacy. The tradition of physics initiated by Galileo and Newton adopts a practice of explanation distinguishing between stipulated initial conditions and timeless laws governing the motions or changes of law-governed phenomena within a configuration space defined by such conditions. Both of us explore the many and connected reasons why this explanatory practice has no legitimate cosmological application: a method useful in exploring parts of the universe misleads us when we apply it to the universe as a whole.

RMU, however, goes on to argue that there is also a second cosmological fallacy. The first cosmological fallacy improperly applies to the whole universe a style of explanation that is legitimately applied only to circumscribed parts of nature. The second cosmological fallacy treats the forms that nature typically exhibits in the cooled-down universe, with its stable, well-differentiated structural elements and regularities, and its clearcut distinctions between general laws, symmetries, or constants and particular phenomena or events, as the only form of nature. For RMU, what cosmology has already discovered about the history of the universe suggests that nature may undergo extreme states, such as those that marked its earliest history or may appear later (for example, in the interior of black holes), in which those characteristics are absent.

RMU reasons that the two cosmological fallacies – the first, one of unwarranted generalization; the second, one of universal anachronism – are closely related. The force of the second can be adequately appreciated only when it is seen in the context of the first. An attempt to rid cosmology of both fallacies helps define a research agenda and requires a break with ways of thinking that are entrenched in physics as well as in cosmology.

In LS's formulation of the argument, the insight RMU attributes to a second cosmological fallacy is implied by the first; given our ignorance of conditions prior to the big bang, he feels it is not necessary to emphasize the point that we are ignorant or raise speculation about conditions so far prior to the era we can observe to the status of a separate principle.

4. *The mathematical infinite and nature.* Debate about the reality of the infinite and about its presence in nature has gone on for several thousand years. In the history of Western philosophy and science, it begins with the pre-Socratics.

RMU holds that it is impossible entirely to avoid engagement with this debate in the development of our argument. He believes that our ideas about the singular existence of the universe, the inclusive reality of time, and the selective realism of mathematics can be fully established only within a view that rejects the natural presence of the mathematical infinite (the mathematical infinite because the history of thought includes other conceptions of the infinite).

In the one real, time-drenched universe, everything has a particular history precisely because it is finite, and not part of an infinite array. Moreover, the cosmological use of the infinite serves to mask the failure of a physical theory taken beyond the boundaries of its proper domain of application. The most notable instance is the inference in contemporary cosmology of an infinite initial singularity from the field equations of general relativity. Finally, the admission of the mathematical infinite into natural science effaces the difference, which we emphasize, between nature and mathematics. Nature works in time, with which mathematics has trouble. Mathematics offers, among other things, the infinite, which nature abhors.

It does not follow from the usefulness of real and complex numbers in science that the infinite must exist in nature because the continuum of such numbers is infinite. Our mathematical conceptions may come laden with certain attributes – a circumstance that LS describes with his idea of mathematics as evoking structures with inbuilt content. However, we deny to mathematics, for reasons that both of us explore, the power of opening a shortcut to the understanding of nature and argue that the effectiveness of mathematics in science is reasonable because it is only relative. To accept these conclusions is to lose any reason to suppose that the characteristics of nature must mirror the features of numbers.

LS does not disagree, and would point to the success of loop quantum gravity in showing that quantum effects eliminate cosmological singularities, thus opening up the time before the big bang for description and modeling. But as Penrose has long pointed out, to eliminate the infinite from physical theory one must eliminate any use of the continuum including real and complex numbers. While there are proposals to eliminate the spacetime continuum in favor of discrete sets of events (causal set models), it seems a harder challenge to eliminate the dependence of quantum theory on the continuum of complex numbers.

5. *The eternity of the world.* RMU argues that the thesis of the eternity of the world should have no place among the claims of this book. Eternity is mathematical infinity in time, and no more acceptable to

natural science than is any other variant of the infinite. We could never know the world to be eternal; it does not become eternal by growing older or by having a more ancient history than we earlier supposed. The idea of the eternity of the world is a metaphysical proposition, to which we may mistakenly believe ourselves driven by our trouble with the apparent alternative to it: that the world has an absolute beginning, or that time emerges from something else.

However, there is another alternative, appropriate to the limitations of science, if disappointing to the rationalist metaphysician: that the history of the world extends indefinitely back into the past and forward into the future. Indefinite longevity is not eternity. It nevertheless gives cosmology all that it requires: a vast and open field on which to pursue its task of exploring the structure of nature in the light of the history of nature.

LS agrees that the next step in scientific cosmology is to push back from the first three minutes after the big bang to draw predictions from hypotheses about the three minutes previous to that. Science proceeds step by step. There is no need to try to solve dilemmas of ultimate origins and ultimate beginnings at one blow. All we need to do – and all we can reasonably hope to do – is to expand our knowledge of the causes of things further back in time.

6. *Time, present, past, and future.* For RMU, the standard options of school philosophy about time – such as presentism and eternalism, and the affirmation or denial of real distinctions among present, past, and future – are all inadequate to develop the implications of the inclusive reality of time. According to presentism, only what exists in the present moment is real. According to eternalism, all present, past, and future events (to the limited extent that time is real and that any such distinction can be made) are equally real because they are all necessitated by the structure and regularities of the universe. Both presentism and eternalism fail to do justice to the inclusive reality of time.

On the view developed in this book, nothing is outside time. Every event that has ever happened, or that will ever happen, in the history of the universe can in principle be placed on a single time line, notwithstanding the objections that result from the established understanding of special and general relativity.

It is true that general relativity, under its most influential interpretations, admits a cosmic time. The trouble is that it admits too many of them: in fact, an infinite number. Relativity of simultaneity widens into a freedom to choose the time coordinates on the spacetime manifold. The choice of coordinates in what has come to be called many-fingered time is arbitrary from the perspective of the theory. Thus, there can be no one cosmic time that is also the preferred time of universal history.

For there to be a preferred cosmic time, the universe must have developed in a certain way, and it must possess certain features, which we later consider. *Now* for the ephemeral human being, lost in a corner of the universe, must also be *now* for the cosmos. In the present moment, the whole of the universe, and every part of it, is, in principle, weighed on the scales of reality, although we may be unable to complete the weighing, given the difficulty of retrieving information and of establishing simultaneity among places in the universe. In this sense and only in this sense, the now has special significance for science, not just for human experience. The distinctions among past, present, and future are real rather than just local observations about the relation of something, or of someone, to something else.

The past is not real: what is past no longer exists, albeit recorded in the prodigal vestiges lodged all around us in the universe. The future is largely unknown. That it is largely unknown, however, need not mean that it is entirely undetermined and unknowable. The future of the universe is a proper subject for investigation by science; it is not a subject that we must abandon to metaphysical speculation. We may be encouraged in the hope of gradually increasing our insight into what the future can or must hold.

There may be room for chance, novelty, and surprise, especially in the movement from one period of universal history to the next, or from one universe, or from one state of the universe, to another. Moreover, in the remote future, we may be able to influence the history of the universe as well as the history of our planet, if we are not extinguished before the strengthening of our powers enables us to intervene against some of the dangers that may beset us.

The primacy of historical over structural explanation, for which we both argue in this book, when taken together with the idea that everything in nature changes sooner or later, condemns any simple determinism, including what has come to be known as the block-universe conception. However, it fails to underwrite the idea of an open future, in any sense that we can analogize to openness in human life.

If eternalism is not to be understood as the trivial claim that the past influences the future, it must mean that there is a degree of determination of all events, presumably by unchanging laws, symmetries, and constants, in a world the structural elements of which remain always the same. The block-universe interpretation of general relativity amounts to a variant of such a view. It is a view that cannot be right if the ideas and arguments of this book are well founded.

These arguments and ideas also resist the demand of presentism to organize science around the specialness that the present moment enjoys in our experience. The subject matter of science and in the highest measure of the most general science, cosmology, is the becoming, and ceasing to be, of everything in nature. Everything, including the fundamental structures and regularities of nature and all the kinds of things that exist, is ceasing to be, or becoming something else, more slowly or more quickly. Change itself is changing.

Becoming is more real than anything else, yet it cannot be confined within the now, for the now is instantaneous. The subject matter of science is not the world viewed, as we cannot help experiencing it, from the perspective of now.

Our experience is such that all each of us ever has is the present moment, extended through the many-sided work of memory. In this sense, we are entrapped in the now. In its struggle to understand the world, however, science seeks to loosen, although it cannot wholly overcome, the constraints imposed by the way in which we experience reality.

The now has no unique value in science. Whether the ultimate subject matter of science is time-drenched becoming (as RMU believes) or timeless being (as the dominant tradition of modern physics supposes), it is not the now. Denial of the reality of time, however, widens the gulf between science and consciousness to such an extent that it renders all our experience questionable, including the perceptual experience into which we must translate, and on the basis of which we must make, our scientific discoveries.

For RMU, the argument of both parts of this book should be understood as a criticism of our pre-scientific experience as well as of certain central ideas and longstanding practices in cosmology and physics, not as a defense of pre-scientific experience against science. A claim for the special significance of the present moment not only lacks a basis in science, it also gives cause for a misreading of our argument. Moreover, it plays into the hand of attitudes influential in the contemporary philosophy of science that are contrary to the aims and ideas of this book. Among such attitudes are an almost limitless deference to whatever scientific theories currently prevail, a disposition to reconcile those theories with the phenomenology of experience whenever possible and to reject whatever part of our pre-scientific sense of reality cannot be reconciled with established science, and an antagonism to ways of thinking that defy these conservative presumptions.

LS would first of all agree that we are not rerunning the old presentist versus eternalist debate, because the key issue for us is the nature of laws on a cosmological scale, with respect to time. What we deny is that it is useful to conceive of the history of the universe as a timeless dynamical law acting on a fixed, timeless, state space. Instead, the laws are changeable as the distinction between laws and states breaks down.

While the exact degree to which the future is open is not resolved by the present arguments, we want to reserve that possibility.

So we need an objective distinction among past, present, and future to be able to assert that there are no certain facts of the matter about the future, which would seem to imply that the future is not presently real. This requirement does not contradict the obvious truth that there are many predictions we can make about the future which are reasonably but not absolutely reliable. Regarding the past, LS proposes it be regarded as also not real, but as having been real; there are present facts of the matter about the past. These can be evaluated on present evidence. This claim leaves the present as real.

The need for an objective distinction among past, present, and future is mainly motivated by the need to let the distinction between law and state break down in a way that allows at least the possibility that the future is not completely determined. This implies the need to deny the argument for eternalism of Putnam, which in turn implies that the relativity of simultaneity – as well confirmed as it is – give way to a preferred global time. This imperative is demanding but is supported by our ability to reformulate general relativity so as to admit such a global time.

7. *The empirical hard core of general relativity and its metaphysical gloss.* We agree in seeing the prevailing interpretations of general relativity as a major challenge to our ideas about the inclusive reality of time. We differ, more in emphasis than in substance, in our views of how to meet this challenge.

For RMU, the decisive element in such a response is the dissolution of the marriage between the empirically validated kernel of general relativity and the metaphysical conception of spacetime as a four-dimensional semi-Riemannian manifold that can be sliced in an infinite number of ways by alternative global spacetime coordinates. RMU argues that, contrary to what LS maintains, the ample empirical support for general relativity need not and should not be interpreted to validate that conception. Riemannian spacetime has served the larger project of spatializing time: of representing time as an aspect of the disposition of matter and motion in the universe. To identify such marriages between empirical and experimental science and supra-empirical ontologies and to propose that they be dissolved, when convenient to the progress of science, must, to his mind, be one of the major resources and tasks of natural philosophy. Resort to this method is one of the features distinguishing natural philosophy from philosophy of science, as the latter is now understood and practiced.

In the pursuit of this task, with regard to general relativity, it is useful to understand that we can translate the insights of general relativity into alternative vocabularies, free from the taint of the spatialization of time. Today, one of these vocabularies belongs to shape dynamics. Nothing in the argument, however, should turn on the truth of shape dynamics or on its fecundity as a research agenda.

It is no conclusive vindication of the metaphysical element in the leading formulations of general relativity that general relativity has been successful in its core domain of study any more than it was a confirmation of Newton's ontology of forces and bodies for classical mechanics to succeed at the work that it undertook, in the realm that it addressed. The ideas now prevailing in cosmology and fundamental physics reach their limit of successful application when we broaden their subject matter to include the universe and its history and attempt to complete the transformation of cosmology into a historical science.

LS would insist that while the concept of spacetime may be emergent, and give a coarse-grained and incomplete description of reality, we can still evaluate claims about its properties – just as one can evaluate claims about other emergent observables like temperature and pressure. And one thing that is well supported by experiment is that, within the realm of validity of general relativity, spacetime is Lorentzian. Thus the empirical success of general relativity supports strongly the conclusion that within the domain in which spacetime is a meaningful description that spacetime is Lorentzian.

There are several definitions of a preferred global time within general relativity, of which the constant-mean curvature slicings singled out by shape dynamics is only one. The hypothesis that each is physically preferred is a hypothesis about the form and content of the deeper theory that will replace general relativity. Shape dynamics is a particularly interesting hypothesis about global time because it is based on trading the many-fingered time gauge invariance of general relativity with a different gauge principle: that of local changes of scale.

8. *The conundrum of the meta-laws.* The thesis of the inclusive reality of time suggests that the laws, symmetries, and supposed constants of nature may be mutable, as they are within time rather than outside it: moved protagonists in the history of the universe rather than unmoved authorities. The conjecture of the mutability of the laws and other regularities of nature generates what we call the conundrum of the meta-laws. It seems unacceptable either to say that the joint evolution of the structures and laws of nature is caused by higher-order, meta-laws or to say that it is uncaused. The former option generates an infinite regress that we can halt only by exempting the higher-order regularities from the reach of time. The latter alternative throws science into explanatory paralysis.

For RMU, the best hope for beginning to solve the conundrum of the meta-laws lies in the combination of several ideas discussed in this book. Among these ideas are the primitive character of causation; the wide range of forms that causal connection takes in nature, most often recurrent and law-like but sometimes, in extreme states of nature, singular and lawless; the susceptibility of everything in the universe to change sooner or later, including the most fundamental structures or constituents of nature and change itself; the association of change of structure, as it becomes more radical, with change of laws, symmetries, and supposed constants; the differential readiness of natural regularities to change, so that the more fundamental ones, which we call principles, may change only more slowly and more rarely, in the most extreme circumstances of nature; and the powerful role of sequence or path dependence in a universe that is historical. He argues that similar problems have not prevented progress in the life and earth sciences or in social and historical studies. These disciplines have grown in insight by freeing themselves from the pretense that higher-order, immutable laws command the co-evolution of the laws and the structures that they study.

Such ideas do not provide a solution to the conundrum of the meta-laws, which we might take to be the holy grail of cosmology. However, they mark out the imaginative space in which it seems most promising to look for a solution. They form pieces of a view that can generate empirical claims, open to challenge and confirmation, precisely because such a view is historical and because the history of the universe leaves records and vestiges in nature.

It is, for RMU, a mistake to seek a special and permanent mechanism of change entrenched against the effects of time. If it is not a mistake, it nevertheless contradicts the spirit of our argument. The idea of cosmological natural selection would give a central and permanent role, in the history of the universe, to one mode of change: the one described by Darwin. It misstates the distinctive character of natural selection to reduce it to the combination of variation with the law of large numbers. Natural selection is a unique variant of functionalist explanation: an explanation converting an effect into a cause thanks to a mechanism of change that ensures the conversion. For Darwinian theory, this mechanism is differential reproduction in a population. Cosmological natural selection seeks to explain features of the universe by an analogy to such differential reproduction: a universe with more black holes generates more universes to succeed it.

If our argument about the inclusive reality of time is correct, there can be no such unchanging master mechanism of change. Moreover, the range of variation on which cosmological natural selection operates must be severely limited if the thesis of the singular existence of the universe, consistent with branching universes as well as with a succession of universes or of states of the universe but not with a vast array of universes (the multiverse), is to be upheld. If, on the contrary, we are to take cosmological natural selection as just one passing mechanism of change among many, we can make sense of it only by understanding its place in relation to all the others, including mechanisms radically different in character, such as the one that LS labels the principle of precedence.

Similar objections apply to this principle of precedence, according to which nature selects stochastically a state of affairs out of the pool of previous states of affairs. It may well be that some changes in some parts of nature at some times conform to this model. However, if stochastic selection from precedent were the pre-eminent procedure of change in the universe, we would once again have conceded to a feature of natural reality an exemption from the reach of time. We would have done so with the additional disadvantage of drastically confining, with meager empirical reason to do so, the room for novelty in universal history, produced through the co-evolution of structures and regularities. Novelty would appear diminished, as blind chance, picking futures from a fixed allotment of fates.

Moreover, the distinction between states of affairs with and without precedent makes sense only in the context of a theory such as quantum mechanics (from which the distinction is drawn) that deals with closed systems and limited patches of the universe. It loses sense when extrapolated to the entire universe. For the universe, there is always a candidate for precedent, drawn from its entire previous history. As applied to a universe in which the new happens and everything changes sooner or later and once reduced to being only one of alternative types of transformation (novelty as selection from precedent versus novelty beyond precedent), the message of the principle of precedent becomes muddled and useless. It tells us that everything changes according to precedent until unprecedented novelty takes place. It casts no light on when, how, or why events conform to the first of these two varieties of change or to the second.

Our own argument lays bare the root of the mistake. The conundrum of the meta-laws presents a cosmological problem. We cannot progress in solving it by appealing to modes of thought that, in the manner of the principle of precedence, require as their subject matter a closed subsystem within nature. Such an appeal exemplifies what LS calls the cosmological fallacy and RMU the first cosmological fallacy: it applies to the whole universe a way of thinking that is licit only when applied to part of it, and even then only to the extent that the part can be understood without regard to its relations to other parts and to the whole.

We do need to search for solutions to the conundrum of the meta-laws that can face empirical challenge or benefit from empirical confirmation. We shall not succeed in our quest by resort to speculative conjectures exhibiting the errors that we denounce.

LS would emphasize that our arguments, if they are to succeed, must generate scientific hypotheses that can be developed and tested by the usual means within science. So if we propose that laws on cosmological scales evolve, it is necessary to put forward hypotheses as to mechanisms by which the effective laws that govern observable phenomena could have evolved. Moreover, these hypotheses must have implications for doable observations that render them testable and preferably falsifiable.

Two hypotheses of this kind are cosmological natural selection and the principle of precedence. As described previously and mentioned in the text here, cosmological natural selection has generated genuinely falsifiable predictions that have so far, over the last twenty years or so, stood up to tests.

It is not necessary to object that "the idea of cosmological natural selection would give a central and permanent role, in the history of the universe, to one mode of change: the one described by Darwin." The mechanism of natural selection is not proscribed or directly coded anywhere in nature, either in biology or in cosmology. All that is coded is reproduction with small variations – and this much exists because physics and chemistry allows it to and the environment is hospitable to it. The rest, the effects of fitness and differential success, is logic, an aspect of the law of large numbers. So natural selection is not one "mode of change," different from others; it is a description of the logic by which a large class of systems may change. There are many mechanisms of natural selection which operate in biology on different scales and in different contexts and there is every reason to expect that new ones may come into play from time to time. Indeed, natural selection is not only compatible with the general notion that "laws evolve with the systems they describe," it is the general logic making such evolution possible.

The principle of precedence is likewise a general suggestion for how laws may have evolved. It suggests testable hypotheses and may apply to a large number of distinct mechanisms.

9. *The relation of mathematics to nature.* One of the three central ideas in this work concerns mathematics and its relation to nature and to science. It is intimately connected with our other core claims about the singular existence of the universe and the inclusive reality of time.

We deny that nature is mathematical: the universe is not homologous to a mathematical object, much less is it such an object. The effectiveness of mathematics in science is reasonable because it is relative. Mathematics is good for some uses in science and bad for others. It has trouble representing some features of nature, especially those features that have to do with time. We agree that neither views seeing mathematics as the discovery of a distinct realm of mathematical objects nor views presenting mathematics as pure invention or convention can account for the effectiveness of mathematics in science and for the relativity of this effectiveness.

We develop the thesis of the selective realism of mathematics in different forms. Our developments are largely but not entirely compatible and complementary. For RMU, mathematics is an exploration of a simulacrum of the world, robbed of time and of phenomenal particularity. Having begun in the representation of aspects of nature – number and space, structured wholes and bundles of relations – it soon soars above the world. It finds its primary inspiration in itself and its secondary inspiration in the problems that science presents to it.

The selectiveness of mathematics is the source both of its power and of its limits. Mathematics accommodates only to a limited extent what in nature is fundamental: time. It offers many conceptions – such as the mathematical conception of the infinite – that have no presence in nature. It empowers science. In so doing, however, it offers science a poisoned chalice: the idea of timeless laws of nature, written in the language of mathematics.

RMU doubts that LS's idea of mathematics as evoking structures that prove to have a predetermined content does justice to these features of the relation of mathematics to nature and to science. LS's view of mathematics seems to him not to serve as an example of an alternative to the familiar contrasting approaches to mathematics as discovery and as invention or convention. Instead, he takes it to exemplify the moderate conventionalism that has historically been the main form of conventionalist accounts of mathematics. Only the most radical conventionalists (e.g. late Wittgenstein) have taken the position that the play of convention is unconstrained other than by collective practice. Mathematical conventionalism in any form, moderate or extreme, is unable to account for the traits that render mathematics so useful in dealing with some aspects of nature and so hobbled in addressing others.

LS and RMU agree in rejecting a Platonic account of mathematics according to which mathematical objects exist in a timeless realm separate from the universe. We also agree in rejecting a purely conventionalist alternative according to which mathematical facts are true by convention or agreement. The challenge we face is then to formulate an account of mathematics that explains its success in physics without requiring belief in anything other than our time-bound single universe. This is a difficult challenge, and LS and RMU each put forward new proposals to frame such an account.

LS's tentative view of mathematical truth as evoked relies on the belief in the reality of time expressed as the objective reality of the present moment and the denial of truths outside of time. The key idea is that structures can be invented – by nature and by human beings acting, if you will, as nature's agents of creation, which did not exist before, but which from that point on have stable, objective properties. This is a form neither of conventionalism nor of Platonism.

10. *Nature as a process of singular events*. Nature, for RMU, is not always and forever composed of either singular or non-singular events.

The extent to which events are singular or non-singular varies, as does everything else, in the history of the universe.

In the cooled-down universe in which we find ourselves (but that coexists with records of an earlier state of the universe), the overwhelming preponderance of events is repetitious. The recurrence of such events exhibits stable laws, symmetries, and constants of nature. They are trivially singular in the sense that each event has a distinct spatial and temporal connection with other events. However, these differences of spatial and temporal placement do not prevent occurring phenomena from being in other respects the same, as only an unqualified relationalism would deny.

In other, extreme states of nature, such as those that may have existed at the formative moments of the present universe and may exist as well, in other manifestations, later in its history, events and their causal connections may fail to exhibit such recurrence, captured by laws, symmetries, and constants. Then, and only then, are they strictly singular.

A disadvantage of the view of nature as a process solely of singular events is that it fails to make room for these variations. It is not just the structure of nature that changes – as does everything – in the course of time; the kind of structure that there is changes as well.

This is again a case in which the goal of translating our general principles into science requires that we tentatively put forward hypotheses and play with models. To specify a model you have to lay down an ontology. An ontology of discrete events provides a simple framework for investigating and representing hypotheses in which causality is prior to law as well as prior to spacetime. So we have used it in recent work with Marina Cortes that is described below.

In this work we were able to focus on some striking but previously unappreciated consequences of relationalism, particularly the insistence that laws refer to events only through relational properties as well as respect for the principle of the identity of indiscernibles. The combination of these ideas implies that elementary events be uniquely specifiable and distinguishable from all other events by their relational properties. This condition implies that laws applying to individual fundamental events cannot be both general and simple. Simple laws can emerge only when applied to large classes of events. This condition limits the reach of the Newtonian paradigm for fundamental, microscopic events and complements the limits from above resulting from the argument about the cosmological fallacy and dilemma.

11. *The openness of the history of the universe.* Some may read this book as implying that the history of the universe is open, and thus as well hospitable to us, human beings, and to our individual and collective plans. The history of natural philosophy is filled with the influence on our views of nature of attempts to devise feel-good philosophical systems.

In this reckoning of points of divergence, RMU earlier remarked that we do not know to what extent the history of the universe is open, much less that nature is favorable to us and our aims. It does not follow from the historical character of the universe, the inclusive reality of time, and the mutability of the laws of nature that nature is on our side.

One way in which nature might be on our side is if mental phenomena were prefigured in the pre-human natural order: qualia, or what we experience in ourselves as consciousness, would form part of physical events, as panpsychism proposes. There is, for RMU, nothing in our ideas and arguments to support or even to suggest this or any other number of claims about nature that we might find encouraging.

We know for a certainty that nature is on our side in one respect: it gave us life. We also know for sure that nature is against us in another regard: it will soon crush each of us. It may later annihilate humankind. For the most part, nature is neither for us nor against us. It is simply indifferent, and organized on a scale unfathomably disproportionate to our concerns.

It should form no part of the program of natural philosophy to inspire or to justify reverence for the cosmos. Reverence for the universe is power worship, under the disguise of pantheistic piety, theistic gratitude, or philosophical wonder. It is unworthy of free men and women and dangerous to their humanity.

It is crucial to the integrity of natural philosophy that it not cast itself in the role of bearer of good news. If there is good news to be heard, it must come from somewhere else.

LS disagrees; not the least because having a reason to challenge the dominant paradigm of strong artificial intelligence, while staying within natural science, is good news for human beings. It is good to know that neither we nor the universe we inhabit can be fully captured by the computational metaphor. Temporal naturalism, as introduced here, is also a form of naturalism that admits qualia as intrinsic properties of physical events, which means that the basic fact that we are living beings who experience qualia need not alienate us from full membership within nature, as conceived by a naturalist.

Index